INTERNATIONAL SYMPOSIUM ON ELECTRON BEAM ION SOURCES AND THEIR APPLICATIONS

CONFERENCE PROCEEDINGS NO. **188**

PARTICLES AND FIELDS SERIES 38

RITA G. LERNER
SERIES EDITOR

INTERNATIONAL SYMPOSIUM ON ELECTRON BEAM ION SOURCES AND THEIR APPLICATIONS

UPTON, NY 1988

EDITOR:

ADY HERSHCOVITCH
BROOKHAVEN NATIONAL
LABORATORY

American Institute of Physics   New York

Authorization to photocopy items for internal or personal use, beyond the free copying permitted under the 1978 US Copyright Law (see statement below), is granted by the American Insitute of Physics for users registered with the Copyright Clearance Center (CCC) Transactional Reporting Service, provided that the base fee of $3.00 per copy is paid directly to CCC, 27 Congress St., Salem, MA 01970. For those organizations that have been granted a photocopy license by CCC, a separate system of payment has been arranged. The fee code for users of the Transactional Reporting Service is: 0094-243X/87 $3.00.

Copyright 1989 American Institute of Physics.

Individual readers of this volume and non-profit libraries, acting for them, are permitted to make fair use of the material in it, such as copying an article for use in teaching or research. Permission is granted to quote from this volume in scientific work with the customary acknowledgment of the source. To reprint a figure, table or other excerpt requires the consent of one of the original authors and notification to AIP. Republication or systematic or multiple reproduction of any material in this volume is permitted only under license from AIP. Address inquiries to Series Editor, AIP Conference Proceedings, AIP, 335 E. 45th St., New York, NY 10017.

L.C. Catalog Card No. 89-084343
ISBN 0-88318-388-9
DOE CONF 881154

Printed in the United States of America.

CONTENTS

Preface ix

EBIS DEVICES

Review of EBIS Devices 3
    J. Faure
Progress Report on the Frankfurt EBIS 15
    M. Kleinod, R. Becker, and H. Klein
**Status Report on the Stockholm Cryogenic Electron
Beam Ion Source** 27
    L. Liljeby and A. Engstrom
Performance of Mini-EBIS Cooled by Liquid Nitrogen 33
    K. Okuno
The Sandia EBIS Program 45
    R. W. Schmieder, C. L. Bisson, S. Haney, N. Toly,
    A. R. Van Hook, and J. Weeks
The Cornell Superconducting Solenoid Cryogenic EBIS 65
    V. O. Kostroun
EBIT: Electron Beam Ion Trap 82
    M. A. Levine, R. E. Marrs, C. L. Bennett, J. R. Henderson,
    D. A. Knapp, and M. B. Schneider
DIONE Status Report 102
    J. Faure, P. Antoine, J. C. Ciret, L. Degueurce, P. Gros,
    A. Courtois, B. Gastineau, R. Gobin, P. Leaux,
    P. A. Leroy, and J. P. Penicaud
The KSU-CRYEBIS Program 114
    M. P. Stockli, C. L. Cocke, and P. Richard
The Reliable Operation of CRYEBIS 1 127
    J. Faure

EBIS PHYSICS

Electron Beam Dynamics in EBIS Sources 131
    M. Tagger
**Computer Predictions of "Evaporative" Cooling
of Highly Charged Ions in EBIT** 145
    B. M. Penetrante, M. A. Levine, and J. N. Bardsley
**Evaporative Cooling of Highly Charged Ions in EBIT:
An Experimental Realization** 158
    M. B. Schneider, M. A. Levine, C. L. Bennett,
    J. R. Henderson, D. A. Knapp, and R. E. Marrs
**Evidence for Ion Cooling and an Observation
of Ion Heating in Cornell EBIS I** 166
    E. N. Beebe
**Panel Session: e-Beam Plasma Interactions in an EBIS Trap;
EBIS Physics; Diagnostics** 179
    N. Rostoker, V. Kostroun, R. Marrs, R. Schmieder, and
    H. Tawara

ELECTRON BEAMS

Thermionic Sources for Hi-Brightness Electron Beams 191
    R. E. Thomas
Advances in *e*-Beam Formation, Focusing, and Collection 219
    R. True
The Effect of Small Transverse Magnetic Fields
on Electron Beam Transmission 233
    K. Amboss
Characteristics of Typical Pierce Guns for PPM Focused TWTs 245
    R. Harper and M. P. Puri
A Novel Electron Source for EBIS Machines 259
    R. True
Extraction of a Steady State Electron Beam
from HCD Plasmas for EBIS Applications 271
    A. Hershcovitch, V. Kovarik, and K. Prelec

RELATED TECHNOLOGIES; PRIMARY IONS

Magnetic Precision Alignment of a Long Horizontal
Ultra-Straight Solenoid 281
    M. P. Stockli, C. L. Cocke, J. A. Good, and P. Wilkins
Brief Comments on Test Results of Both Metal
and Elastomer Vacuum Seals 303
    K. M. Welch, G. T. McIntyre, J. E. Tuozzolo, and D. J. Pate
Primary Ion Sources for EBIS Devices 314
    R. Keller

PROPOSED EBIS BASED SYSTEMS

The Relativistic Heavy Ion Collider (RHIC) Project
at Brookhaven 327
    T. W. Ludlam
Source Options for RHIC 341
    K. Prelec
RFQ for an EBIS-Based RHIC Injector 347
    J. Staples
The Feasibility of an EBIS for the CERN Lead Project 359
    R. Becker
Panel Session: Future Prospects and Limits 371
    J. Faure, R. Becker, C. Herrlander, M. Kleinod, and
    M. Levine

OTHER DEVICES

CRYRING—A Heavy-Ion Storage and Synchrotron Ring 379
    C. J. Herrlander
RFQ Injector for CRYRING 388
    A. Schempp, H. Deitinghoff, H. Klein, A. Kallberg,
    A. Soltan, and C. J. Herrlander
ECR Ion Source Status 1988 398
    T. A. Antaya and C. M. Lyneis
Experiments with a Synchrotron X-Ray Source
and Conventional, ECR, and Storage-Ring Ion Sources 412
    K. W. Jones, B. M. Johnson, and M. Meron

ATOMIC PHYSICS

Atomic Physics Research Using Highly Charged
Ions from EBIS 427
    H. Tawara
Atomic Physics Measurements in an Electron Beam Ion Trap 445
    R. E. Marrs, P. Beiersdorfer, C. Bennett, M. H. Chen,
    T. Cowan, D. Dietrich, J. R. Henderson, D. A. Knapp,
    A. Osterheld, M. B. Schneider, J. H. Scofield, and
    M. A. Levine
Concluding Remarks 459
    R. Becker

APPENDICES

Appendix I: List of Participants 465
Appendix II: List of Authors 469
Appendix III: Symposium Program 470

Preface

It was a most rewarding experience for me to be Chairman and Proceedings Editor for the International Symposium on Electron Beam Ion Sources and Their Applications. Krsto Prelec suggested having such a symposium at BNL after the two of us served on an interdepartmental committee assigned to examine various heavy ion preinjectors for RHIC. Among the Committee's findings was that for the long term (up to 25 years), an EBIS with a capacity of 7.5×10^{11} charges could be a suitable heavy ion source for a RHIC preinjector. Consequently, we wanted to learn more about EBIS devices. The most effective way was to have a meeting at BNL involving most of the top researchers in the field. Three prior EBIS workshops had tracked the progress of EBIS devices until 1985. This symposium was more formal and much broader in scope than the previous workshops. In addition, status reports regarding the state-of-the-art of EBIS devices and research accomplishments, experts in plasma and atomic physics, as well as in e-beam, magnet, and vacuum technologies, participated and shared their expertise with the EBIS community. Hopefully, this interdisciplinary interaction will accelerate progress in this field. Regardless of the symposium impact on EBIS research, a number of BNL staff members, in general, and myself, in particular, found this meeting to be a great learning experience.

Thanks to the AGS Department at Brookhaven for its generous support of this symposium. I am thankful to my Department Head, Derek Lowenstein, and to my Division Head, Theo Sluyters, for making this support possible. Additional funding was provided by the BNL Accelerator Development Department, thanks to Eric Forsyth. Associated Universities, Inc. hosted the wine and cheese reception and provided the coffee during breaks with Jerry Hudis making that possible. Members of the Program Committee provided many helpful comments and suggestions. Special thanks to Jean Faure, an EBIS veteran from Saclay, for his help as Co-Chairman of the symposium. His ties to the EBIS community and his knowledge of the field contributed greatly to the success of the symposium. The experience the Local Organizing Committee had with the BNL Negative Ion Symposia made my job much easier. Ron Clipperton's help as Symposium Coordinator was invaluable. Ron and his assistant, Barbara Cox, extended themselves well beyond the call of duty. Among other things, Barbara suggested and Ron arranged the highlight of the week—the banquet cruise. Proceedings Secretary, Lenore Dudzick, very skillfully handled the difficult task of transcribing all the taped discussions following each talk. Special thanks to Bob Schmieder from Sandia for having his secretary transcribe the panel sessions and for editing them. Their help is greatly appreciated and it also reduced our secretarial cost. Again, I would like to thank Krsto Prelec for proposing this symposium.

Marion Heimerle was an excellent Symposium Secretary. Her experience, forethought, and expertise contributed greatly to the success of this symposium. It is important to note that her invaluable contributions were made during a period marred by a tremendous personal tragedy; her husband Bob's untimely death a few hours after the symposium ended. I was fortunate to have known him as to the wonderful person he was. It is a great loss to all those who knew him.

When editing the transcribed discussion tapes I felt compelled to strike a balance between improving spontaneous oral language into well-written text, and ensuring that spoken content and intent remained unaltered. I apologize if I strayed either way. Finally, I would like to thank the participants for their cooperation and for making it a truly productive and enjoyable symposium.

Ady Hershcovitch
Brookhaven National Laboratory

LOCAL ORGANIZING COMMITTEE

Ady Hershcovitch, Chairman
Jim Alessi
Ron Clipperton, Symposium Coordinator
Barbara Cox, Assistant Coordinator
Ahovi Kponou
Charlie Meitzler
Krsto Prelec

INTERNATIONAL PROGRAM COMMITTEE

A. Hershcovitch	Brookhaven National Laboratory, U.S.A., Chairman
J. Faure	Centre d'Etudes Nucleaires de Saclay, France, Co-Chairman
K. Amboss	Hughes Electron Dynamics Division, U.S.A.
R. Becker	Institute für Angewandte Physik, Frankfurt, F.R.G.
E. D. Donets	Joint Institute for Nuclear Research, Dubna, U.S.S.R.
V. O. Kostroun	Cornell University, U.S.A.
M. A. Levine	Lawrence Berkeley Laboratory, U.S.A.
K. Prelec	Brookhaven National Laboratory, U.S.A.

SYMPOSIUM SECRETARY
Marion V. Heimerle

PROCEEDINGS SECRETARY
Lenore Dudzick

EBIS DEVICES

REVIEW OF EBIS DEVICES

J.Faure
Laboratoire National SATURNE
91191 Gif/Yvette France

INTRODUCTION

In this paper we introduce the EBIS device although it has been described many times before. I went into this business in 1980 and I discovered, making some historical research for writing this paper , that EBIS appeared in the sixties. The first words were said in russian in DONETZ laboratories.

The process consists in step by step ionization of ions using electrons bombardment.

GENERAL ARRANGEMENT

If we don't want to enter into too many details one can describe the sources presently in operation in the following way :

Electron beam is provided by an electron gun and travels under vacuum up to a collector. During this travel, the electrons are focused by solenoidal magnetic field.

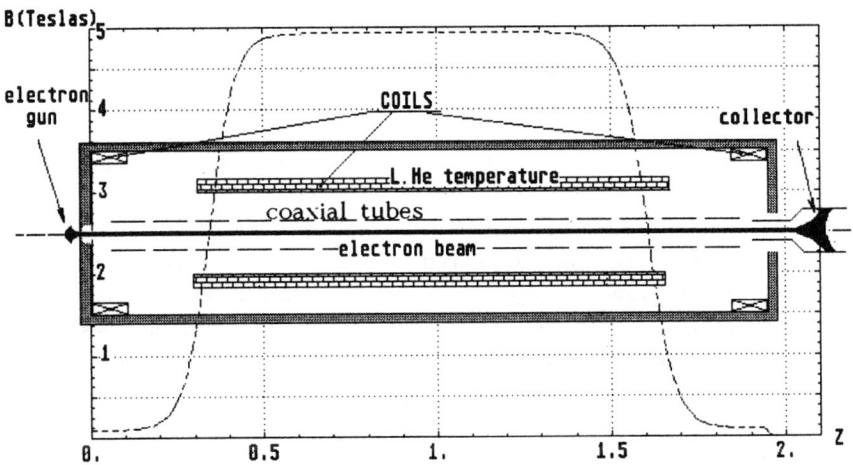

Fig 1 Electron beam inside magnetic configuration

© 1989 American Institute of Physics

The figure 1 shows this arrangement. The coaxial tubes are used to determine a potential distribution along the electrons path.

The low charge state ions are produced inside the electron beam or injected from an external ion source. Radially, the ions are maintained by means of electron beam space charge and longitudinaly trapped by the longitudinal drift tubes potential arrangement.

The ions stay there a time long enough to get stripped of the maximum of electrons.

DC MODE

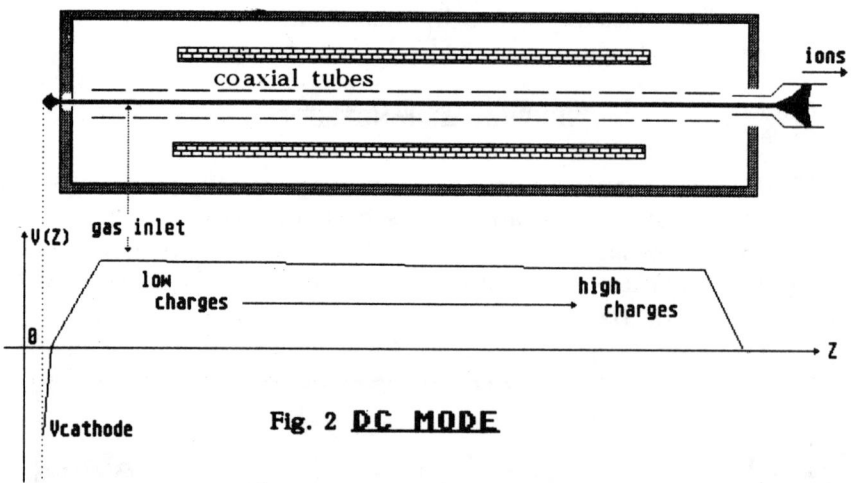

Fig. 2 **DC MODE**

For a DC source, the potential distribution is constant. The ions are produced near the electron gun, ionized during their travel path along the source axis inside the electron beam and extracted through the collector. This regime can be compared with an ECR source. As the ion speed has to be low, charge compensation limits the ions intensity around $10^7/10^3$ ions per second according to the ion species (C^{6+}... Ne^{10+}.... Ar^{16+}.... Kr^{26+}.... Xe^{36+}) ; (Frankfurt Group [1], Tokyo University [2] ....)

PULSED MODE

For the other operating modes, the potential distribution is cyclic, there are 3 successive steps :

1) - Injection 2) - Containment 3) - Expulsion (or extraction)

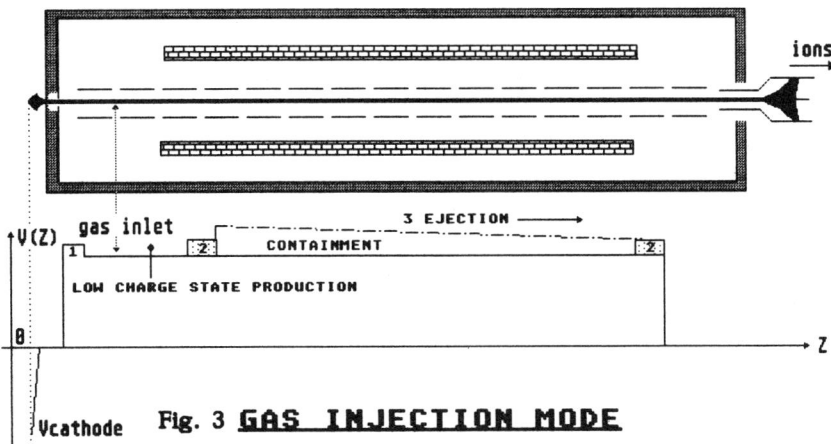

Fig. 3 **GAS INJECTION MODE**

Gas injection

The first configuration corresponds to the creation of low charge state (N^+ or Ar^+ for instance) ions in a "high" pressure part of the source where gas is injected. This mode is used in almost all the sources.

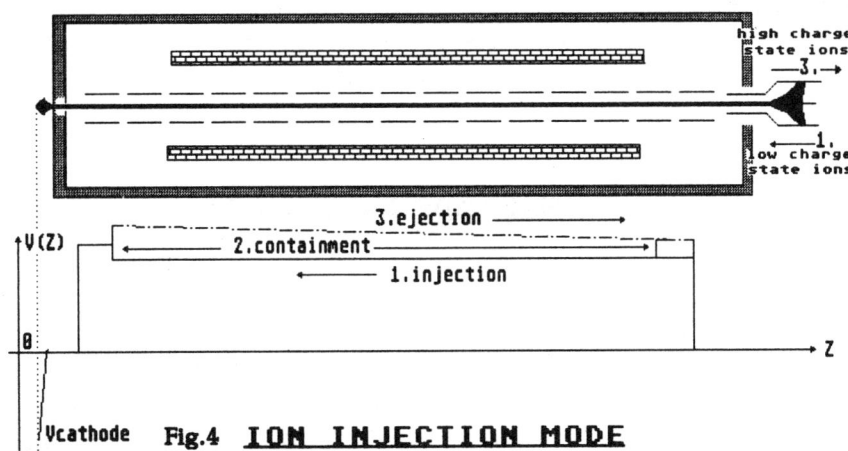

Fig.4 **ION INJECTION MODE**

Ions injection from an external source

Another possibility tested and regularly used in Saclay (CRYEBIS 1, DIONÉ is to inject low charge state ions in the electron beam through the collector. They are slowed down to a few eV. This procedure allows a more flexible control and the possibility to ionize any kind of ions provided that the low charge ions source is available.

IONIZATION PROCESS
and concequences

It is confirmed by experiments that step by step ionization is the basic process involved.

In order to avoid recombinations and charge exchanges the residual pressure has to be as low as possible. That means 10^{-11} torr or less and cryogenic pumping system distributed along the containment region. The ion injection mode looks more attractive from this point of view than a continuous gas injection, because the exact quantity is injected.

The behaviour of the electrons created during the ionization process is not clear. They look like disappearing. Indeed recombination effect seems negligible and the ions emittance calculated taking into account the only primary electron beam is in agreement with the measurements. They should be fast heated, travelling towards the source ends, and accelerated by the longitudinal space charge.

The important parameter for the source is the product ($j\tau$).

j electron beam current density

τ confinement time

The probability to ionize a given particle increases if the flow of electrons per area unit (j) increases and if the time of exposure (τ) increases too.

We have introduced the main parameters of this type of source:

-**pressure or residual gas**
-**electron beam density**

PUMPING SYSTEM-RESIDUAL GAS

Most sources are liquid Helium cryopumped.

The exceptions are:

-CEBIS 1 built by V.KOSTROUN and pumped by an ingenious integrated ion pump system.
-mini-EBIS of TOKYO UNIVERSITY built by K.OKUNO using a liquid nitrogen cryopumping system.
-The Brown-Feinberg EBIS was ion pumped at the origin but has been rebuilt by R.Schmieder with cryogenic pannels.

Now we have two types of liquid Helium cryopumping system:
1) The pumping system cryostat is separated from cryogenic magnet coil cryostat:
   - CRYEBIS 1 built by ARIANER & al. and modified by the SACLAY group.
   - CEBIS 2 under tests in CORNELL UNIVERSITY

2) There is a common cryostat;
   - CRYEBIS 2 built by ARIANER & al.
   - KSU–CRYEBIS built by P.STOCKLI & al. and under tests.
   - CRYSIS the STOCKHOLM EBIS operating and under development.
   - FRANKFORT CRYEBIS built by R. BECKER & al. and running for atomic physics.
   - DUBNA EBIS KRION 1 ,2 and 3 built by DONETS running and under development.
   - SUPER EBIS under construction by R. SCHMIEDER in S.N.L.
   - DIONE built by J.FAURE & al. in SACLAY and in operation at L.N.S.

In DIONE the insulating cryogenic vessel is separated from the confinement region. This option for two separated vacuum vessels comes from the CRYEBIS 1 behaviour. When the liquid helium level went down the surface temperature of the upper part increased leading to residual gas outgassing(H^+ appearance on T.o.f.). Another advantage is, to avoid Helium contamination from occasional Helium circuit leak.

This option has been taken at K.S.U. and at S.N.L.

In DIONE we have set up two pannels with the possibility to pump on them so that their temperature goes down to 2K. They are charcoal platted too. It has been proved that this device is efficient. But if we make a comparison with CRYEBIS 1, we have to point out that the residual gas level is larger in DIONE in spite of all the precautions mentionned above. An explaination could be that the overall pumping speed was larger in CRYEBIS 1 and ,especially, the liquid nitrogen screen more efficient.

Nevertheless DONETZ sources run without residual gas effect, long containment times....and there is not any vacuum separation... but the electron beam intensity is smaller.

ELECTRON BEAM DENSITY

It is obvious that EBIS owner try to reach the highest density in order to decrease the containment time.

Fig. 5 shows the densities obtained on DIONÉ and CRYEBIS 1

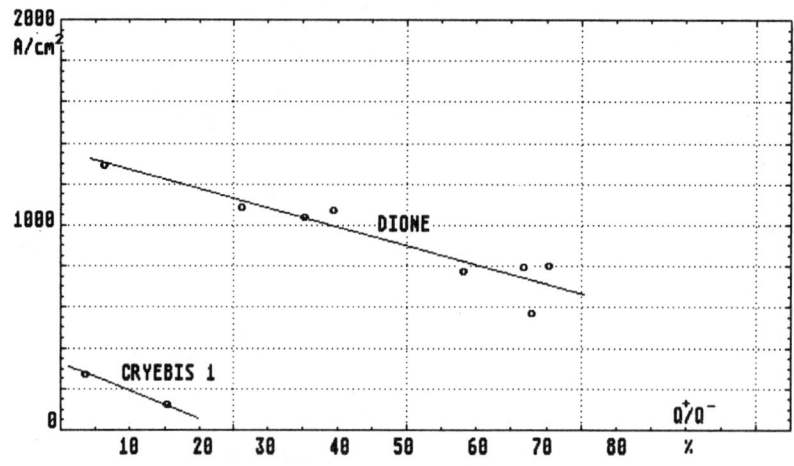

Fig. 5 DIONÉ and CRYEBIS 1 densities versus electron space charge neutralisation

The other figures are:
- KRION-2   400 A/cm^2
- CEBIS   1000/2000 A/cm^2
- FRANCFORT CRYEBIS 1000 A/cm^2
- CRYSIS   100 A/cm^2

Therefore the effective range is 100 A/cm^2-1000 A/cm^2 and the results of Fig.5 show that the density depends on neutralization. Another figure could be added: ECR (Q^+/Q^- = 100%) density is about 100 A/cm^2.

The questions are:
- Is it possible to increase the density?
- Are these values large enough for future projects?

One goal of this Symposium is to answer to these questions!

As soon as the discussion will start two important features will be considered:
- electron guns
- electron beam compression

Electron guns.

The electron guns are bought from companies like HUGHES or LITTON Systems or calculated using Herrmannsfeldt code and dispenser cathodes bought from companies like THOMSON. DONETZ uses LaB_6 emitting surface cathodes bought from electron welding devices manufacturers.

A.HERSHCOVITCH proposes to use hollow cathode discharge.

Electron beam compression.

If we assume that the gun is optically perfect:
    the electron distribution is uniform
    the cathode temperature is low
    there is no magnetic flux at the cathode
and that the magnetic field varies so that we follow adiabatically the Brillouin equilibrium , the density should be as high as 10^5 A/cm^2 !

But non uniformity of the electrons distribution , geometric imperfections and several instabilities lead to more modest performances.

QUANTITIES OF IONS

The quantity of charges that we can expect depends on the maximum rate of neutralization that the electron beam is able to withstand.
Examples:

DONETZ. KRION-2 : $2 \cdot 10^8$ charges of Ar^{18+} represent 4% of neutralization of 100 mA at 30kV

SACLAY. DIONÉ produces $1.5 \cdot 10^{10}$ charges with a 360 mA 10kV electron beam that is 50% of neutralization

But in the Saclay conditions , the density is such that only $7 \cdot 10^9$ charges of N^{7+} are obtained or 10^9 charges of Ar^{16+} with a containment time of 30 ms. If we increase this time the relative quantity of high charges states increases but the the total intensity decreases so that the Ar^{16+} quantity remains the same.

To overcome these ions losses DONETZ proceeds to ion cooling by cold ions injected periodicaly during the containment process. The confinement time is 1.5-2 s. and the periodicity of cooling is 15 ms.

This procedure is probably the future for long confinement time and high charge states.

The ion heating/cooling effects has been recently observed and calculated at SANDIA NATIONAL LABORATORIES.

EBIT SOURCES

This source type is used for atomic physics but the phenomena studied are interesting for EBIS understanding. It is a very short EBIS where the containment region can be investigated. The cooling process has been used to maintain Au^{69+} during few hours at LLNL.

EDITORIAL SUPPLEMENT

- EBIS LABORATORIES -

LOCATION/LAB	KEY PERSONNEL	MACHINE	PURPOSE	IONS PRODUCED	REMARKS
Dubna/JINR	Donets	KRION-1	Synchrophasotron injector	Bare C,N,O & Ne	Converted to KRION-S
	Pikin	KRION-2	Highly charged ions	Bare Ne,Ar,Kr & Xe	Highest charge state ever produced
	Vadeev	KRION-3	Synchrophasotron injector	10^9 charges of Bare Ar/pulse	Plan to inject 5×10^8 Ar^{18+}/pulse
	Ousyannikov	KRION-S	EBIS development	Bare Ar, U^{40+}	Average U charge distribution (U) from TOF measurements
Orsay	Arianer	CRYEBIS-I	EBIS development	Kr^{34+}, Xe^{44+}	Transferred to Saclay
Saclay/LNS	Faure	CRYEBIS-I	SATURNE injector	Bare C,N,Ne as injector;	Retired after 3 years of very reliable operation as an injector
	Faure	DIONE	SATURNE/MIMAS injector	Bare C,N,Ne,Ar^{16+} $17+$ $18+$, Kr^{30+}	Up to 2×10^{10} charges/micropulse. Up to 6 micropulses injected into MIMAS per cycle. Has had 2000 hours of reliable operation.
Frankfurt/IAP	Becker/Kleinod	Frankfurt-EBIS	Development and physics	$Ne^{10+}, Ar^{16+}, Kr^{26+}, Xe^{36+}$	Steady state; single passage ionization
Nagoya/IPP	Tawara	NICE	Naked ion collisions	$B^{5+},C^{6+},N^{7+},O^{8+},F^{9+},Ne^{9+}$, Kr^{25+},I^{42+}	
Tokyo Metro U.	Okuno	Mini-EBIS	Naked ion collisions	Bare C,N,O,Ne^{9+},Ar^{16+}	
Stockholm/MSI	Liljeby/Herrlander	CRYSIS	CRYRING injector	Ne^{10+}, Ar^{16+} $17+$ $18+$, Xe^{35+} $40+$ $44+$	Has been operational for a short time. Maximum output 3×10^9 charges/pulse; designed for 3×10^{11} charges/pulse
Ithaca/Cornell	Kostroun	CEBIS-I	Development/atomic physics	$C^{5+},N^{6+},O^{7+},Xe^{28+}$	
		CEBIS-II	Ion physics		Currently undergoing tests
Livermore/Sandia	Schmieder	BEBIS	Training/development/physics	Bare C,N,O,Xe^{40+}	Construction close to completion
		Super-EBIS	Ion collision physics	Plan to produce ions up to U^{82+}	
Livermore/LLNL	Marrs/Levine	EBIT	Spectroscopy	Bare Ti & Ni, Ba^{46+}, Au^{69+}, Pb^{2+}	
Manhattan/KSU	Cocke/Stockli	KSU CRYEBIS	Ion collision physics	Just started operation with argon. No extraction yet.	

Some of this information was provided by Martin Stockli (KSU). His paper NSCL Report #MSUCP-47, Dec. 87 and Oct. 23-30 1988 trip report have been very helpful.

DISCUSSION

Stockli: I am aware of Donets' publications in which he showed that he made bare krypton and helium-like xenon, however I have not seen any proof of bare xenon which you mentioned. Have you seen any proof of this claim?
Faure: Donets came to Saturne in France last year, and I have something I can give to you.
Stockli: Are you referring to the MIMAS meeting?
Faure: Yes.
Stockli: But he showed only bare krypton in the proceedings.
Faure: Yes.
Stockli: It does not mention anything like bare or hydrogen-like xenon.
Faure: No, it is only krypton.
Stockli: That is right. But do you know anything about xenon with an ionized K-shell?
Faure: I do not know. I have not seen anything.
Stockli: When I was in Dubna, I was told that there is a JINR-report which shows hydrogen-like and bare xenon, but Dr. Pikin could not find the report at that time.
Kostroun: Three weeks ago when I was in Stockholm, Donets showed x-ray spectra from ions incident on a copper target and I think he had seen Lyman radiation of bare xenon. This would be clear proof. It is very difficult to separate highly charged ions by time-of-flight or magnetic analysis, so that is a way of doing it.
Levine: Could you justify your requirement that the vacuum be $\approx 10^{-11}$ Torr or better?
Faure: If you consider the hydrogen cross section based on the calculations I showed, you see that you need at least 10^{-10} if you do not want to be adversely affected by protons. That is why I said 10^{-11} Torr.
Levine: You mean at higher pressures the trap would fill with protons?
Faure: Yes.
Becker: In the usual operation of an EBIS with no ion cooling, a considerable amount of ions from the background can be collected during ionization. And ionization times of seconds mean that pressure must be low.
Levine: I would argue that that is too stringent a requirement when one considers effects similar to those presented at this conference in the theory of "evaporative cooling".
Hershcovitch: I have a comment about the table that was presented. Some critics of EBIS claim that you only use noble gas ions. But there have been many metallic ions produced by EBIT. Could either Jean or Mort expand on what ions were achieved in EBIT?

Levine: EBIT currently uses a Metal Vapor Vacuum Arc source (MEVVA) for injection.

Faure: I can say that if you have an external source injecting in EBIS you can have any kind of ions you want.

Kleinod: Could you expand on the cooling Donets uses? You only showed it briefly but I know you have more information.

Becker: I suppose the difficult thing which is being revealed with this kind of ion cooling is that most of us have started out doing EBIS work by saying, "Now I want to build a source which is fully understood". Instead research is moving into the field of ion sourcery, isn't it?

Penetrante: You mentioned that you can essentially achieve any charge that you want from EBIS. I think that statement is only true if you can cool it as efficiently as you can.

Faure: Yes, but I said that we have to cool it if you want to go as far as possible.

Penetrante: Right, correct me if I am wrong. I thought I heard you say that you can achieve almost any charge state that you want, but I think that has to be taken in the light that you could only do that if you can cool very efficiently.

Faure: I do not know. You can have bare ions but in small quantities as Donets did for Ar^{18+}, for example. If you want to increase their quantity, cooling is needed. Anyway, he has Ar^{18+} or Kr^{36+} without cooling. When you cool, you increase the quantity, is what I understood. Now, for xenon or uranium, I cannot answer.

Antaya: I would like to come back to the question of the residual pressure again. I still think that 10^{-11} is too low, if even for hydrogen, because once you have ionized hydrogen, why is it a problem in the discharge?

Faure: The problem is that we want to maximize the number of ions for injection, so if you have 50% proton and 50% argon, few ions can reach the synchrotron. Regarding recombinations, I do not know.

Antaya: What is the total ion density inside the discharge?

Faure: It is not a discharge. It is not a plasma.

Antaya: Inside the electron column.

Faure: How do we measure this density is what we have to say. We know from the time-of-flight the relative quantities in each of the species. We come back to the curve I showed previously and we say this is the way we measure the density. So it is an effective density. I do not know if it is the real density.

Antaya: You quote numbers like 10^8 particles per second or per pulse coming out. That is an effective pressure like 10^{-8} Torr, or something like that. Still, why do you need 3 orders of magnitude lower for residuals?

Faure: We hope to have only one species inside the source. We do not want to do atomic physics and to say what happened or what we do with these species. We want to inject in the synchrotron. Therefore, we want to get as many ions as we can. Maybe another reason is recombination and loss of high charges, I do not know.

Becker: If you have an EBIS, you have negative charges by the electron beam and you can accumulate as many positive charges as you have negative ones. Unfortunately, we do not know what is happening with the secondaries. Therefore, sometimes the secondaries help, and sometimes they are starting to kill the operation because they cause instabilities. So, a reasonable upper limit for an EBIS is to accumulate about as many positive charges as negative charges. Of course you do not want residual gas atoms participating in the compensation of the electron beam.

Levine: I would still think that the only restriction is the charge exchange restriction that the ion-ion collisions between the protons and your highly ionized gas, say Ar^{18+}, are rapid enough so that the hydrogen is heated to high enough temperatures so it can no longer be contained in your system and will be lost from the system almost as rapidly as you make it.

Faure: When we look with your time-of-flight spectrometer, if we inject argon, for instance, when the vacuum is good enough, we see Ar^{16+} with other charge states. When the vacuum is not very good we have H^+ nitrogen, or carbon, or fully stripped light ions.

Levine: We have done similar time-of-flight and what we have observed in our system is that, as you turn it on, the amount of hydrogen increases quite rapidly. Then it decreases and you see ionization of the N_2 or O_2 or so forth, and then that decreases. Finally, we get barium and we go up to Ba^{46+}. But what happens in our system is that the high charge states pump out low charge states and you end up with the high charge states.

Becker: I expect that this will be the topic of your talk and we are looking forward to it. I think that this is something which is newly coming into the field and that you can have a mixture of gases to cool the heavy species. But, of course, you will not expect as many ions due to the participation of lower masses in the negative container.

Antaya: When the vacuum is bad, what sort of pressure is that?

Faure: I do not know. We do not measure the pressure inside the EBIS. That occurs, for instance, when we have something going wrong inside, like sparking or some other problem. We come back on the air after a few hours, but during this time the experimental physicists are not happy. This is the reason why we try to have as good a vacuum as we can.

Becker: I think Jean Faure has give a nice overview about EBIS and I thank you very much.

PROGRESS REPORT ON THE FRANKFURT EBIS[*]

M. Kleinod, R. Becker, and H. Klein
Institut für Angewandte Physik der Universität Frankfurt am Main
D-6000 Frankfurt am Main 1, FRG

ABSTRACT

The Frankfurt superconducting EBIS has been operated mainly in the dc mode with continuous ion extraction, yielding ion beams of excellent quality with charge states up to Ne^{10+}, Ar^{16+}, Kr^{26+} and Xe^{36+}. These results are achieved by bombarding the ions during their single passage through the source with high power electron beams Brillouin focused to densities of about 1000 A/cm^2.

Simpler ways to obtain these high densities were examined, running the superconducting EBIS with a hairpin gun in an unusual cathode operation or omitting the solenoid by employing periodic focusing with rare earth permanent magnets. An approach without the use of any focusing field is in preparation resembling the welding gun operation, allowing to extend the beam energies for the production of extremely high charge states. The design of an improved analyzer for internal and external experiments for ionization and charge exchange measurements is presented.

INTRODUCTION

Any electron beam ion source development is charaterized by the careful design and construction of electron guns as well as focusing magnetic fields. The high accuracy in the winding of our 5 T superconducting solenoid and the ability to adjust the source structure relative to the magnetic field axis have proven to be essential. The successful operation of the Frankfurt EBIS is to an important part the result of a high standard being reached in the use of electron beam optics computer codes including magnetic fields[1]. This experience being demonstrated here in the design and the development of a variety of electron guns or focusing devices for EBIS use has also resulted in the construction of similar electron beam devices like an electron target for the study of dielectronic recombination in a merged beams experiment at the GSI[2].

[*] Supported by the BMFT, FKZ 06 OF 851 I

THE SUPERCONDUCTING EBIS

The emphasis of our work on the superconducting EBIS in the last years has been laid on the continuous operation of the electron as well as the ion beam, since in this mode the ion beam is matched better to the demands of the detectors in most of the atomic physics experiments than in the pulsed containment mode. Since the ions have to be ionized during their single passage through the source, the electron beam density has to be of the order of 1000 A/cm^2 to shift the maximum of the charge state distribution to e.g. Ne^{8+} for the production of bare nuclei of neon^3. This has been achieved now routinely, delivering highly charged ion beams of high quality - defined by two slits of 0.5 mm separated by 0.6 m - with rates reaching up to a few 10^5 s^{-1}. With an electron beam of 3 keV/80 mA, e.g. Ar^{16+} has been produced, with the charge state spectrum peaking at Ar^{10+}. For Kr and Xe the spectra peak at higher charge states of about 15+ to 20+, Kr^{26+} and Xe^{36+} clearly being detected (s. fig. 1). The applied electron energy was not sufficient to expell e.g. the 1 s-electrons of Ar. Unfortunately the necessary increase to more than 5 keV was of no success, since the working conditions of the source were changed. With the given perveance of $2 \cdot 10^{-6}$ A/V$^{3/2}$ of the electron gun, the increase in beam power led to intolerable desorption in the collector region, while also the loss currents to the ion source structure exceeded the normally achieved level of about 1 µA.

To extend to higher energies, a new electron gun with adjustable beam properties is being developed. A second anode will allow to vary the perveance, thus decreasing the load to the collector. To fulfil Brillouin focusing conditions, new bucking coils and soft iron shields will be employed to keep the cathode free of flux. To get high accuracy near the cathode improved calculations of the magnetic field distributions were performed by the use of conformal mapping^4 in combination with the TRIM code^5. Beam formation was calculated using the magnetic field data in the SLAC electron trajectory code^6, taking special care for sufficient mesh resolution in splitting up the computations for the gun region and the high compressed beam area.

A rather simple approach to achieve high density Brillouin beams at quite low electron currents of a few mA had been shown by us with our first normal conducting TOFEBIS^7, delivering a few 10^{-8} A of moderately charged Ar and Xe ions. It uses an unusual cathode process, where the emission density is increased orders of magnitude by lowering the work function through backstreaming residual ions in the gun

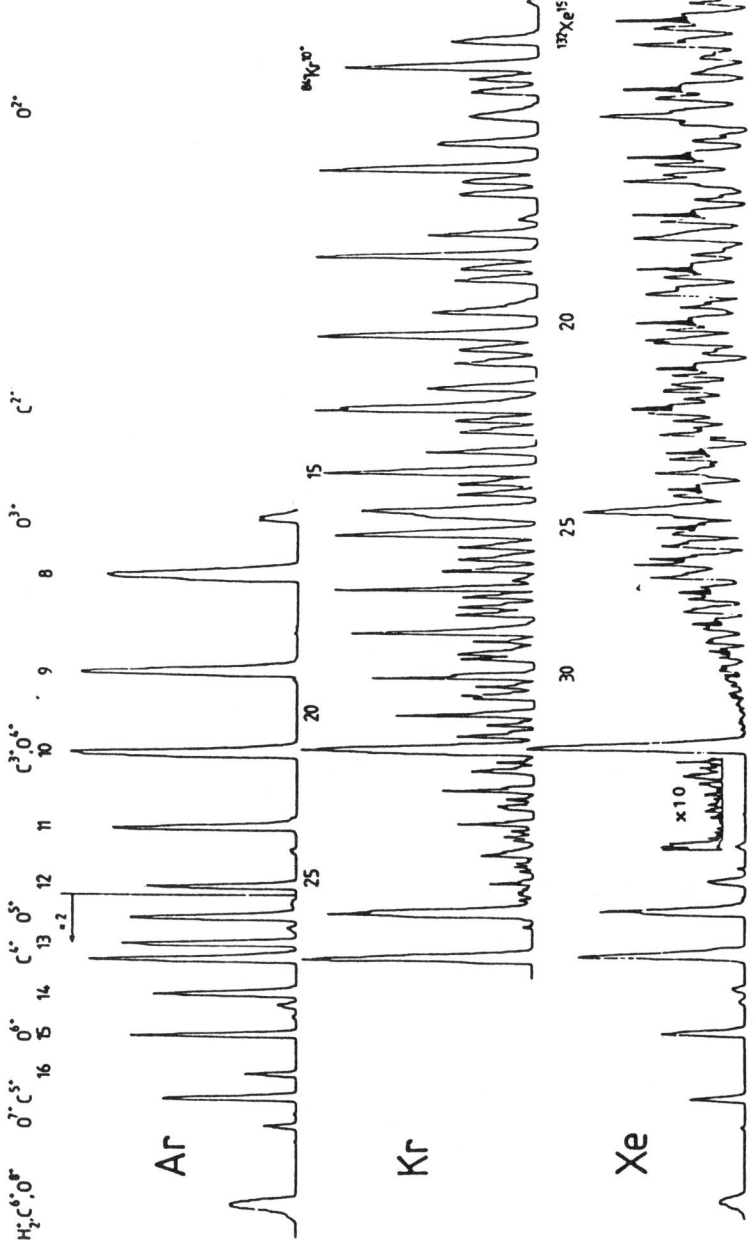

Fig. 1. Ar-, Kr-, and Xe-spectra obtained in continuous extraction mode with the Frankfurt EBIS ($E_e = 3$ keV, $I_e = 90$ mA, $j_e \sim 1000$ A/cm^2)

region. At specific parameters of pressure, magnetic field, voltage, current and dimensions these ions get focused to a spot, resulting in current densities of more than 10^3 A/cm^2. When launching inside an axial magnetic field, the beam will only blow up to the ideal Brillouin value even at remarkable field strengths at the cathode, since the spot is negligible compared to the focused beam cross-section, containing almost no flux.

Since high charge states had suffered from the bad vacuum conditions in the long small diameter tube of the warm solenoid of the TOFEBIS, it seemed worthwhile to test this cathode behaviour in our cryopumped EBIS by simply exchanging the electron gun. With the same working conditions as in the TOFEBIS (B = 0.4 T, E_e = 1.3 keV), the source could easily be operated to produce spectra peaked at Ne^{5+}, confirming the formerly observed Brillouin current density of ~ 240 A/cm^2. At higher magnetic fields, the spectra didn't improve much further, necessitating the redesign of other parameters like gas feed, electrode structure etc. Nevertheless it seems to us, that because of the simplicity of the electron gun a very dense electron beam at very high energy could be reached using this kind of operation. The low electron current releases the collector problem and results in a low potential depression in the thin electron beam, thus the beams of highly charged ions should be of excellent emittance.

PPM-FOCUSED EBIS

The high complexity of the superconducting EBIS is to some kind prohibitive for its use at small laboratories. Hence there is a need to study more simple approaches, which omit the superconducting solenoid. Using the periodic permanent magnet focusing principle, a test device for a dense electron beam of 20 cm length had been constructed consisting of 80 SmCo$_5$-rings of 1.6 mm inner diameter[8]. Spacing, thickness, and outer diameter were optimized to 0.8, 1.6 and 8 mm resp. for a maximum axial field of 0.59 T as well as sufficient vacuum transparency. A magnetically shielded converging gun (cathode diameter = 17.4 mm) with a perveance of 1.7 · 10^6 A/V$^{3/2}$ launched the electrons into the focusing structure on a distance of 20.5 mm (s. fig. 2). The final beam diameter of 0.53 mm is reached at 6 keV, which should correspond to a calculated current density of 560 A/cm^2.

The measurements were carried out with a pulsed power supply, the voltage decreased from 7 to 1.5 kV in 14 msec. The design values were reached, fig. 3 shows the beam transmission during the discharge,

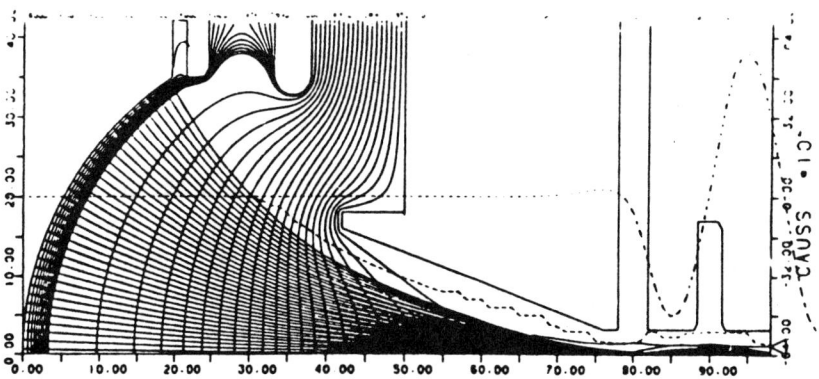

Fig. 2. Matching of the shielded electron gun to the periodic permanent magnet structure

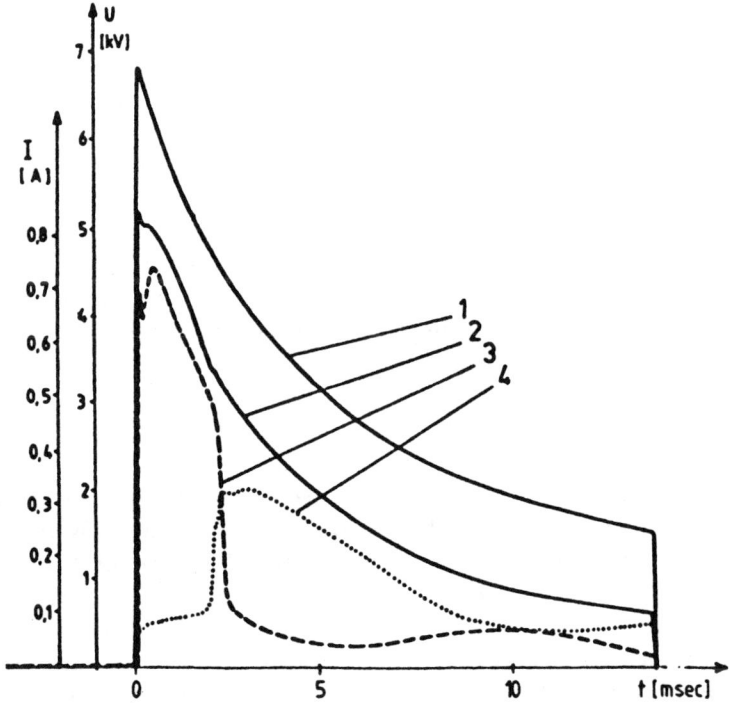

Fig. 3. Pulsed electron beam transmission test of the focusing structure
1 - anode voltage, 2 - total current, 3 - transported beam,
4 - beam losses to structure

the passband with losses of only 6 % ranging from 6.3 to 4.8 kV. Below this voltage the transported current decreases rapidly as expected from periodic focusing theory.

With these encouraging results, a complete ion source is now under development, where the injection of the electron beam into the focusing field will be optimized for a better fit to the periodic passbands and a dc operation of the electron beam as well as ion extraction are planned.

CROSS-OVER EBIS

An even more simple device is a source, where the electron beam is not focused by any magnetic fields but only by kinematics. It consists of only a short high density region near the beam cross-over, - we therefore call it a "cross-over" EBIS - , but the extracted ion currents can easily reach values of µA. By chosing the beam energy to a few hundred keV as in welding guns, the repelling forces of the electrons will decrease, and assisted by space charge compensation with ions, one can expect interesting source length of about 20 cm with current densities of more than 10^3 A/cm^2.

To show the principle an experiment using an electron gun of perveance $0.1 \cdot 10^6$ A/V$^{3/2}$ with a 4 mm cathode diameter (spherical radius = 16.6 mm) is in preparation, where a 20 kV/300 mA beam is launched to form a smooth cross-over[9]. The high density region is calculated to have a length of 37 mm with at least 10 A/cm^2 (40 A/cm^2 maximum). With compensation by ions, the maximum current density increases to 140 A/cm^2. Ion extraction will be pulsed at variable frequency to observe the evolution of the charge state distribution as a function of containment time.

MULTIPASSAGE MAGNETIC SPECTROMETER

Our magnetic spectrometer will be extended to a multipassage system, each deflection angle being 90^0 at each passage. By positioning four electrostatic mirrors, the ions extracted from the EBIS can thus travel the four passages several times, improving the resolution of the analysis. The well-defined ions in charge state and energy can also be re-injected into the emptied EBIS, e.g. allowing to investigate the ionization of highly charged ions in the electron target of the EBIS of known radial profile when using an immersed gun. The high density beam will lead to sufficiently high reaction rates even at extremely high charge states, allowing to decompose the effective cross-sections of Donets[10] into its individual contributions. Finally, the injection of metal ions from the microwave ion source to be installed can also be handled easily with this switchyard.

The magnetic chamber consisting of the electrostatic mirrors and lenses was calculated using the SLAC electron trajectory code. Special care was taken for the accumulative action of aberrations. Ions starting with different energies will arrive at different transverse positions after the passage of one quarter of the analyzer. This will be counteracted by a lens, which influences the trajectories in such a way, that the ions seem to come from the center of the magnet (s. fig. 4). An additional electrostatic lens in front of each mirror will focus the ions to its center to reduce the influence of aberrations in the region, where the ions are slow. Fig. 5 shows the combined action of lens and mirror on ions with different energy or angle.

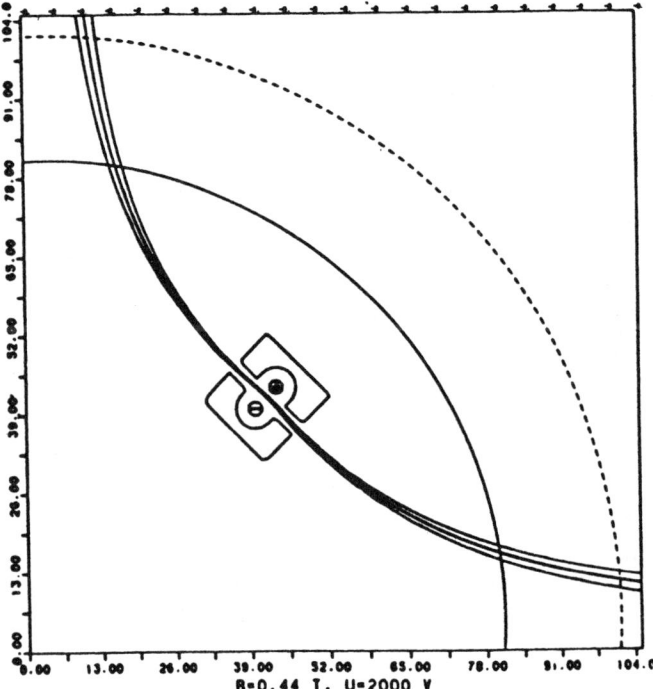

Fig. 4. Correction lens for adjustable reduction of the dispersive action of the analyzing magnet
(ion trajectories with $\Delta E_i/E_i$ = -1, 0, +1 %, starting at 3 different angles each)

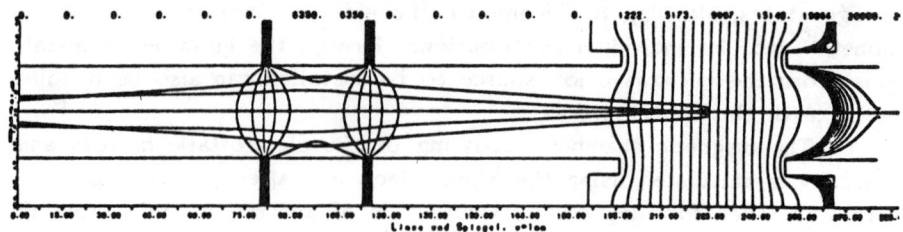

Fig. 5. Electrostatically focused and reflected ion beams in one mirror of the multipassage magnetic analyser

CONCLUSIONS AND OUTLOOK

The EBIS development has by far not reached saturation, which is expressed by the rising number of new superconducting EBIS sources. On the other hand the construction of more simple devices, which sometimes renounce on extremely high charge states or yield, but compare well with the charge states from ECR sources must not be lost from sight. Promising results can be expected from devices using the focusing action of periodic permanent magnets or even only the extended cross-over of an electron gun.

Continuous ion beams of lower rates but excellent quality were extracted from our EBIS with charge states up to Ar^{16+}, Kr^{26+}, and Xe^{36+} using a 3 keV/80 mA beam. Higher dc rates should be possible by splitting up the source into an ionization region and into a container for slow quasi-continuous extraction.

The first EBIS was constructed to work with a cyclic accelerator. This symposium at BNL offers the excellent possibility for a combined effort for the construction of a highly developed EBIS for such a machine.

ACKNOWLEDGEMENTS

We would like to thank Elke Jennewein for her assistance in calculating the improved EBIS electron gun and Irmgard Langbein for the calculation and design of the multipassage magnetic spectrometer.

REFERENCES

1. R. Becker, this conference
2. R. Becker et al., Proc. Europ. Part. Acc. Conf., Rome, Italy, June 6 - 11, 1988
3. R. Becker, M. Kleinod, H. Klein, Nucl. Instr. Meth., B 24/25, 838 (1987)
4. G. Barzik, diploma thesis, Inst. f. Angew. Physik, Universität Frankfurt, (1983) unpublished
5. A.M. Winslow, J. Comp. Phys. 2, 194 (1972)
6. W.B. Herrmannsfeldt, SLAC-226 (1979)
7. M. Kleinod, R. Becker, H. Klein, W. Schmidt, IEEE-NS-23, 1023 (1976)
8. R. Bikowski, diploma thesis, Inst. f. Angew. Physik, Universität Frankfurt (1985), unpublished
9. A. Fallah, diploma thesis, Inst. f. Angew. Physik, Universität Frankfurt (1988), unpublished
10. E.D. Donets et al., JINR R7-12095 (1979)

DISCUSSION

Antaya: In the crossover EBIS, what is the electron compression mechanism?

Kleinod: It is by kinematics. You start with a spherical cathode and you pull the electrons from the cathode through the hole of an anode by an electrostatic field.

Antaya: In your last transparency you quoted an expectation of 10^9 particles per pulse for Pb^{54+}. Why do you quote such numbers when other numbers I was hearing for EBIS sources this morning was sort of 10^7 particles? What makes you think that you can reach it?

Kleinod: Simple. We talk about 10^9 particles all the time. This is in containment mode, of course. Sorry if I did not point this out carefully enough. This is an accelerator asking only every second for a pulse and we have enough time.

Faure: About the number of particles, if you take 54 and multiply by 10^9, and multiply by 3.7, you have a little bit more than 10^{11}. In DIONE we do 10^{10} so we are within order of magnitude.

Kleinod: It is a demand that we show that an EBIS really runs like we always talk about and that we are using the electron container completely. And one has to be sure that the cross sections are OK.

Faure: I want to say that on the side of the emittance, if you decrease the electron current density too much, you increase the emittance and you can lose on the acceptance of any machine.

Becker: But we have only 1 kG; that is a very low magnetic field.

Faure: No, you have two contributions to emittance from the magnetic field and from the energy of the electron inside the electron beam. I was speaking of the last one.

Beebe: On one of your earlier viewgraphs, you showed the perveance of the electron gun. I think you said it was 2 μperv. What was the actual voltage on the gun and what was the energy of the electron beam inside the trap region?

Kleinod: In the superconducting source, it is 3 kV.

Beebe: That 3 kV, is that the beam energy inside the trap?

Becker: Yes.

Beebe: OK, what is the actual voltage on the gun?

Kleinod: The gun is operated at −1 kV below ground potential and the ionization region is at 2 kV, so altogether it is 3 kV. We have the possibility to adjust a little bit with our anode which was operated at a few hundred volts above ground potential. We are restricted to ground potential because our iron shields are on ground potential where we have to go through, coming from the cathode.

Beebe: Are you running the gun in the space charge limit then?

Becker: Of course.

Beebe: Another question I have is on the calculation that you showed for the compensated beam current density. How is that calculation made? How/where do you compensate?
Kleinod: We compensate for the space charge.
Becker: We have a very thorough method of introducing compensation by looking on the actual distribution of the potential. Whenever we find a potential valley, then we start to eliminate the space charge, but not right to the border of the valley, just one mesh less. Then through the cycles of calculation, a very stable procedure develops to take compensation into account. We have had two experiments where we could state that this procedure is good — one is a famous ionizer which is used in Giessen for crossed beam at atomic physics experiments and the other one was a special apparatus of Elke Jennewein having a 60% depression.
Beebe: OK, so then you are compensating the beam only in a region where you know the ions are?
Becker: No, it does not take into account the temperature of the compensating species. It is a compensation by very cold ions which is the high pressure limit.
Beebe: In other words, you are not making an approximation back at the gun. You are doing it right in the region of interest.
Becker: Oh, yes. Through influences you get more current because the depression goes up.
Beebe: So you are actually starting with ions in one region and not after many interations?
Becker: No, the ions are not treated realistically, but artificially. You just fill up potential valleys on axis.
Marrs: I did not understand what you said about the hairpin cathode immersed in the magnetic field. Did you say that you had something like 200 A/cm^2 without any compression of the beam?
Kleinod: At the cathode we expect a much higher current density due to lowering of the work function by the backstreaming ions to a spot. The spot is necessary for the focussing conditions here. That is not easy to be reached and, obviously, this time we were not so lucky. We did not change anything else to put a hairpin gun in there. With the TOFEBIS we had the possibility of adjusting all geometrical dimensions, gas feed and so on, which is really necessary to reach this very difficult, stable situation where you focus your residual gas ions or the ions you want to ionize back to cathode on one spot. If you reach that, the beam will explode. In this case it will only expand to the diameter which is given by the Brillouin focussing conditions.
Marrs: And the work function is lower because there is this positive charge of ions right next to the surface of the electron sample?
Becker: There is no explanation on how the work function is lowered. However, you can imagine that you no longer have the surface of tantalum or tungsten, but a mixture of cathode material atoms and of the back focussed ions.

Kleinod: This has been observed by others and can be found in literature.
Stockli: What is the gap between the magnets of your periodic focussed EBIS?
Kleinod: It is 0.8 mm.
Stockli: And is the bore 0.6 mm?
Becker: The bore is 0.4 mm in radius.
Stockli: Do you have any idea how uniform these magnets are?
Becker: Yes, quite uniform. We tried to measure them from the outside because it is difficult to measure them from the inside. They are reasonably uniform. And for us this was a test: Can we just order magnets made to industry precision (which means grinding to 0.01 mm and magnetized in a standard way) and put them into grooves made in our workshop with precision limits of 0.02 mm and does this work in focussing? It turned out, it works.
Stockli: Were these custom-made magnets which were first ground and then magnetized?
Becker: Yes.

STATUS REPORT ON THE STOCKHOLM CRYOGENIC ELECTRON BEAM ION SOURCE

L. Liljeby and Å. Engström
Manne Siegbahn Institute, S-104 05 Stockholm, Sweden

INTRODUCTION

The EBIS project in Stockholm started as a collaboration between MSI and IPN in Orsay, France. Two almost identical cryogenic EBIS sources were constructed at IPN: CRYEBIS II for IPN and CRYSIS for MSI. The main difference between the two sources was that CRYEBIS II was equipped with a 50 keV electron gun as compared to 10 keV for CRYSIS. Both sources were planned to be dedicated to atomic physics experiments but during the construction plans to use CRYSIS as an injector for a storage ring evolved. Both sources were completed in the beginning of 1984 and after initial tests, CRYSIS was moved to MSI in December 1984 and installed in a temporary laboratory.

DESCRIPTION OF CRYSIS

CRYSIS consists of a 1.66m long superconductive solenoid which is cooled by a KPS 1430 He liquefier connected in a closed circuit. Thermal insulation is achieved by two screens, one at 20 k and the other at 80 K, are cooled by return gas from the cryostat. The Helium consumption is 3 liters per hour. An external electron gun produces an electron beam which is compressed electrostatically and magnetically and passed through a system of 33 cylindrical tube electrodes (9 mm inner diameter) inside the cold bore of the solenoid. After expansion in the fringe field the electron is stopped by a water cooled collector. The potential distribution on the tube electrodes is produced by 5 fast voltage amplifiers (5 kV) and a resistive divider chain. The whole source including liquefier and control electronics is mounted on a 50 kV platform to allow it to be used as a injector to CRYRING.

STATUS OF THE PROJECT

When the CRYEBIS project in Orsay was terminated in the end of 1987, the electron gun with power supplies was moved to Stockholm and installed on CRYSIS. This gun (NC4) has a smaller cathode, 12 mm, a perviance of 0.45 uP and allows post acceleration of the electron beam up to 50 KeV. Nc4 proved to be an improvement over NC1, giving higher current densities and smaller losses on the electrode system. With this gun we were able to extend the charge state range up to 18+ for Ar (see fig. 1) and 44+ for Xenon using electron energies up to 12 KeV. The electron beam has successfully been retarded down to 500 eV before collection.

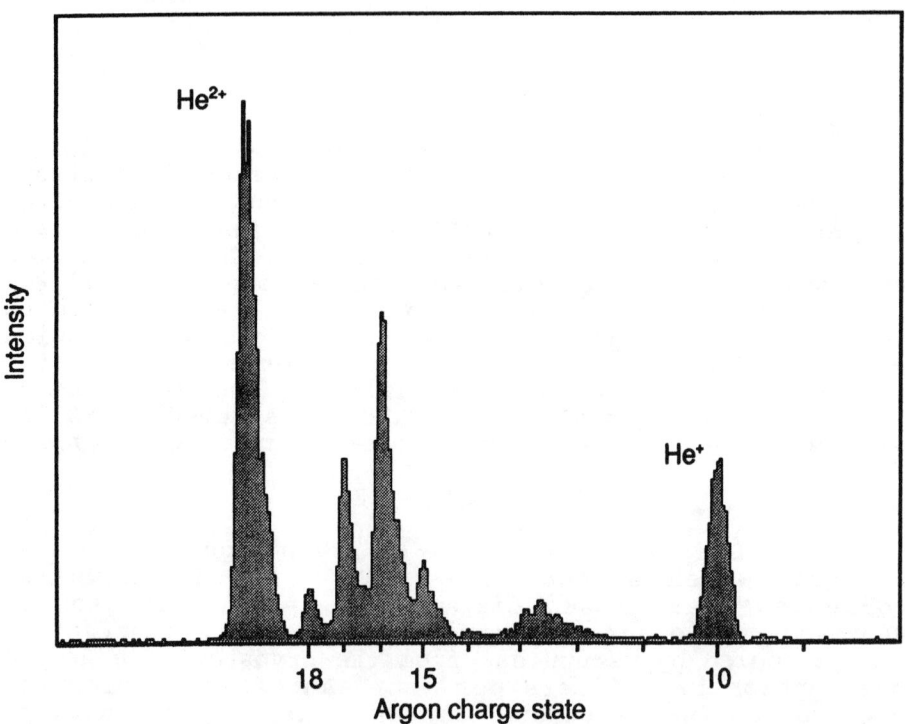

Figure 1. Time-of-flight spectrum for Argon. E_e=12kV, t=1s

Table 1 shows a summary of achieved source parameters compared with design values.

Table 1. Source parameters

	Achieved	Design value
Electron energy (KeV)	12	50
Electron current (A)	0.1	3.5
Electron density (A/cm^2)	100	1000-10000
Magnetic field (T) (during ion production)	1-3	5
Intensity (Charges/pulse)	3×10^9	3×10^{11}
Pulse length (ms)	.05-20	
Emittance (π.mm.mrad)	50	
Energy spread (q.eV)	8	

During 1987 and 1988 CRYSIS was used for low energy collision experiments[1,2] using Xe-beams with charge states up to 35. Sufficient beam intensities were obtained despite a very tight collimation (0.03°) and energy selection ($dE/E=10^{-4}$). The operation of the source has been very stable and reproducible.

Table 2. Beam intensities

Element	Charge state	Intensity (Charges)	Rep.rate. (Hz)
Neon	≤ 10	10^9	5
Argon	≤ 16	10^9	2-5
	17	2.10^8	1
	18	5.10^7	0.5
Xenon	≤ 35	5.10^8	3
	≤ 40	2.10^8	1
	≤ 44	a few	0.5

THE NEW CRYSIS LABORATORY

In the end of October 1988 we started moving CRYSIS to its final location at CRYRING. Figure 2 shows a layout of the new laboratory.

A small isotope separator will be connected to CRYSIS to be used for injection of ions using the scheme developed and successfully used in Saclay. This injection method has several advantages over the present gas injection:

1. The gas load on the system decreases leading to better vacuum (less recombination).
2. Increased range of elements (earlier limited to gaseous elements).
3. Isotopically pure injection which improves the intensity for some elements.

The injector is a traditional isotope separator using a uno-plasmatron ion source (similar to Danfysik 910). This type of source can produce ions of virtually any element using either injection of gas or solid materials evaporated in a oven. The ions are accelerated to 30 q Kev and mass selected by a doubly focussing 50 cm radius, 90° sector magnet. before entering CRYSIS the ions are retarded so that they will have an energy of 10-20 eV when they are trapped. The injector is designed with several differential pumping stages reducing the pressure, from $\sim 2.10^{-6}$ torr near its ion source, down to $\sim 10^{-9}$ in order to minimize pollution of CRYSIS. The mass resolution of separator is >600.

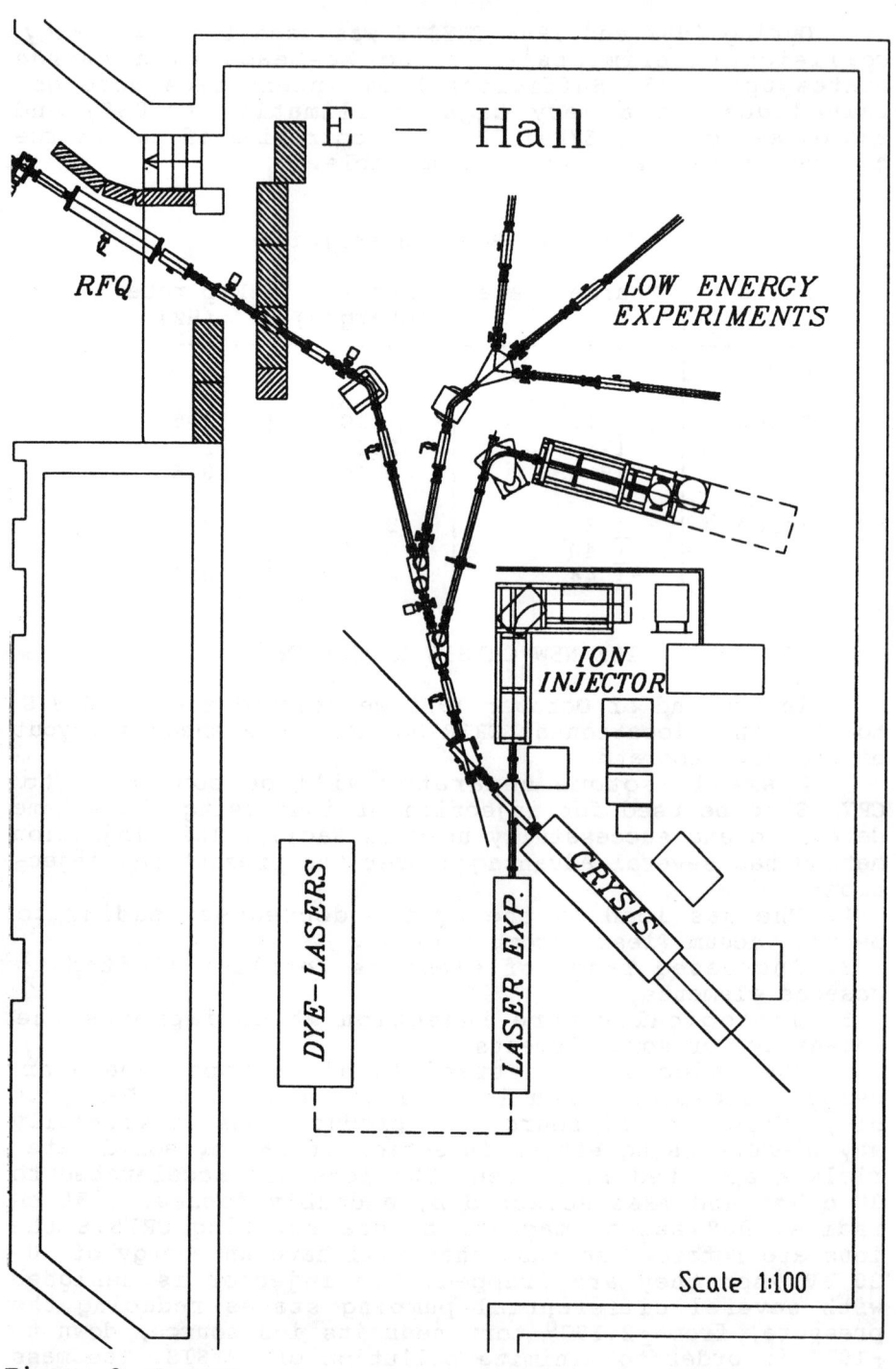

Figure 2. Layout

Four beam lines for experiments using low energy (0-50 q KeV) CRYSIS beams are being built. One of them is designed for high energy resolution using a doubly focussing 50 cm radius analyzing magnet. The other three will have moderate energy definition but higher transmission and will be used mainly for spectroscopy and surface physics.

As indicated in figure 2, the ion injector will also deliver beams of low charge states (energy ≤ 80 q Kev) for laser experiments.

CRYSIS and the high resolution beamline should be in operation spring or early summer 1989.

REFERENCES

1. H. Andersson, G. Astner and H Cederquist,
   J. Phys. B 21(1988)L187

2. H. Cederquist, H. Andersson, G. Astner, P. Hvelplund and J.O.P. Pedersen,
   submitted to Phys. Rev. Lett.

DISCUSSION

<u>Stockli</u>: Did your calculation of the magnetic field, which you showed, include the field of bucking coil?
<u>Liljeby</u>: Yes. It is included in our calculations. The sketch I showed you was without any current through the coil, but we have been varying that also and trying to adjust it.
<u>Stockli</u>: Did you try to shape the magnetic field with magnetic shims?
<u>Liljeby</u>: We have not tried that. We have not had time since these calculations were done to make any modifications on the source. But that is an alternative for sure.

PERFORMANCE OF MINI-EBIS COOLED BY LIQUID NITROGEN

Kazuhiko Okuno
Department of Physics, Tokyo Metropolitan University,
2-1-1 Fukasawa, Setagaya-ku, Tokyo 158, Japan

ABSTRACT

A MINI-EBIS cooled by liquid nitrogen has been developed for the low energy collision experiments. An idea of cooling magnetic solenoid coil in liquid nitrogen can was very effective not only for size down of the solenoid coil and its power supply but also for the achievement of very low pressure. All parts of electron gun, ion drift tube, magnetic solenoid coil, electron collector and ion extraction lens are housed in the vacuum envelope of 150 mmϕ diameter and 500 mm length. In the DC ion extraction mode with the electron current of 15 mA at 2 keV, beam intensities of C^{6+}, N^{7+}, O^{8+}, Ne^{9+} and Ar^{16+} were obtained at least of 5×10^4, 3×10^4, 1×10^4, 5×10^3 and 3×10^3 ions per second, respectively.

INTRODUCTION

The electron beam ion source (EBIS) and the electron cyclotron ion source (ECRIS) are the alternatives to produce multiply charged ions and their developments are progressed in aim at the production of fully stripped uranium ions. For the time being, EBIS is at an advantage for production of fully charged ions and ECRIS is at an advantage for production of intense beam of multiply charged ions.

According to the development of these ion sources, the collision experiments with highly and fully charged ions have been propagated extensively. However, these ion source devices are not so handy, or rather large scale and expensive, so most of such devices are located at some large laboratories.

Previously, we had constructed a classical EBIS "PROTO-NICE"[1,2] and a cryogenic EBIS "NICE-1"[3] in order to perform Naked Ion Collision Experiments at IPP of Nagoya University. Operating NICE-1 in a DC ion extraction mode, we investigated state selective electron capture processes of highly and fully charged ions at several keV x q energies, systematically. Such a mode, in which ion trapping potentials are supplied but not altered, favors to extract a narrow energy spread beam of only moderate intensities for the low energy collision experiments.

As the growth of interest in highly charged ions, the development of a low cost and compact ion source is urgently required. In this paper, a new trial of cooling the magnetic solenoid in liquid nitrogen can is tested to make EBIS compact. A NINI-EBIS developed for the atomic collision research is reported.

DESCRIPTION OF THE MINI-EBIS

The use of cryogenic superconducting magnet solenoid has made to improve the power of EBIS, while it brings rather the size-up and cost-up of EBIS until the high temperature superconducting magnet becomes utilizable. In EBIS, the strong magnetic field is not necessary for the production of highly and fully charged low-Z ions. In order to make EBIS compact, the size down of magnetic field device is most essential.

As is well known, when the copper coil is cooled down from the room temperature to the liquid nitrogen temperature, its electric resistance becomes small to about one tenth of the room temperature one. Usually, a conventional magnetic solenoid generates a great quantity of heat even at the operation of 1 kG. Without the cooling, the coil temperature raises up more with increasing the coil resistance. To cool the hot coil with the liquid nitrogen seems to be an impracticable plan since too amount of liquid nitrogen is consumed. However, the calorification of the solenoid coil cooled at 77 kelvin is very small and hence the low coil temperature can be kept with a little consumption of the liquid nitrogen. The simple idea of cooling the solenoid coil in a liquid nitrogen can is very effectual not only for the size down of the magnetic solenoid and its power supply but also for the achievement of very low pressure.

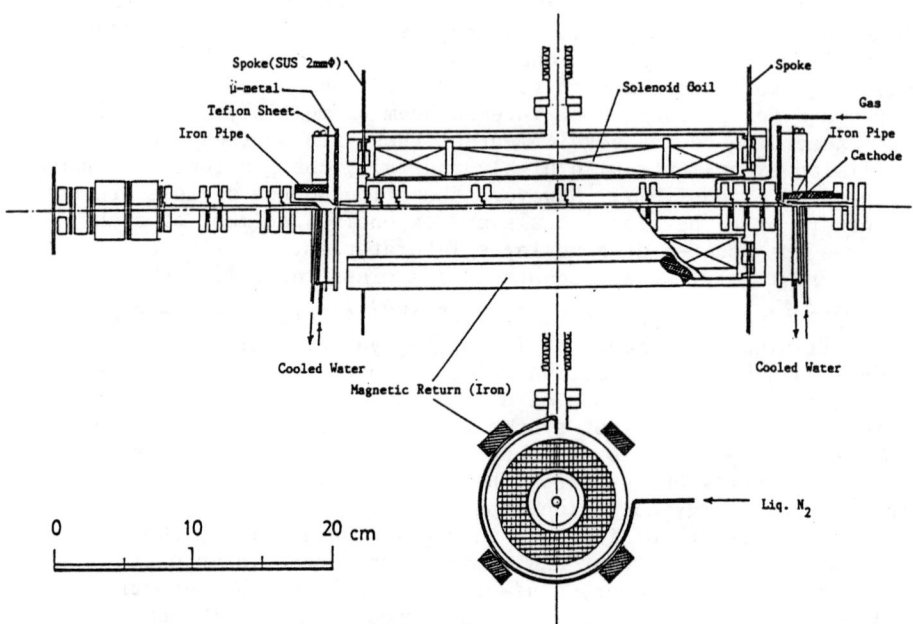

Fig.1. Schematic diagram of the MINI-EBIS.

The MINI-EBIS constructed is schematically shown in Fig.1. All parts of a magnetic solenoid, magnetic shields, an electron gun, ion drift tubes, an electron collector and an ion extraction lens system are housed in the vacuum envelope of 150 mmϕ diameter and 500 mm length. Parameters of the MINI-EBIS are summarized in Table 1.

1. Magnetic solenoid.

Solenoid coil of 252 mm length, of which Cu wire of 0.7 mmϕ diameter is wound by ten thousand turns into three sections, was set within the liquid nitrogen can of 40 mmϕ inner diameter, 100 mmϕ

Table 1. Parameters of the MINI-EBIS

1. Magnetic solenoid	
Length of solenoid	25.2 cm
Inner diameter of bobbin	40 mmϕ
Diameter of winding Cu wire	0.7 mmϕ
Total winding number	10000 turns
Electric resistance of coil	
at the room temperature	89.8 Ω
at 77 K	11.6 Ω
For generation of 1 kG magnetic field at 77 K	
Calorific value of coil	49.5 Watts/hour
Evaporation of liquid nitrogen	1.14 liters/hour
2. Electron gun and drift tube	
Species of cathode	BaO
Diameter of cathode	2 mmϕ
Indirect heating power	2 Watts
Anode	Water cooling
Anode aperture	2 mmϕ
Perveance	0.17-0.34 μ A/V$^{3/2}$
Species of anode, drift tube and electron collector	Al
Minimum inner diameter of drift tube	4 mmϕ
Collecting efficiency of electron collector	more than 95 %

outer diameter and 290 mm length. in order to cut off the thermal conduction from the vacuum envelope, both ends of the bobbin are hung by four spokes of 2 mmφ diameter. The position of the magnetic solenoid in the vacuum envelope can be controlled externally. The liquid nitrogen is automatically replenished from the cold trap of 8 liters reservoir to the solenoid can under the control by a level sensor in the solenoid can and its level can be kept at the desired one. Of course, the charge of liquid nitrogen to the cold trap is free at any times. The whole electric resistance of solenoid coil is 89.9 ohm at the room temperature and it is reduced to 11.6 ohm by the liquid nitrogen cooling. At the temperature of 77 kelvin, the heat power of the coil is slightly 49.5 watts per one hour for generation of 1 kG and requires the evaporation of only 1.14 liters liquid nitrogen one hour.

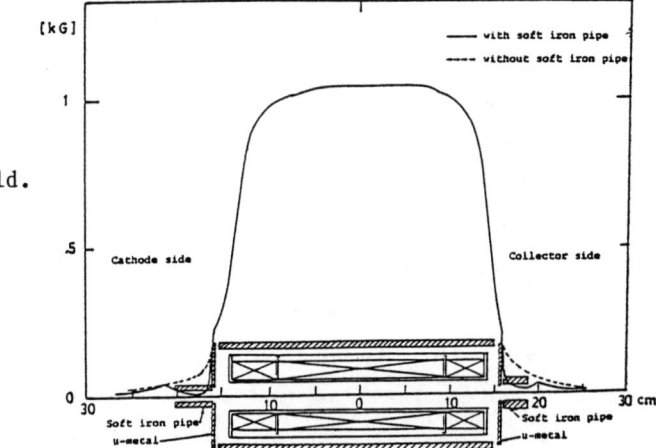

Fig.2. Magnetic field.

As shown in Fig.1, the solenoid has four magnetic return yoke of soft iron, and the electron gun and the electron collector are magnetically shielded by μ-metals and soft iron pipes. Shape of the magnetic field can be variable in the independent operation of three sectionalized coils as shown in Fig.2. The soft iron pipes act to minimize the field strength at electron cathode and collector and to make the sharp rising at both ends of magnetic field.

2. Electron gun and ion drift tube.

The electron gun is placed outside the solenoid induction and also shielded by μ-metal and the soft iron tube. The magnetic field strength at the cathode is less than 0.5 % of the maximum one. Barium oxide cathode of 2 mmφ diameter, which is commercial, is heated indirectly by the power of two watts. Anode with a 2 mmφ

aperture and electron collector are cooled by the water. All
material of anode, ion drift tubes, electron collector and ion
extraction lens are of aluminum. Minimum inner diameter of ion
drift tubes is 4 mmφ.

In the MINI-EBIS, the electron gun assembly gives a low
perviance of 0.17–0.34 $\mu A/V^{3/2}$. Electrons emitted from the cathode
are compressed by the rapid increasing of magnetic field and the
high current density beam is formed in the condition of the
brillouin flow. Lifetime of the cathode is longer than several
hundred hours in the low pressure operation at less than 1×10^{-8}
torr. The transmission rate of electron beam between the cathode
and the electron collector is always more than 95 %.

OPERATION OF THE MINI-EBIS

The MINI-EBIS is connected with the tandem mass spectrometer
having an octopole beam guide (OPIG)[4] placed at the middle as
shown in Fig.3 and tested. The MINI-EBIS is pumped by a turbo
molecular pump of 300 l/sec through the liquid nitrogen cold trap.
The ion source and the pumping system are constructed on a movable
box-type rack (50 x 50 cm^2 and 120 cm height) in which all control
power supplies can be set. Usually, the background pressure is
achieved of a few 10^{-9} torr when the magnetic solenoid is cooled by
the liquid nitrogen. Sample gases are directly introduced into the

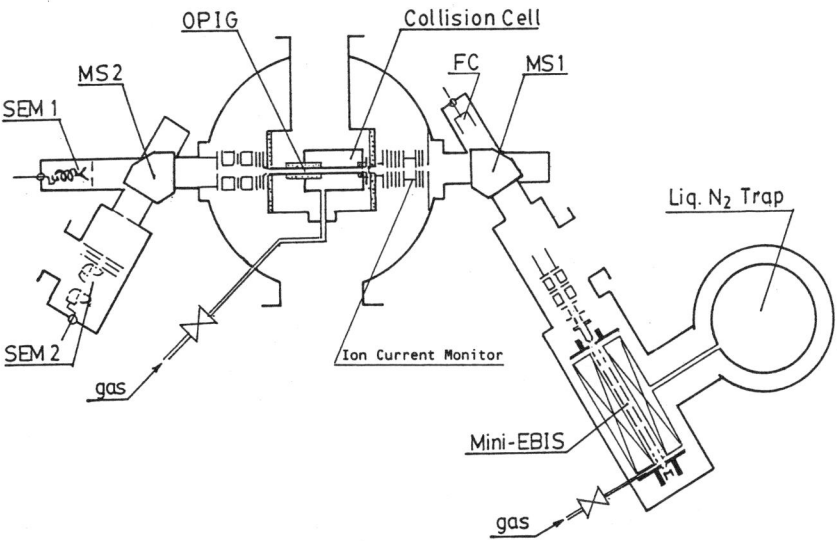

Fig.3. Schematic diagram of the OPIG tandem mass spectrometer.

ion drift region through a pneumatic control valve which is available for the pulsed and continuous small amount gas puff. The MINI-EBIS is operated at several 10^{-9} torr during the gas feeding.

The extracted ions from the ion drift region are reformed in the rectangle beam of 4 x 0.5 mm^2 by a doublet of electrostatic quadurupole lens. The total ion currents can be measured at the Faraday cup. After mass analyzed by MS1, most of ions are collected by the ion current monitor which is an entrance slit of the beam collimator with a small aperture of 0.5 mmφ. The small parts of ions pass through the collimator and are measured by the secondary electron multiplier at SEM-1 in Fig.3.

In the MINI-EBIS, the potential of the cathode is fixed at the ground and ion trapping potentials are applied at both ends of ion drift region, however, the potential barrier at an exit is very low and tuned at a few volts. Usually, sample gases are directly and continuously introduced into the ionization region and ion extraction from the ionization region is performed in the DC mode. The magnetic field of about 1 kG is applied for the electron beam of 15 mA at 2 keV.

In Fig.4, typical mass spectrum of background ions extracted from the MINI-EBIS at 5.8×10^{-9} Torr is shown. Electron beam is of 15 mA at 2 keV. This spectrum is measured by SEM-1 and its current scale are converted into that at the position of the ion current monitor. Hydrogen like ions of C, N and O can be observed clearly. Background ions seem to be caused mainly from the CO absorbed at

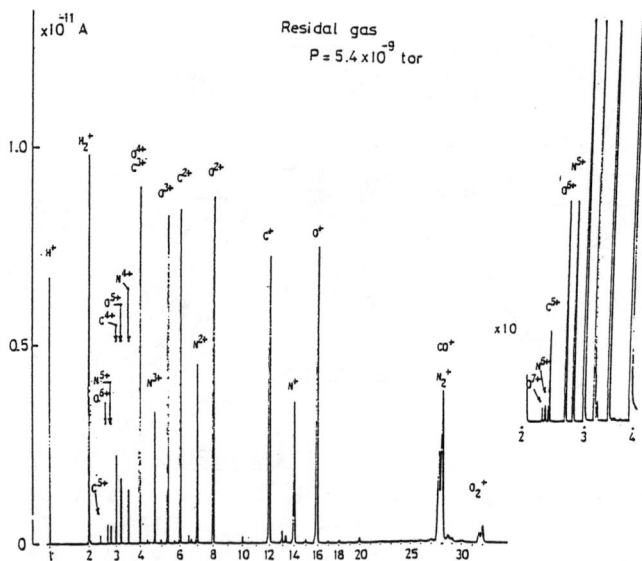

Fig.4. Background ion spectrum by 15 mA electron beam at 2 keV.

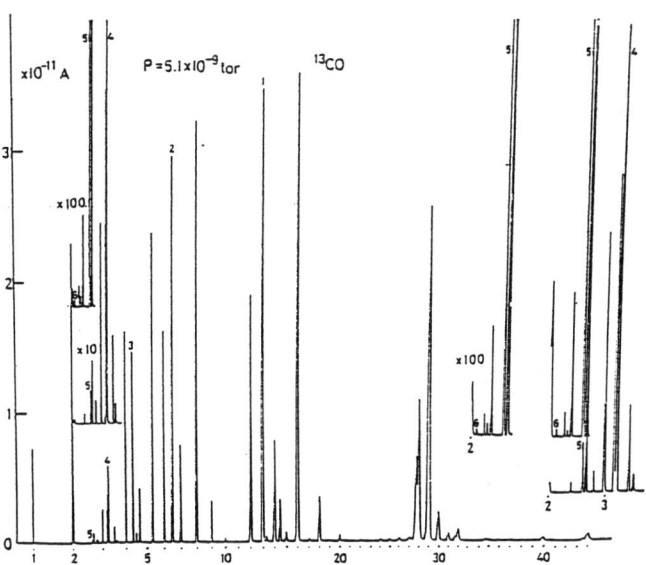

Fig.5. Mass spectra of $^{13}C^{q+}$ ions by electron current of 15 mA at 2 keV.

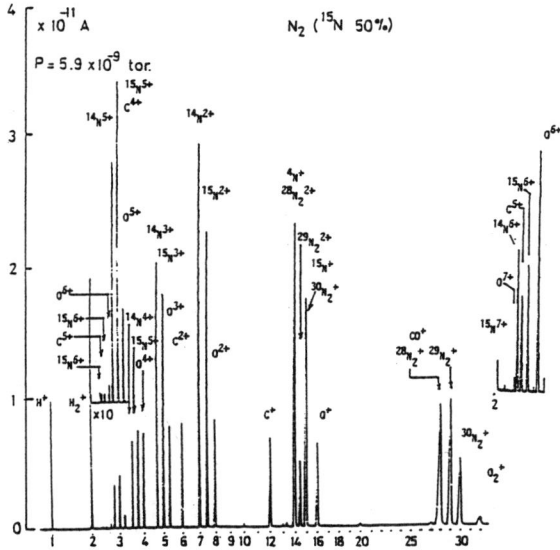

Fig.6. Mass spectra of $^{15}N^{q+}$ ions by electron current of 15 mA at 2 keV.

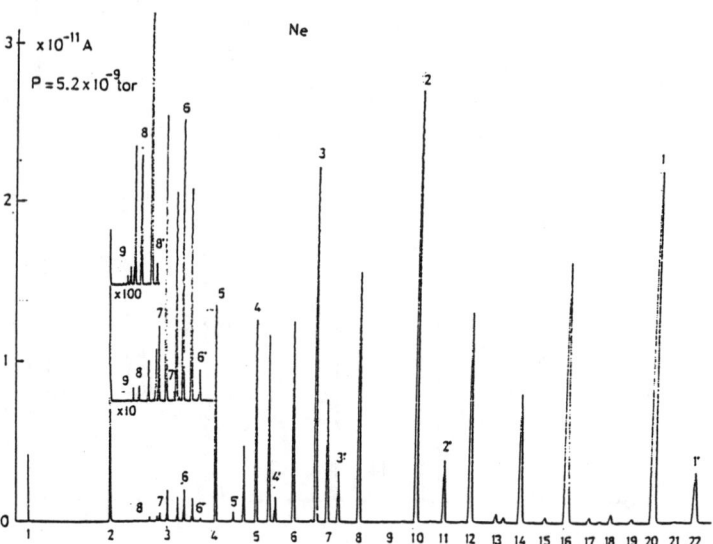

Fig.7. Mass spectra of Ne^{q+} ions by electron current of 15 mA at 2 keV.

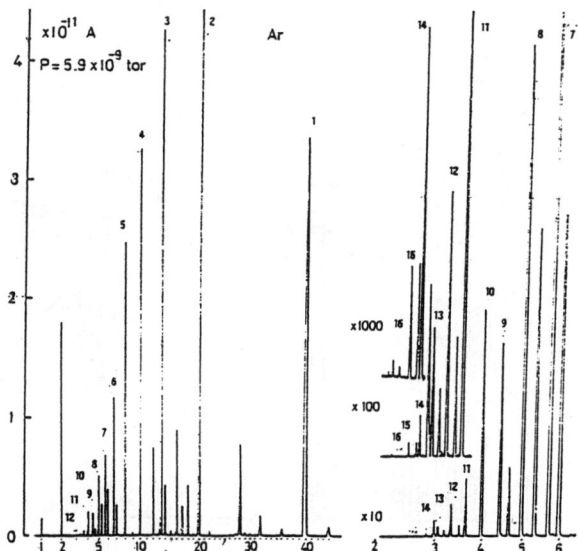

Fig.8. Mass spectra of Ar^{q+} ions by electron current of 15 mA at 2 keV.

the anode and the electron collector due to using BaO cathode and also from N_2 and O_2 due to the small vacuum leak.

Mass spectra shown in Figs.5, 6, 7, and 8 are obtained by using the electron beam of 15 mA at 2 keV for CO (^{13}C rich CO gases), N_2 (50 % ^{15}N in atomic percent), Ne and Ar gases, respectively. As seen in these spectra, peak shapes of highly charged ions are very sharp, but their charge distributions are not so good and their intensities are not so high. It agrees with our previous experiences in the NICE experiments that the DC ion extraction mode yields a narrow energy spread but only moderate intensities.

Obtainable intensities of fully charged ions of C^{6+}, N^{7+} and O^{8+} are at least 5×10^4, 4×10^4 and 1×10^4 ions per second, respectively. The highest charge state ions of Ne^{9+} and Ar^{16+} are obtained of 5×10^3 and 4×10^3 ions per second, respectively. Even these intensities can be used enough for the low energy collision experiments. These performance of the MINI-EBIS compares almost favorably with that of the cryogenic EBIS "NICE-1".

CHARGE DISTRIBUTION

Charge distribution of multiply charged ions produced in EBIS can be described by means of a balance model. If multiply charged ions are produced by only single ionization processes due to successive electron impact, the number density n_q of ions in charge q follows the time-dependence

$$\frac{dn_q}{dt} = J_e n_{q-1} \sigma_{q-1,q} - J_e n_q \sigma_{q,q+1} - \frac{n_q}{t_q} \qquad (1)$$

where J_e is electron current density, $\sigma_{q,q+1}$ is cross section for ionization from charge state q to q+1 and t_q is mean ion confinement time for ions in charge state q. With increase the total charge $\Sigma q\, n_q$ of produced ions, ion trapping effect by the space charge of electron beam reduces. When the total charge excesses the electron density n_e, ions diffuse from the ion trapping region. The limit of plasma quasineutrality is

$$n_e = \sum_{q=1}^{z} q\, n_q . \qquad (2)$$

If charge distributions of ions extracted from the ionization region accord with balanced solutions of coupled differential equations

(1), we can estimate effective diffusion coefficients $\tau_q = J_e t_q$ for respective ionic charge q in steady state of $dn_q/dt = 0$ from the relation

$$\tau_q = \frac{1.6 \times 10^{-19} n_q}{n_{q-1}\sigma_{q-1,q} - n_q\sigma_{q,q+1}} \quad [\text{A cm}^{-2} \text{ s}] \quad (3)$$

using the experimental values of n_q and cross sections $\sigma_{q,q+1}$ by Lotz's formula[5].

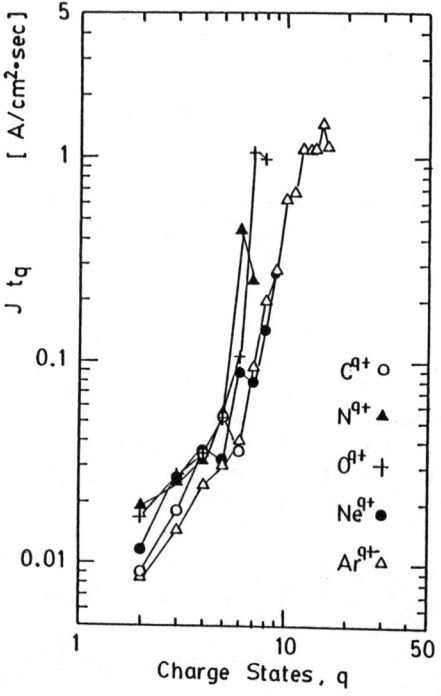

Fig.9. Effective diffusion coefficients for various ions.

As shown in Fig.9, the estimated values of τ_q increase rapidly with increase of charge states. In the presented MINI-EBIS, since the electron beam current densities used for various ion productions were almost the same of about $Je \simeq 10$ A/cm^2, it is found from Fig.9 that the confinement times in the DC ion extraction mode depend almost on only ionic charge charge state, a few msec for doubly charged ions but longer than 100 msec for highly charged ions.

In this experiments, the ion source is run in the DC mode, in which sample gases are continuously introduced into ion trapping region and ions are continuously extracted over the small potential barrier from ion trapping region. In this DC operation mode, since space charge of electron beam is partially or almost all neutralized by residual gases before gas puff, the ion trapping effect should depend on the grade of neutralization of space charge. In results from the time-dependences (1) with using estimated values of τ_q in Fig.9, it was found that experimental charge distribution can be reproduced by the equilibrium charge state distribution in only a condition in which the diffusion term contribute always from t=0.

CONCLUSION

A MINI-EBIS cooled by liquid nitrogen has been developed for the low energy collision experiments. An idea of cooling magnetic solenoid coil in liquid nitrogen can was very effective not only for size down of the solenoid coil and its power supply but also for the achievement of very low pressure. The constructed "MINI-EBIS" is open to further improvement, but it can produce fully stripped low Z ions in the DC mode operation. Although the DC mode dose not use the ion trapping effect due to the space charge of high density electron beam effectively, such a mode is convenient for the low energy collision experiments since it yields a narrow energy spread but only moderate intensities.

REFERENCES

1). H. Imamura, Y. Kaneko, T. Iwai, S. Ohtani, K. Okuno, N. Kobayashi, S. Tsurubuchi, M. Kimura, and H. Tawara, Nucl. Instrum. and Meth. 188, 233 (1981).
2). T. Mizokawa, Y. Awaya, Y. Itoh, T. Kambara, Y. Kanai, T. Koizumi, S. Ohtani, K. Okuno, H. Shibata, S. Takagi, and S. Tsurubuchi, Mass Spectroscopy, 35, 1 (1987).
3). N. Kobayashi, S. Ohtani, Y. Kaneko, T. Iwai, K. Okuno, S. Tsurubuchi, M. Kimura, H. Tawara and T. Hino , IPP report, IPPJ-DT-84 (1981).
4). K. Okuno, J. Phys. Soc. Japan, 55, 1504 (1986).
5). W. Lotz, Z. Physik, 232, 101 (1970).

DISCUSSION

<u>Kleinod</u>: Your results look very much the same as those done by the Salzborn group in Giessen. He also made Xe^{15+} and things like that with a thousand ions per second and we had a lot of contact with them, of course. So we asked, "How did you run your source?" Again, the gun was underheated and was operated at a very special temperature, well below the temperature you need to heat tungsten for a current density of about 1 A/cm^2. So we guessed that you have the same situation as in our case. How did you operate your gun? Here it is an impregnated one. I am not quite sure, but that might be the same situation.

<u>Okuno</u>: Our electron gun uses a commercial cathode on which barium oxide is coated on a nickel plate. This cathode is able to emit electrons of some amperes per square centimeter with the indirect heating of about 700°C. Here, in order to reduce the damage of the cathode by back shearing ions, our electron gun was operated in low perveance and the electron current density emitted from the cathode was about 0.5 A/cm^2 at 2 kV voltage between cathode and anode. However, the electron beam is magnetically compressed to more than 10 A/cm^2.

<u>Antaya</u>: Do the four yoke bars around the solenoid coils contribute at all to the field inside and, in general, is it important in EBIS sources not to break the azimuthal symmetry of the solenoid field?

<u>Okuno</u>: I think that the azimuthal symmetry of the solenoid field is important in EBIS sources.

<u>Antaya</u>: Is that good or bad?

<u>Okuno</u>: An EBIS source needs the high current density of an electron beam. These yoke bars are magnetic return circuits of μ metal plates and soft iron tubes shielding the cathode and electron collector. They are effective to make sharp magnetic gradients for magnetic compression of electron beams.

<u>Antaya</u>: Do you get a multipole moment inside the radius of the solenoid coil from the iron yoke?

<u>Okuno</u>: I measured the magnetic field directly. I have not seen any evidence of multipole moment inside the solenoid coil. The distribution of the magnetic field along the center axis is measured as shown in the figure.

<u>Antaya</u>: Just along the axis?

<u>Okuno</u>: Along the axis in the whole range from cathode to electron collector.

<u>Faure</u>: How do you know the density of the electron beam?

<u>Okuno</u>: I could not measure the density of the electron beam in the magnetic field directly. I estimate it from the condition of Brillouin flow.

THE SANDIA EBIS PROGRAM

R. W. Schmieder, C. L. Bisson, S. Haney,
N. Toly, A. R. Van Hook, and J. Weeks
Sandia National Laboratories, Livermore, CA 94551

ABSTRACT

Sandia National Laboratories Livermore (SNLL) has been engaged in developing a source of high charge state ions, principally for use in atomic physics experiments. The program began when SNLL acquired the EBIS test stand built at Berkeley (LBL). This source was upgraded by a major rebuild, and currently is producing extracted ions of xenon up to Xe^{40+} at energies of 1-2 keV/Q. Experiments are in progress to study the production of secondary ions and electrons from solid targets. Concurrently, SNLL is constructing a state-of-the-art EBIS which uses a 5 T superconducting magnet of length 1.2 m. Scheduled for completion in early 1989, this source should produce ions up to U^{82+}.

INTRODUCTION

The SNLL EBIS program was described at the Third International EBIS Workshop held at Cornell University in 1985 [1]. The purpose of the program was to develop the EBIS as a source of low-energy high-charge-state ions for studying fundamental processes of importance to keV plasmas such as lasers and reactors. As the program has developed, the interest has broadened to include all aspects of the physics of these ions and their interactions with particles, plasmas, and bulk matter. In this paper we present some of the developments that have occured since the 1985 workshop.

THE UPGRADED LBL EBIS

The LBL EBIS test stand was described by Brown and Feinberg [2]. Some studies of instabilities and ion heating were made on this machine [3]. Upon decomissioning at LBL, it was moved to Livermore and reassembled while major modifications were designed and built. The upgrade included LHe-cooled panels to improve the vacuum, new drift tubes to operate at higher voltage, a new electron gun and collector, new ion extraction and transport optics, a dipole magnet charge state analyzer, and various spectrometers for analyzing secondary electrons, ions, and X-rays. The upgrade was completed in the summer of 1987, and the next year was devoted to studying the behavior of the source. In mid-1988 experiments were begun to study the interaction of the extracted ions with solids.

The source normally is operated at room temperature in DC mode. The ions are extracted continuously by allowing some of them to leak over the potential barrier set by the voltage on the end drift tube. This "Leaky EBIS" configuration is most convenient for tuning up the source and for many experiments. It is found that the tuning for the highest charge states in leaky mode is very critical--variations of a fraction of a volt on the top of a 20-volt barrier cause major changes in the yields. One expects to obtain the highest charge states by confining the ions for very long times (confinement/dump mode); experiments with this EBIS show the expected rise of mean charge with confinement time.

Typical parameters for the source in DC ("Leaky") mode are:

Axial Magnetic field	3	kG
Field at gun	<1	G
Beam voltage	1500-2500	V
Beam current	10-30	mA
Gun perveance	0.28	upervs
Central DT (trap) voltage	1000-1500 = DT	V
Rear DT (barrier) voltage	DT+50	V
Exit DT (barrier) voltage	DT+20	V
Collector voltage	0	V
Extractor electrode voltage	1500-2000	V
Trap length	60	cm
Drift tube inner diameter	1	cm
Drift tube length	5	cm
Pressure in drift tube chamber	5(-10)	Torr
Pressure in transport tubes	1(-9)	Torr
Distance from trap to analyzer	250	cm
Analyzer magnet angle	90	deg
Distance from analyzer to target	150	cm

Extracted ions were identified by separation according to Q/M in a 90° magnetic analyzer. Typical spectra of the residual gas are shown in Fig. 1. These show the normal pattern of ions of C, N, and O, from Q=1 to H-like ions, plus the ions of H and H_2. The purity of these spectra is taken as indication that we are observing only ions produced within the trap.

Ions of xenon are produced by bleeding xenon gas continuously into a hole in one of the drift tubes. Fig. 2 shows typical spectra taken using isotopically enriched Xe gas (90% 136-Xe + 10% 134-Xe), with the EBIS operated at room temperature in leaky mode. Evident in these spectra are the same residual gas ions as in Fig. 1, plus a series of peaks at the expected positions of the Xe ions, some overlapping the former. The sequence of Xe ions begins to fall quickly above Q=26, and ends near Xe^{40+}. The count rate for Xe^{+36} is 10-100 ions/s, but this may be low compared to the actual production rate, due to low efficiency of the transport optics. The yield of high charge state ions was higher than expected, given the confinement time in DC mode, and suggested that some mechanism was enhancing the trapping time.

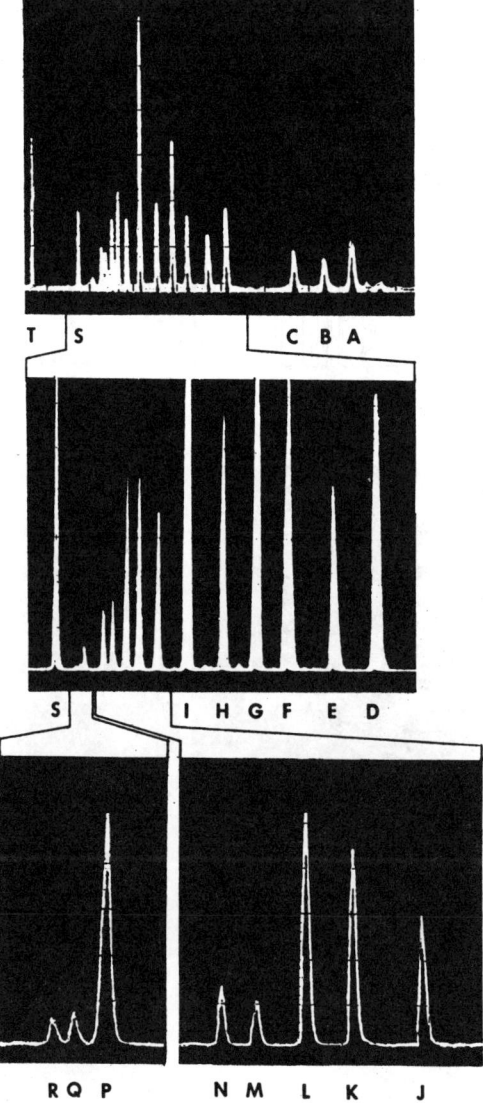

Fig. 1 – Residual gas ion spectrum observed with a 90° magnetic analyzer, observed in DC ("leaky EBIS") mode.

$S = H_2^+$ $M = N^{5+}$ $J = N^{4+}$ $G = O^{3+}$ $D = O^{2+}$ $A = O^{1+}$
$ + C^{6+}$ $N = O^{6+}$ $K = O^{5+}$ $H = N^{3+}$ $E = N^{2+}$ $B = N^{1+}$
$ + N^{7+}$ $P = C^{5+}$ $L = C^{4+}$ $I = C^{3+}$ $F = C^{2+}$ $C = C^{1+}$
$ + O^{8+}$ $Q = N^{6+}$ $ + O^{4+}$
$T = H^+$ $R = O^{5+}$

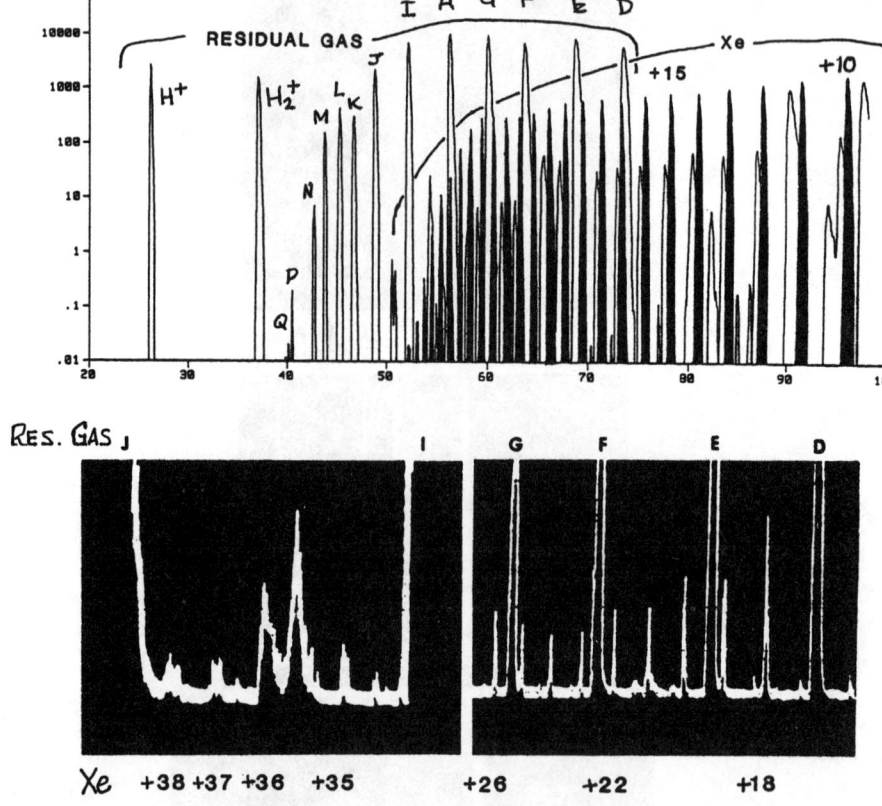

Fig. 2 – Spectra of residual gas plus (90% 136-Xe, 10% 134-Xe), observed under conditions similar to Fig. 1. Superimposed on the residual gas spectrum is the series of Xe charge states.

Fig. 3 – Spectra of residual gas plus Xe (natural isotopic composition) observed by trapping the ions for times from 10 ms to 1 s. These data suggest the light residual gas ions are being heated and lost, while the heavy ions are being cooled and retained in the trap. The residual gas spectrum at the bottom was taken in DC ("Leaky EBIS") mode; these peaks are more nearly monoenergetic.

The EBIS also was operated in a gated mode. Xe gas was bled in continuously, while the end barriers were raised for a specified length of time to trap the ions. Then the downstream barrier was dropped and the ions ejected in a pulse. Fig. 3 shows spectra of Xe plus residual gas as a function of trapping time, from 10 ms to 1 s. The surprising result is that although the residual gas spectrum is constant, the Xe ions apparently have not reached equilibrium, even at 1 s. We interpret this result to indicate that the light residual gas ions are quickly reaching an equilibrium between ionization and loss from the trap, while the heavy Xe ions, which take much longer to reach the high charge states, are not being lost from the trap, but are continuing to evolve toward higher charge states. This is evidence for enhanced trapping of the heavy ions, presumably due to collisional cooling by the light ions.

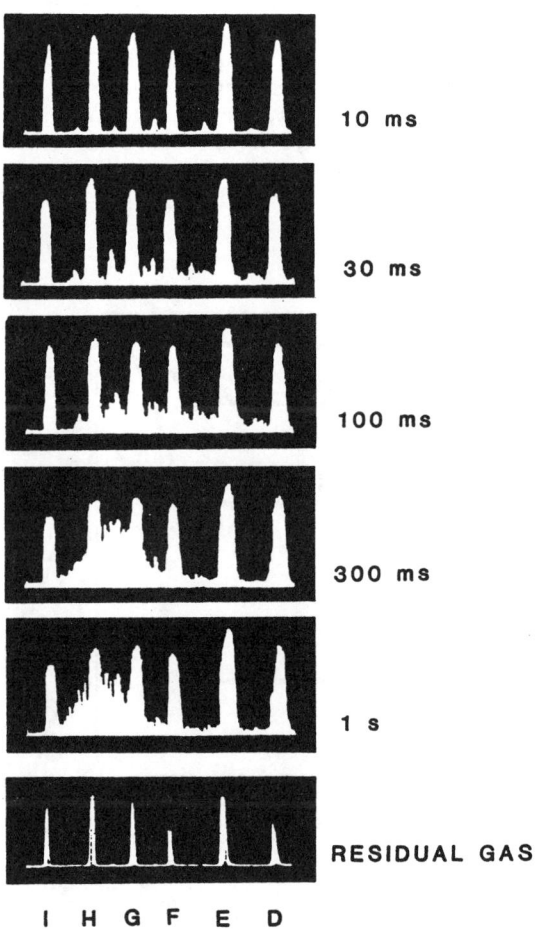

The relative number of residual gas carbon ions extracted as a function of exit drift tube (barrier) voltage is plotted in Fig. 4. This is really a demonstration of the "Leaky EBIS" mode. All ions are created with Q=1 at the trap voltage (1100 V), and when the barrier is less than this all ions simply leak out the exit. The yields of higher charge states are very low because there is insufficient time for the ionization. As the barrier is raised above the trap potential, the ions are trapped for longer times, and the proportion of higher charge states increases. Simultaneously, the ions are heated, and eventually gain sufficient energy to surmount the barrier. Due to longer trapping, these ions have a higher mean charge. As the barrier is increased further, the ions are heated enough to escape through the rear drift tube, biased at 1120 V, so the yield extracted to the detector falls rapidly.

Ions extracted from the leaky EBIS mode are monoenergetic. This is demonstrated for C^{3+} ions in Fig. 5, in which is plotted the current in the analyzing magnet as a function of the exit barrier voltage. The magnet current is proportional to the ion momentum, i.e., the square root of the energy. As before, all ions are created at the trap potential (1100 V), so for barriers less than this, the ions all have the same energy. As the barrier is increased above the trap voltage, the ion energy abruptly increases in exact correspondence to the barrier voltage. This plot is direct evidence of heating of the ions.

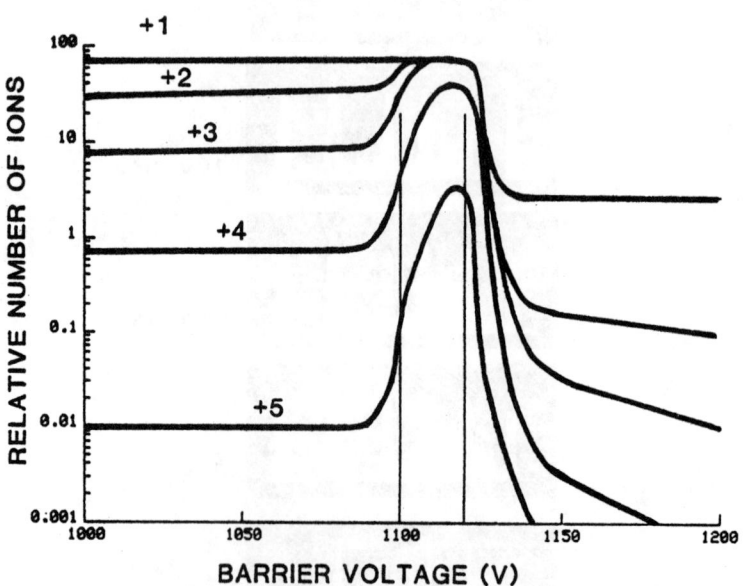

Fig. 4 – Relative number of carbon ions extracted through the last drift tube as a function of the voltage on that tube. The ions are produced in the central drift tubes biased at 1100 V. The rear drift tube was biased at 1120 V.

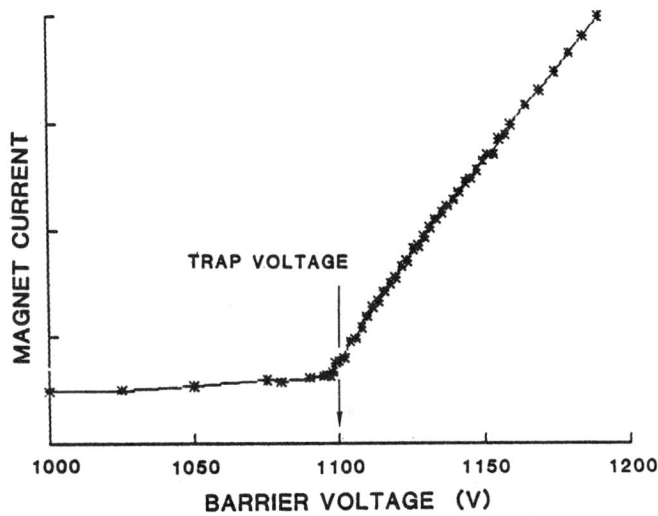

Fig. 5 - Relative energy of ions extracted through the last drift tube, as a function of the voltage on that tube. These data were taken at the same time as those in Fig. 4. The curve shows that ions extracted from the EBIS in "leaky" mode are monoenergetic at the energy of the barrier. This is direct evidence of heating of the ions, since they are created in the central trap at 1100 V.

The observation of light ion heating and the surprisingly high yield of high charge state ions suggested that the trapping time of the heavy Xe ions was being enhanced. It was natural to expect that ion-ion collisions would transfer energy from the high charge state ions to the low ones, thereby cooling the heavy ions and keeping them confined in the beam. Accordingly, a computer model for the system including ion-ion collisional energy transfer was developed. This model, and typical results, is described in a separate paper [4]. The calculations verify the importance of the collisional cooling. We expect the introduction of ion cooling in the EBIS to significantly influence the design of future machines and to enhance their performance [5].

Extracted ions were used to study the production of charged secondary particles from a solid copper target. A retarding potential difference spectrometer [6] was used to detect secondary ions emitted from the target and measure their yields and energy spectra. Fig. 6 shows a scan of the relative yields of secondary ions as a function of analyzer magnet current, i.e., vs incident ion. Only secondary ions with energies $E<20$ eV were accepted by the spectrometer. The incident ion energy was 1400 eV/Q, and the

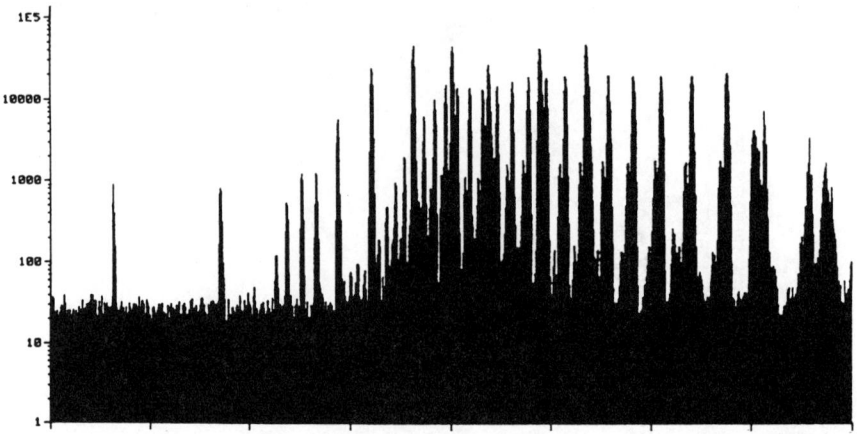

Fig. 6 - Relative yields of secondary ions released from a solid copper target as a function of magnetic analyzer current (i.e., incident ion). Note that the ordinate is plotted logarithmically. Each Xe ion is seen as a pair of peaks, corresponding to the isotopic composition of the beam. This spectrum shows that secondary ions are emitted in rough proportion to the flux of incident ions, but that the secondary yield for incident Xe ions is much higher than for the light residual gas ions (cf. Fig. 2 for the relative residual gas/Xe abundances).

incident ion spectrum was essentially that shown in Fig. 2. The plot clearly shows secondary ions produced by each species of incident ion, including both the residual gas and Xe. Presumably the secondaries were sputtered Cu ions, but they may also include ions from surface contamination by residual gas. A significant aspect of the plot is that the yields of secondary ions produced by Xe ions are enhanced (10-100x) over the yields from residual gas ions. This is consistent with the higher mass of Xe, and with a sputtering process that is independent of the incident charge state. Current experiments are directed at attempts to observe a dependence of the yields on incident charge state.

Secondary electrons were measured in a similar way, using a 20-cm diameter hemispherical electrostatic analyzer [7]. Fig. 7 shows a scan of the relative yields vs analyzer magnet current, comparable to Fig. 6. In contrast with the secondary ions, the secondary electron yields for incident Xe ions are not enhanced over the yields for incident residual gas ions. Work is in progress to see if there is enhancement of the secondary electron yield at high charge states, predicted by models of Auger neutralization [8].

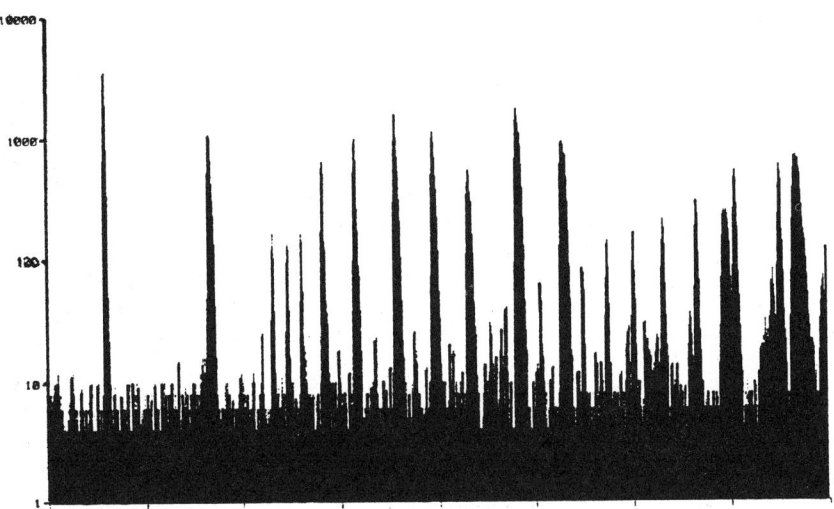

Fig. 7 - Relative yields of secondary electrons released from a solid copper target. This plot was obtained under essentially the same conditions as Fig. 6.

THE SNLL Super-EBIS

The Super-EBIS being developed at SNLL is shown in Figs. 8-9. Construction will be complete in early 1989, and operation by summer is expected.

In order to produce the ions of interest to the program, it was necessary to build an EBIS with high beam voltage, high beam current density, very long confinement times, relatively large volume trap, and to extract the ions to an external target. These very demanding and sometimes conflicting requirements have caused the system to become physically large and expensive.

An early decision was to build the EBIS on a vertical axis to minimize mechanical distortions from gravity. This means, however, considerable inconvenience in working on the source, and the extracted ions must be deflected by 90° to pass into the experimental beam pipe.

The high voltage drives the design to large size. For safety and convenience, the source is kept at ground potential, and the high voltage for the gun and collector is brought into the vacuum chamber through insulators. This limits the voltage to about 75 kV in the present machine. The high voltage is maintained in a screen room, and supplied through cable to the system. The gun and collector are biased at high negative voltage, with the collector about 5 kV positive with respect to the gun. This allows reducing

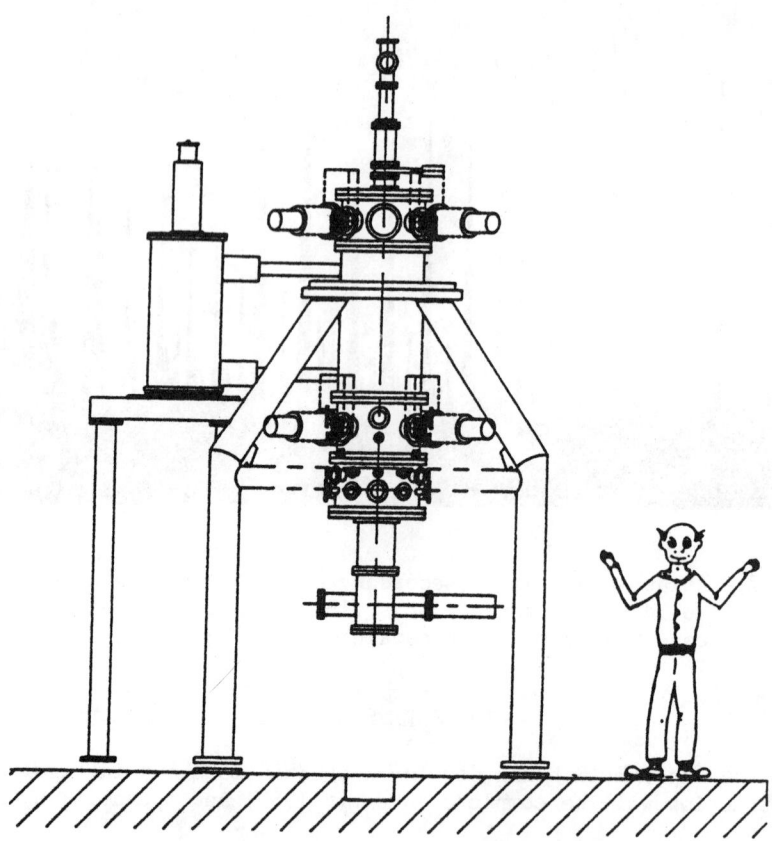

Fig. 8 - Elevation drawing of the SNLL Super-EBIS.

the power delivered to the collector while providing adequately controlled electron trajectories. The drift tube voltage is never more than a few kV; these voltages, and those for transport optics, are supplied from separate rack mounted supplies.

The electron gun is totally immersed in the fringing field of the solenoid, where the field is roughly 0.1 T. This scheme has the advantage of simplicity and provides better beam laminarity, but does not produce as high beam compression as designs using the gun in zero field. In order to attain high current density, a LaB_6 cathode 2 mm in diameter is used. At 10 A/cm^2 extracted from the cathode this produces a beam of 300 mA. Inside the solenoid, this beam is compressed 5T/0.1T=50x to 500 A/cm^2. The electrode configuration was designed to match the gun to the fringing field profile calculated for the solenoids [9].

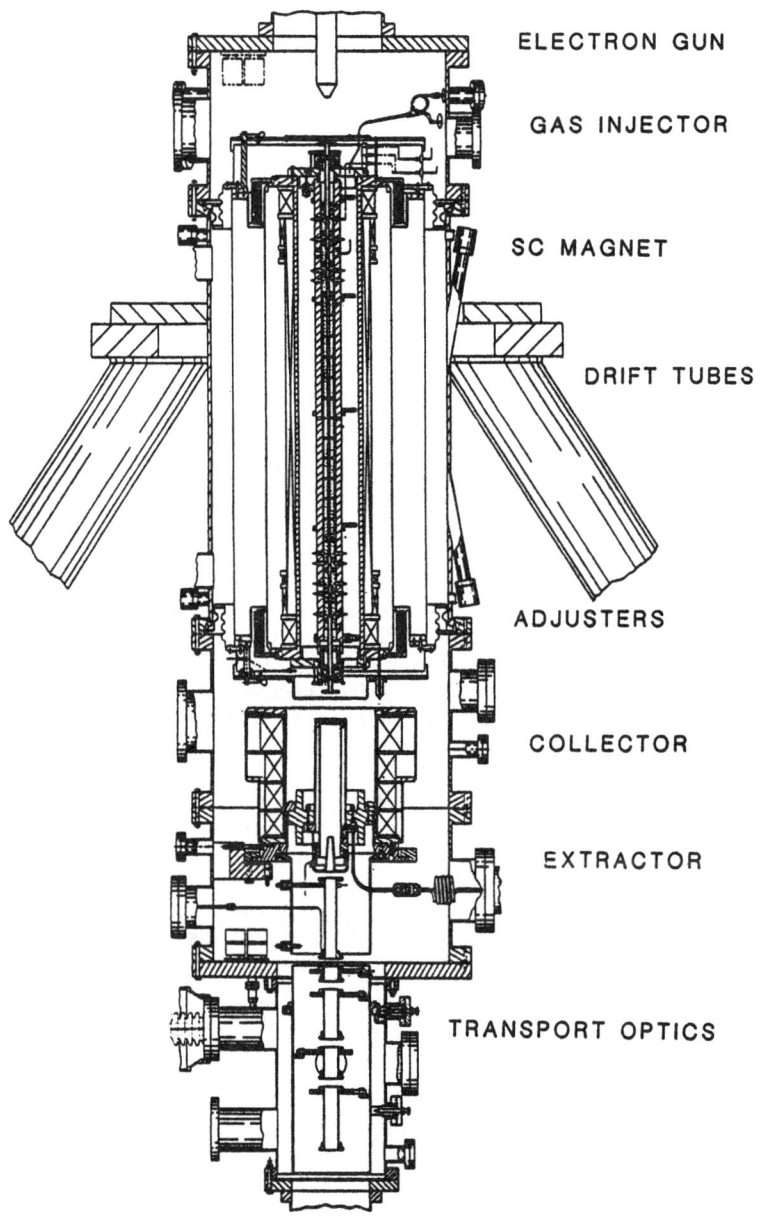

Fig. 9 - Cross section of the Super-EBIS. The entire support assembly is non-magnetic. The superconducting magnet bore is 1.2 m long, and can be adjusted inside the vacuum vessel.

The requirement for high beam compression and straightness dictated the use of a superconducting magnet [10]. A rather large bore, 15 cm, was specified to allow flexibility in the drift tube structure, and possible internal devices such as gas injectors. Another early decision was to adopt a cold bore magnet, and require the magnetic axis to be accurately coaxial with the mechanical axis of the bore. All internal structures such as drift tubes are kept coaxial with the mechanical axis, hence also the magnetic axis by close tolerancing; there are practically no internal alignment adjustments. The field was specified to be 5 T, dictated mostly by the cost of producing the high field in the large bore. The magnet is suspended inside the vacuum cryostat, and can be positioned within it by external radial and axial adjusters.

The magnet has three separate coils: a central coil of length 90 cm, and two end coils; length of the bore is 120 cm. The field was specified to be straight within 1 part in 10000 within the bore, and uniform to within 1 part in 1000 within the central 90 cm (when the end coils are used to flatten the axial profile). If desired, the end coils could be run in opposition to the main coil, nulling the field at the end of the bore at each end. A plan to include radial access ports at the midplane had to be abandoned due to engineering risks and the impossibility of achieving the flat axial profile. The magnet former was cut from aircraft L-111 aluminum heat treated to relieve stresses. The magnet normally runs in persistent mode. Its attached LHe dewar holds 100 liters, and is filled from a 500-liter supply dewar every 3-5 days. The supply dewar lasts about 2 weeks.

The drift tube structure was designed principally to maintain azimuthal symmetry. In order to generate the beam bore (0.625 cm diameter, 100 cm long, straight within 0.002 cm), a two-component construction was devised: a 6.3 cm diameter aluminum rod was drilled and honed to a 3.15 cm diameter, providing the requisite straightness on an ID. The drift tubes are 304L stainless plugs 5 cm long that are precision ground with 0.003 cm clearance to the rod ID. These are drilled with the 0.625 cm ID for the electron beam to high precision, and are supported by ring insulators of alumina. As the structure cools, the aluminum rod contracts onto the insulators and drift tubes, maintaining coaxiality.

The vacuum is generated by a set of 8 in. cryopumps [11] and a set of non-evaporated getter pumps [12], as well as the LHe cold bore. The main vessels are 304L stainless, which was taken though an elaborate heat-treating process to remove dissolved hydrogen. Numerous internal parts such as screws are coated with molybdenum disulfide to prevent vacuum welding. The collector and the bucking coil around it are water cooled; the assembly is demountable using Cajon fittings.

The entire structure sits on an isolated floating concrete pad designed to withstand moderate earthquakes.

Alignment of the system is accomplished as follows: First, the gun and the collector are pointed at each other using their screw adjustments. This establishes the axis for the beam, which is made essentially vertical by screw adjustments (on the legs of the stand) of the entire structure. Next, the magnet bore is mechanically set on this axis by optical sighting. Finally, the beam is run through the drift tube structure. The first and last drift tubes are split quadrants, allowing a final centering of the magnet by observing the capacitively induced transient on the four elements when the beam is modulated.

Working gas is supplied to the trap by an electrically heated capillary to a hole in a drift tube near the end of the magnet, where the magnetic field has a broad peak. Proper biasing of the drift tube assures that the ions are created at precisely the voltage of the main trap in the center section of the magnet. Cooling of the trapped heavy ions is necessary for long confinement. An ironic aspect is that warm-bore sources automatically provide light residual gases for cooling, whereas the vacuum in a cold system is too good, and coolant ions must be supplied. This is accomplished by having a separate gas bleed, either near the center of the main trap or at the end.

Ions are extracted from the trap by lowering the downstream potential barrier, allowing them to move into the collector, where they are separated from the low-energy electron beam. Exiting the collector they return to kinetic energy $E=QV$, where V is the drift tube potential where they were created, and then pass through a series of focussing optics to the spherical-sector electrostatic bend, thence to the analyzing magnet and target chamber. Because the main magnet has no magnetic shielding, a series of rectangular compensating coils nulls the transverse magnetic field everywhere along the ion trajectory.

CONCLUSION

The SNLL Super-EBIS is projected to produce ions up to U^{82+}, with the aid of in-situ ion cooling. It is expected to be operational by mid-1989, and will be used mainly for experiments on the highest charge ions available. In general these experiments will study the processes and fragments resulting from collision of these low-energy ions with neutral targets, including atoms and solids. The general scientific problem is to determine how the enormous potential energy of these ions is diffused into a variety of products. Much of the impetus behind all this is to better understand the processes occuring in keV plasma devices that have the potential for being useful.

ACKNOWLEDGEMENTS

Major contributions to the early stages of this program were made by K. W. Battleson (SNLL), L. Hansen (LBL), V. O. Kostroun (Cornell), M. Levine (LBL), M. Libkind (SNLL), R. Marrs (LLNL), M. Stockli (Kansas), and J. Vitko, Jr. (SNLL). R. Becker (Frankfurt) has provided continuing collaboration on all aspects of this project.

REFERENCES

1. V. O. Kostroun and R. W. Schmieder, Eds., Proceedings of the Third International EBIS Workshop, Cornell University, June, 1985.
2. I. G. Brown and B. Feinberg, Nucl. Inst. Meth. $\underline{220}$, 251 (1984).
3. M. A. Levine, R. Marrs, and R. W. Schmieder, Nucl. Inst. Meth. $\underline{A237}$, 429 (1985).
4. R. W. Schmieder and C. Bisson, "Heating and Cooling of Ions in the EBIS: Model Calculations," this symposium.
5. We respectfully acknowledge the dramatic demonstration and extensive study of ion cooling in the EBIT by the LLNL group (M. Levine, R. Marrs, and others).
6. Designed by R. W. Schmieder and C. L. Bisson.
7. Surface Sciences/VSW Instruments.
8. E. S. Parillis, Dokl. Akad. Nauk. SSR, Ser. Fiz. $\underline{37}$, 2565 (1973).
9. Carried out by Prof. R. Becker.
10. Cryogenic Consultants, Ltd., London.
11. Varian Associates.
12. SAES, Inc.

DISCUSSION

Faure: The experiments you did with xenon, did you do any with argon?

Schmieder: We did some experiments with argon but mostly to see which charge states we would get. I did almost everything with xenon.

Faure: But did you have cooling with argon?

Schmieder: I do not know. I did not make those measurements, or the calculations.

Faure: In your calculation, if I understood, did you take into account coulomb forces between nitrogen and xenon?

Schmieder: That is correct. The collisions are strictly coulomb collisions between two point charges of fixed charge and mass.

Faure: About the hollow beam, do you know the emittance of the ions coming out?

Schmieder: I do not know yet, but it comes right out of the code. The emittance will be a consequence.

Faure: Is this emittance too high?

Schmieder: Certainly I expect the emittance for the light ions to be high because they are going all over the place, out to the walls. Furthermore, we see the conversion of radial kinetic energy, which is driven by the potential energy, increase as the ions increase in charge. We see the conversion of that radial kinetic energy into longitudinal energy. There is nothing clever going on here. We do see that in the code. One result of the code when we get all the parameters right numerically, is the rate at which that energy gets converted. Of course, to some extent, these are things you can look up in books on rate coefficients for plasmas. But remember, this is a non-equilibrium plasma of very peculiar spatial distribution so it is doubtful that you can really look up a quantity that is meaningful. That is why we are calculating.

Schneider: I have two questions about your cooling. One is that the orbit you showed has a minimum of angular momentum. Have you considered other angular momenta?

Schmieder: Yes. What we do is set up the ions initially with random locations within the beam, as if they are being created from gas, and we give them a thermal velocity with random direction. They evolve from there.

Schneider: The second question: In EBIT we think of axial escape as being much easier than radial escape, so we feel a lot of our cooling happens by ions escaping axially. Can you comment on this?

Schmieder: I think that depends on what the voltage of the last drift tube is. If the voltage of the last drift tube is the same as the bottom of the well, sure, they will just leak out; you tip a bottle over and it pours out the end. However, if you put the barriers very high on the ends, the ions have no choice but to go radially to the walls. I think I have some data that shows that as a function of the end barrier (viewgraph). This is the number of ions extracted axially. What you see is that so long as the barrier is low, you essentially extract them all; they all just dribble out. But now as the barrier increases, you are cutting off more and more. Only those that have been heated high enough to get over the barrier get out. Any plasma physicist will recognize this as a characteristic curve of a probe in a plasma.

Levine: That only works if you know that the number of ions in the beam is constant. It seems to me that the argument you have just made depends on the fact that you are assuming the number of ions in the electron beam stays the same, independent of the barrier.

Schmieder: If you like. I will certainly not argue with you.

Levine: Is your code fully kinetic with a Monte Carlo calculation for the radial transport?

Schmieder: The transport is being done at the moment by coulomb collisions.

Levine: Does the code take into account momentum transfer in each collision?

Schmieder: Yes.

Levine: The way I interpreted your graphs, you had most of the high charge states in the beam.

Schmieder: The xenons were all trapped within the beam.

Levine: I infer that you have just N^{7+} on the outside.

Schmieder: It was N^{6+} and N^{7+}.

Levine: OK. Ion-ion collisions generally do not lead to diffusion because of conservation of angular momentum.

Schmieder: Collisions between like ions, that is, nitrogen-nitrogen and xenon-xenon. But collisions between xenon and nitrogen, and nitrogen and xenon, will lead to an exchange of energy. The reason is that the xenons, of course, have higher charge state and higher mass, so they have higher total available energy.

Levine: What I did not understand is, once they get outside the beam, the nitrogen is only colliding with nitrogen in the radial area.

Schmieder: That is correct. The nitrogen is only colliding with nitrogen.

Levine: How do you get the cross field diffusion?

Schmieder: They still go through the beam and can collide in the beam.

Levine: Then you are telling me that the radius of the device is of the same order as the gyro-radius for the nitrogen?

Schmieder: That is hard to say because there is such a distribution. I think we would have to talk about the distributions and momenta of the ions. These ions are created, initially, with random thermal energies. But once they start exchanging energy through collisions, there is no longer one gyro-radius; there are lots of different energies.

Levine: Is the gyro-radius small compared to the size of your tube?

Schmieder: I think not. These calculations, by the way, were done modelling the old machine, 3 kG. So these ions are not magnetized.

Levine: I do not know whether your gyro-radius is small.

Schmieder: It does not matter because the trajectory is determined by the electric and magnetic fields in the volume and the differential equation solver just follows that trajectory, whatever it is. It is in a self-consistent field. We then generate radial distributions of ions according to their residence time versus radius. That is why a single ion, which spends most of its time near its apogee, generates a distribution of ions in the code that is peaked out at its apogee. We do not care whether they are magnetized or not, or what their gyro-radius is. The code takes care of it all.

Levine: And that apogee is large compared to the drift tube radius?

Schmieder: Yes, apparently, unless we made a coding error, which I will not claim we have not. The ions were getting out to the walls and we started losing ions on the walls. I have to say, of course, that is the whole purpose of modelling. We know we can plug the ends and force them to go to the walls. They cannot get through the ends if you plug it hard enough.

Levine: My confusion is that in EBIT the ion orbits are very small compared to the drift tube radius.

Schneider: In our machine the drift tube end barriers are 300 volts at the most. So, for radial escape, an orbit that would go from the axis to the drift tube wall would need at least a thousand volts of energy, which is why they escape axially.

Schmieder: Well, that is OK except that in EBIT you may not be able to plug the ends strongly enough. In the model I can certainly plug them, and in general, you can imagine a plug. Certainly in our machine we can plug the end. You know that if you put a 1000 volt barrier on the ends and you have neutral gas in the system that continually gets ionized and produces new ions, sooner or later it is going to have a radial distribution that reaches to the walls. In equilibrium, ions will be adsorbed or absorbed on the walls and new ions will be produced in the beam. That has to be the equilibrium; nothing else can go on.

Becker: I suppose, Bob, this was a convincing spectrum and I even see another point. Your residual gas peaks are always broader than your injected gas peaks. As we see in our spectrum and have already shown at the last Denton meeting two years ago, the colder ions survive at the expense of the hotter ones. Therefore, we see, for instance, enrichment of the ^{22}Ne isotope against ^{20}Ne isotope by a factor of three in the dc output for the most abundant charge state.

Schmieder: Yes, thank you. That is a very important point to make. Here is the residual gas spectrum without the xenon. These background peaks are much narrower in dc mode than in pulsed mode.

Beebe: I have a question relating to the last viewgraph you had up, the one where you made the measurement of the energy of the ions. Was that for ions coming out over the barrier?

Schmieder: Yes. In dc mode. What I did was set the barrier and increase it in small steps and then go to the steering magnet and measure the energy of the ions coming out.

Beebe: How well do you know the constant from the analyzer? What kind of analyzer are you using here?

Schmieder: It is a 90° steering magnet and we know its calibration from the background gas ions.

Beebe: Do you know what the space charge depression is on the barrier electrode as compared to the trap electrodes? Such a depression can enable ions to leak out of the source at a lower energy than that suggested by the applied potentials.

Schmieder: No, I do not, but it is obviously a good question.

Beebe: If you use a low current beam, then the depression is not much, but if you use a high current beam then you have much more depression.

Schmieder: Yes, it is a good question. I cannot answer it. Our beam current is 15 mA.

Beebe: OK, so that is not very much. I imagine it is only about 50 volts.

Schmieder: I do not think it enters into this, except that I do know that the drift tube voltage here was 1100 volts. Essentially, I started seeing ion heating at the drift tube voltage. In other words, if the end barrier was too low, I saw exactly the same energy ions coming out, namely 1100 volts. But as soon as it got a few volts above the drift tube voltage, I started seeing the heating.

Beebe: I am doing a similar measurement and am wondering if some of what we might be calling heating might also be accounted for because of the difference of the depression in the ionization region as opposed to the last barrier.

Schmieder: Well, I have shown you some data of ion energy versus end barrier that is actually the nicer data. I had some others that had some peculiar steps. I cannot explain those.

Beebe: Do you have different voltages on your drift tubes?
Schmieder: I tried it at several voltages: 1100, 1200, 1400.
Beebe: Was your well flat?
Schmieder: The well was flat, all drift tubes were at the same voltage. It was 50 volts high on the backside and then variable on the downstream side. But on some of them, and repeatably, it showed a kind of funny behavior here and I had no way to account for it.
Beebe: Were you ever able to plug the end?
Schmieder: Sure. The number of ions decreased very rapidly, exponentially, with barrier height.
Antaya: A comment first. This discussion of the radial diffusion and coulomb collisions - you can write down a diffusion coefficient for coulomb collisions that gives you a diffusion mechanism which does not include anything like the radial field profile from the solenoid field. Also, I did not understand your comment that coulomb collisions will conserve angular momentum and that that will not lead to radial diffusion. I do not think that is right.
Hershcovitch: To first order, like particle collisions do not lead to cross-field diffusion.
Antaya: Well the solenoid field does not give you any mechanism that suppresses radial diffusion through coulomb collisions. My question, Bob, is what was the electron energy and the barrier height for that spectrum that showed the Xe^{37+} and Xe^{38+}?
Schmieder: I will show the spectrum again. The electron gun was at about 2300 volts and the drift tubes were at about 1200 or 1400 volts. So it was a little over 3 kV, 3 and something kilovolts, total beam energy. Those of you I have talked to as this has gone on know that, at first, I thought I was seeing Xe^{44+}. I think Reinard was the first to point out I did not have enough beam energy to see it and I certainly agree. There was just some structure there. That was when I was using natural xenon. I did not get to do it with the separated isotope.
Beebe: I had one more question on the sputtering data that you had, the one that you showed for the different xenon impacts. You said you normalize that per particle for incident xenon?
Schmieder: No. That is why I did not claim anything about the amplitudes because it is just a scanned spectrum. The system is rather stable, that is, I could scan this several times and get something that looked pretty much the same. But to get the yields will take more measurements.
Beebe: Were you monitoring the ion current hitting the target?

Schmieder: No, the target is on a plunger. I have to plunge it in and plunge it out and that is too much trouble. So all I did was scan the magnet to sweep the incident charge states. Some were background gas, some were xenon. This measured the yield as a function of scanning magnet current. Now what I should do and what I assume I will get to do is, for this particular peak, operate the plunger and see how many ions I have incident and how many secondaries I am getting and get a relative yield. Then, essentially, replot this kind of data or make a table of more precise relative yields. I just did not feel confident in reading them off this plot.

Beebe: You could put an electrical lead on the target. Since you do not really have a proper Faraday cup, it is also difficult to know how many secondary ions are produced.

Schmieder: Yes, and I just elected to ground the target so I knew it was at zero voltage relative to the system and go back and operate the plunger when I had time.

True: A comment. I was looking at Okuno's paper before and he showed a magnetic return structure that was wide open. I wonder if that would be a way of shielding your magnet?

Schmieder: It is late in the game. Is it a big plate or something?

True: Bars. Instead of using a complete shield around the outside, he used four longitudinal bars. It is so wide open that it might be something that you could put on your system fairly easily.

Schmieder: Indeed, yes. Thank you.

THE CORNELL SUPERCONDUCTING SOLENOID CRYOGENIC EBIS

V. O. Kostroun

Nuclear Science and Engineering Program, Ward Laboratory
Cornell University, Ithaca, NY 14853, USA

ABSTRACT

Over the past four years, a superconducting solenoid, cryogenic electron beam ion source, EBIS, has been designed, constructed and tested at Cornell University. The source is specifically intended for atomic physics experiments requiring low energy, very highly charged ion beams. The source and its operating characteristics are described.

INTRODUCTION

In recent years, new sources of multiply charged ions have been developed, particularly in the Soviet Union and France, to extend the range of kinetic energies from existing charged particle accelerators used in medium energy nuclear physics research. The two types of source available are the electron cyclotron resonance ion source, ECRIS, and the electron beam ion source EBIS. The ECRIS can produce fairly intense beams of highly charged ions, including bare nuclei of light Z elements[1], while the EBIS has produced the highest extractable charge state of any source[2], and can, in principle, produce bare, hydrogen and helium like ions of almost all the elements.

The basic operation of an EBIS is as follows. Neutral atoms or singly charged ions are injected into an energetic, high current density electron beam, propagating in ultra high vacuum and focused by a solenoidal magnetic field. The ions formed are trapped radially by the electron beam space charge potential and magnetic field lines, and axially by a potential distribution applied to a series of drift tubes concentric with the beam. Highly charged ions are formed by sequential electron impact ionization of the ions trapped in the beam. The highly charged ions thus formed are extracted by dropping the potential barrier at one end of the axial trap after a predetermined confinement time, and are accelerated by the potential applied to the drift tubes.

In 1980, our laboratory embarked on a project to construct a small EBIS, called CEBIS I. CEBIS I is a relatively simple, reliable source of multiply charged ions whose magnetic field and vacuum in the ionization region limit the charge states that can be obtained[3]. In order to create very highly charged ions of medium to high Z elements, electron beam energies in the 10-20 keV range or more are required, and ions have to be trapped for hundreds of milliseconds in electron beam current densities of 10^3 A/cm^2 or more. Such current densities can be obtained by adiabatic magnetic compression of an electrostatically focused beam from an electron gun injected into a magnetic field of

© 1989 American Institute of Physics

several Tesla. Very long ion confinement times are possible if the electron beam propagates in a vacuum of 10^{-10} Torr or better. Although the principle of an EBIS is deceptively simple, the actual realization of a working source is beset by a number of technical difficulties.

In this paper I discuss the design and evolution of a superconducting solenoid, cryogenic EBIS specifically intended for atomic physics experiments with low energy, very highly charged ions of medium to high elements.

DESIGN CONSIDERATIONS

Experiments with extracted ion beams from an EBIS are limited by the magnitude of ion currents produced, and by the inherently low duty factor of these sources. Increasing either the ion output or the duty factor or both is therefore highly desirable in extending the usefulness of these sources. The ion output is proportional to $PV_c^{3/2}L / (V_c+V_{dt})^{1/2}$, where P is the electron gun perveance, V_c and V_{dt} are the gun cathode and drift tube potentials respectively, and L is the length of the ionization region. The duty factor depends on the electron current density in a more complicated manner[4], but the greater the current density, the shorter the confinement time and hence the greater the duty factor. It was decided at the outset to try to design a source with $PV_c^{3/2}$ (electron current) and current density as high as possible at electron energies that would produce the highest possible charge states of medium to high Z elements.

The highest current density electron beam that can be produced by magnetic focusing occurs under the Brillouin flow condition, i.e. when the beam is launched from a magnetically shielded cathode and the beam space charge electric field and the Lorentz force compensate the centrifugal force[5]. Accordingly, external gun beam injection into a magnetic field was chosen.

The original design of the source was dictated by the ready availability of a 3 T, iron shielded, 16.5 cm diameter, warm bore, 0.96 m long, superconducting solenoid with cryostat, originally intended as one of the compensating magnets for the interaction region of the Cornell 8 GeV electron-positron storage ring CESR. In particular, the warm bore, fringe field profile and apparent lack of coaxiality of the mechanical and magnetic axes of the solenoid were important points of design consideration. The warm bore meant that the cold mandrel of the solenoid could not be used as a cryopumping suface and to cool the EBIS structure to cryogenic temperature as is commonly done in most EBIS sources. A separate cryopump with attendant refrigerator or liquid helium supply and radiation shields was therefore required. Such a system had already been used on CRYEBIS I at Orsay[6].

The shape of the fringing field and its slope of 20 T/m at 3 T, (determined by the dimensions of the magnet), led us to adopt an external injection scheme that in retrospect has turned out to work remarkably well. Proper injection of an electron beam into a magnetic field requires that the beam envelope radius and slope have the correct values at a preselected entrance point. These entrance conditions are most easily found by numerical integration of a differential equation which describes the motion of electrons under conditions that very closely approximate the problem. That is, one assumes that the electrons are moving with perfect Brillouin flow in a region of constant magnetic field strength, and then follows them back through the magnetic field towards the gun anode[7]. Unfortunately, this standard method for setting up Brillouin flow in travelling wave tubes[8] does not work for EBIS sources which utilize several Tesla fields to confine the beam because the electron beam size produced by available electron guns is incompatible with beam diameters focused by such high magnetic fields. One could of course shape the fringing magnetic field in such a way that the magnitude and slope over the build up range are consistent with electron trajectories from the electron gun. However, such a solution is highly unsatisfactory in that it ties the shape of the magnetic field rise to the properties of a specific electron gun. In our design, we have adopted a solution that circumvents this difficulty[9]. We launch the electron beam into a conventional, low field solenoid and establish more or less Brillouin flow using standard techniques. Once the beam is formed, we compress the beam adiabatically using the main superconducting solenoid. The advantage of this scheme is that it decouples the launch of the electron beam into a magnetic field from adiabatic compression, and therefore can be used with any external electron gun.

The basic idea of the method is as follows. In the paraxial theory, ($\dot{r} << \dot{z}$), the equation of motion for the edge of a laminar beam is[10]

$$\frac{d^2 r}{dz^2} = \frac{\eta I}{2\pi \varepsilon_0 u_z^3 r} - \frac{\eta^2 B^2(z)}{4 u_z^2} r \tag{1}$$

where $\eta = e/m$, I is the electron beam current, u_z the electron axial velocity and B(z) the axial magnetic field. For

$$B r = \left(\frac{2 I}{\pi \eta \varepsilon_0 u_z}\right)^{1/2} = \text{constant},$$

$$\frac{dr}{dz} = -\left(\frac{2 I}{\pi \eta \varepsilon_0 u_z B^2}\right)^{1/2} \frac{1}{B} \frac{dB}{dz}.$$

Since paraxial theory requires that $dr/dz << 1$, the axial variation of the adiabatically compressing magnetic field also has to satisfy

$$\frac{dB}{dz} << -\left(\frac{\pi \eta \varepsilon_0 u_z B^2}{2 I}\right)^{1/2} B = \frac{B}{b(z)}.$$

an easily met condition. Following reference 10, we assume that adiabatic compression creates a ripple of amplitude ρ around the Brillouin radius $b(z) = \left(\frac{2 I}{\pi \eta \varepsilon_0 u_z B(z)^2}\right)^{\frac{1}{2}}$ so that $r = b + \rho$. If $\rho << b$, then substituting r into (1) and keeping only zero and first order terms in ρ gives

$$\frac{d^2 \rho}{dz^2} + \frac{\eta^2 B^2}{2 u_z^2} \rho = -\frac{d^2 b}{dz^2}$$

The solution of the reduced equation, i.e. with the right hand side equal to zero, can be obtained using the JWKB [11] method.

$$\rho = \sqrt{\lambda_L} \left\{ \alpha \cos\left(\sqrt{\lambda_L} \int^z (2\pi/\lambda_L) dz'\right) + \beta \sin\left(\sqrt{\lambda_L} \int^z (2\pi/\lambda_L) dz'\right) \right\} \quad (2)$$

where α and β are constants, $\lambda_L = 2\pi u_z/\Omega_L$, and Ω_L is the Larmor frequency. This solution describes any initial ripple present in the beam. Since the amplitude of the initial ripple varies as $\sqrt{\lambda_L}$ and b varies as λ_L, the fractional ripple in the beam, ρ/b varies as $1/\sqrt{\lambda_L}$ and increases as the beam is compressed. It is therefore advantageous to start compression with a ripple free beam. The solution of the general equation (2) represents ripple introduced by compression. To describe compression of the beam in a magnetic field B= B(z), we let r(z)=b(z) R(z) and substitute into equation (1) which has to be solved numerically for R(z). The resulting equation is,

$$\frac{d^2R}{dz^2} - \frac{2}{B}\frac{dB}{dz}\frac{dR}{dz} + \left\{\frac{2}{B^2}\left(\frac{dB}{dz}\right)^2 - \frac{1}{B}\frac{d^2B}{dz^2}\right\} R = \frac{\eta^2 B^2 (1-R^2)}{4 u_z^2} \frac{1}{R}$$

Numerical integration of this equation, using the measured magnetic field, in both the entrance and main solenoids, gives an electron beam envelope that agrees remarkably well with the observed electron beam profile measured with a beam scanner[12].

The above approach says nothing about what happens when ions are present in the beam, a much more complicated situation. In a related problem, Yao[13] has considered the neutralization of the electron beam space charge by ions trapped in the beam. A simple fluid model shows that the eam collapses as it is neutralized. The phenomenon of "ion focusing" due to stepwise ionization plays an important role in this collapse. The results obtained with the simple fluid model were also verified by a 1 D (r-only) electrostatic particle code written to simulate this collapse.

Consider a non-relativistic electron beam propagating along a uniform external magnetic field $\mathbf{B} = B_o \mathbf{e}_z$ in a uniform-density ion background. We assume that the system is azimuthally and axially symmetric, i.e. $(\partial/\partial\theta = 0)$ and $(\partial/\partial z = 0)$. The fluid equations describing the behavior of the electron beam are then:

$$\frac{\partial n_e}{\partial t} + \frac{1}{r}\frac{\partial}{\partial r}(r n_e v_r) = 0,$$

$$\frac{\partial v_r}{\partial t} + v_r \frac{\partial}{\partial r} v_r - \frac{v_\theta^2}{r} = -\frac{e}{m_e} E_r - \Omega_e v_r,$$

$$\frac{\partial v_\theta}{\partial t} + v_r \frac{\partial}{\partial r} v_\theta + \frac{v_r v_\theta}{r} = \Omega_e v_r,$$

$$\frac{\partial v_z}{\partial t} + v_r \frac{\partial v_z}{\partial r} = 0,$$

$$\frac{1}{r}\frac{\partial}{\partial r}(rE_r) = 4\pi e(n_i - n_e)$$

where m_e is the electron mass, n_e is the electron fluid density, n_i is the ion density, v_r, v_θ and v_z are the components of the electron fluid velocity, E_r is the radial electric field, and $\Omega_e = eB_0/m_e c$ is the electron gyrofrequency. (For simplicity, we have assumed the ions to be singly charged.)

If we also assume that the electron beam is launched with no rotational velocity ($v_\theta = 0$) from a shielded cathode, and that the beam is initially in equilibrium ($v_r = 0$), then it can be shown that the electron beam radius r(t) evolves according to the equation

$$\frac{d^2 r}{dt^2} = \frac{1}{r} - (1+n_i)\, r,$$

where r,t and n_i are dimensionless quantities. (r is normalized to its initial value, t to 1/2 π times the electron gyroperiod, and n_i to the initial electron density.) We can assume quite generally that the characteristic time scale t_N for neutralization of the electron beam space charge is much longer than the electron gyroperiod, and hence write n_i as

$$n_i = n_i(\varepsilon t)$$

where $\varepsilon = O(1/\Omega_e \tau_N) \ll 1$. Using the method of multiple scales, we then find, to order ε^2,

$$r(t) = (1 + n_i)^{-1/2} + \varepsilon \frac{n_i'(0)}{2\sqrt{2}} (1 + n_i)^{-1/4} \sin(\sqrt{2} t^*) \qquad (3)$$

where

$$t^* = (1 + \varepsilon^2 \omega_2 + ...)\,\tilde{t}$$

$$\tilde{t} = \int_0^t dt\,(1 + n_i)^{1/2}$$

(ω_2 is a constant). According to (3) as the ion density n_i increases, the electron beam radius decreases. To get the correct electron beam launch condition with ions in the beam, the equilibrium beam in the main solenoid would have to be followed out to the gun anode as described above. However, the 1D nature of this simple code and the complexity of a more realistic calculation preclude this. In any case, we assume that the formation of an ion free, ripple free, quiet beam, propagating from cathode to collector with essentially 100% transmission is a necessary condition for the proper operation of an EBIS.

In order to assure coaxiality of the electron beam with the magnetic axes of the launch and main solenoids and the mechanical axis of the drift tube structure, the following solution was adopted. The entrance solenoid, which together with its magnetic shim defines the initial electron axis, is attached to the main vacuum chamber, whose outer diameter is somewhat smaller than the warm bore of the main solenoid. The whole vacuum chamber, together with entrance solenoid, pumps, etc. is supported at each end by a cradle which can be accurately translated both vertically and horizontally. Thus it is possible to align the electron beam axis in the entrance solenoid with the main solenoid axis. The drift tube structure inside the main vacuum chamber is also moveable. It is suspended at each end by four positioning rods which can be adjusted from the outside. The positioning rods are used to align the drift tube structure with respect to the electron beam.

Finally for ease of maintenance and operation, it was decided that the entire drift tube structure, electron collector-ion extractor, beam forming optics, and all electrical and water connections should be removeable from the source as a single unit. Overall, every attempt was made to design a source mechanically and electrically as simple as possible in order to ensure reliability and the minimum of maintenance.

THE CORNELL SUPERCONDUCTING SOLENOID, CRYOGENIC EBIS, CEBIS II

The design of CEBIS II drew heavily on operating experience with CEBIS I, our original, EBIS. In CEBIS I [3], the axial magnetic field is generated by a conventional, water cooled solenoid which produces a uniform field region 50 cm long at 4.2 kG. Ultra high vacuum in the ionization region ($\sim 10^{-9}$ Torr) is produced by a distributed sputter ion pump whose configuration is consistent with an axial magnetic field. The electron beam is launched into a small entrance solenoid from an external electron gun whose electrode structure is a scaled version of a Hughes 112-2B modified Pierce type gun [14]. The beam from the entrance solenoid then enters the main magnetic field. The feed gas is injected continuously, and the source produces useable beam currents of such ions as C^{5+}, N^{6+}, O^{7+}, Ar^{16+} and Xe^{28+}.[15]

CEBIS II has undergone several modifications over the past three years. An overall schematic of the second version is shown in fig.1.

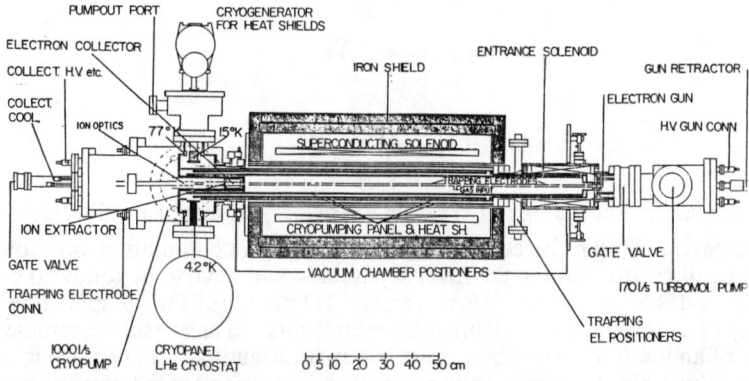

Figure 1. Schematic of the Cornell superconducting solenoid, cryogenic EBIS, CEBIS II.

The main components of the source are: gun vacuum chamber, electron gun, entrance solenoid, main vacuum chamber, cryopumping panel array cooled to ~5 K, drift electrode structure, electron collector, ion extractor and focusing optics. The electron gun vacuum chamber, gate valve, entrance solenoid, main vacuum chamber and main vacuum chamber pump are all supported as one unit by two positioning assemblies. As mentioned above, the positioning assemblies permit vertical and horizontal translation of each end of the source, thereby
allowing precise alignment of the electron axis in the entrance solenoid with the main solenoid magnetic axis. In addition, the drift tube structure can be adjusted with respect to the electron beam by external adjustment of four rods at each end.

The electron collector and ion extractor as well as the ion optics form an integral part of the drift tube structure assembly to ensure alignment of all of the main components of the EBIS. The drift tube structure and support assembly are attached to a 25 cm diameter copper gasket sealed flange which also carries some of the electrical, cooling water and feed gas connections. Electrical connection to individual drift tubes is through high voltage connectors attached to the main 33.7 cm diameter copper gasket sealed flange, and Be-Cu spring contacts between the high voltage connectors and the drift tube structure assembly. This arrangement allows one to withdraw the entire drift tube structure assembly, electron collector and ion optics as one unit through a 20 cm opening in the main flange.

Ultra high vacuum in the ionization region of the source is created by cryopumping on a cryopanel cooled to ~5 K by a copper cold finger connected to an external liquid helium reservoir. The cryopanel is an oxygen free high conductivity copper tube surrounded on the outside by two heat shields kept at 20 and 80 K respectively by a two stage refrigerator (expander-compressor) from a VARIAN Cryostack-8 cryopump. Nominally, the first stage has a 35 W heat removing capacity at 77 K, and the second, 5 W at 15 K. On the inside of the copper tube is a slotted aluminum tube, also maintained at 20 K, that surrounds the drift tube structure support assembly. The slots allow residual and feed gas molecules to escape the ionization region and be pumped by the copper tube cryopump. The support assembly and drift tubes are maintained at 20 K through copper tipped spring contact with the slotted aluminum tube. At 20 K, cryopumping by drift tubes is avoided, and at the same time, outgassing of species from the drift tubes and support structure is eliminated by cooling the support material below the desorption activation energy of most gases 16.

As in CEBIS I, the drift tubes are made from stainless steel mesh to improve pumping in the ionization region. The mesh grid is about one tenth of the 8 mm drift tube diameter. One of the drift tubes in the transition region between the entrance and main solenoid is segmented into quadrants to facilitate beam alignment. Feed gas to the ionization region is supplied by a 1.6 mm diameter, 0.15 mm thick wall, stainless steel tube. The tube is grounded at the feed end, and electrically insulated at the other end. A small current through the tube heats it slightly and prevents feed gas from freezing out.

An external gun with a well shielded cathode at negative potential and grounded anode, injects an electron beam into a constant magnetic field of ~0.15 Tesla to form a solid, ripple free beam with a current density of 40 A/cm^2. The beam is allowed to propagate for some distance (20 cm in this version) before it enters the main solenoid magnetic field through a slowly varying (6-20 T/m) transition region which adiabatically compresses the beam to the required current density. The beam is 1.25 m long, of which 0.4 m forms the ionization region of the source in the flat portion of the main magnetic field. Upon exiting the solenoid, the merged electron and ion beams enter a magnetically shielded collector-extractor where the electrons are collected and the ions extracted. The electron collector and ion extractor are electrically insulated to 25 kV and the collector can be operated in the depressed mode.

The electronics required by the source is relatively straight forward and kept at a minimum. The two major pieces of electronics developed for the source were a 5 kV, 1A, variable pulse width and duty cycle, electron gun cathode supply, and an eight channel, 100 state CMOS memory (battery backed), Apple Macintosh programmable 0 - 6.5 kV high voltage drift tube supply. The cathode supply is built around a commercial, 5 kV, 1A, DC power supply (Hipotronics, Inc., Brewster, NY) into which we incorporated the pulser which is essentially a SCR series switch[17]. The pulse duration can be varied from 10 μsec to 100 msec, and the pulse period from .001 to 100 sec. If desired, the pulser can be operated in an external trigger mode, or completely bypassed to provide a DC beam.

RESULTS

Initial source studies were carried out with our inhouse built version of a Hughes 112B electron gun. With this gun, the source could be operated either in the pulsed or DC electron beam mode. In the pulsed mode we ran with a 0.6 A, 8 keV electron beam in the ionization region (-4.5 kV on the gun cathode, +3.5 kV on the drift tubes), while in the DC mode the current was typically 0.2 A and the energy 5.5 keV (-2.0 kV on the the cathode, +3.5 kV on the drift tubes). In the pulsed beam mode typical pulse lengths were 15 msec at a repetition rate of 10 pulses/sec. In both cases, the average power dissipated in the collector was 400 W. The best transmission from cathode to collector obtained for a 8.5 keV, 0.6 A electron beam at 1.5 T was 98 %. Electron beam current densities in the range 1000-2000 A/cm^2 at 1.5 T have been inferred from the extracted ion charge state distributions for different confinement times. These current densities are consistent with electron beam scanner measurements obtained earlier[12] and numerical calculations of the electron beam envelope. Measured total ion currents 0.75 m from the extractor were typically a few microamperes electric.

In order to operate the source, it first has to be aligned. That is, the electron beam and the drift tube structure have to be made coaxial with the magnetic axes. Initially, we had planned to align the source with the help of an ultra-high vacuum compatible beam scanner[12]. The idea was to remove the assembly supporting the drift tube structure and collector-extractor, replace it by the beam scanner, and use the electron beam, probed by the beam scanner, to define the alignment. Once aligned, the beam scanner was to be removed, the drift tube structure re-inserted and its alignment fine tuned by external adjustment. However, the beam scanner did not posses sufficient mechanical accuracy to align the beam. Nevertheless, it provided valuable information on the formation, propagation and adiabatic compression of the electron beam. In particular, beam scanner measurements showed that electron beam envelope calculations for a uniform density beam with space charge, combined with experimentally measured magnetic fields, agree very well with observation.

In the end, we found that the source could be easily aligned in two to three hours with the drift tube structure in place. (The above mentioned segmented drift tube in the transition region between the entrance and main solenoids is indispensable for this alignment.) Our criterion of alignment is that electron beam transmission through the source from cathode to collector remains constant at its maximum value as the main solenoid magnetic field is varied from 0.3 to 2 T. In addition, it is possible to vary the individual drift tube potentials from 0 to 5.5 kV with little or no effect on beam transmission. When the source is aligned, the measured electron current to all the drift tubes is less than 10 microamperes. Another feature of an aligned electron beam is that it is very quiet, i.e. the beam shows no sign of instabilities or noise growth. Typically, the peak to peak amplitude of the high frequency hash present on the beam is less than 0.1% of the beam signal measured across a 10 ohm resistor connected between the collector and ground.

With the proper amount of feed gas injected into the ionization region and an appropriate trapping potential distribution, the source produced all charge states of C, N and O, including bare nuclei, and all charge states of argon up to and including Ar^{11+} in a 1 millisecond confinement time. After 1 millisecond, the charge state distribution did not evolve further with time, even though the inferred electron beam current density from short confinement times indicated that bare argon should be produced in 20 msec. In searching for possible causes of this behavior, our inhouse built Hughes 112-2B electron gun was replaced by a commercial Litton M-707 gun, for which the transmission obtained at 8.5 keV, and 1.5 T is 100% within experimental error. The M-707 perveance is 0.26 compared to 2.2 microperv of the 112B, and the improved transmission is most likely due to better laminarity of the Litton gun at the lower perveance.

Unfortunately, good alignment and excellent transmission of the electron beam through the source did not improve its performance significantly over that observed with the Hughes gun. After an extensive search for the cause or causes of this behavior, we have obtained direct evidence from both CEBIS I and CEBIS II that the major problem was poor vacuum in the electron collector-ion extractor. Because of the very restricted pumping speed in the collector region, the collector could not be properly conditioned and each beam pulse produced a burst of desorbed gases, (mostly CO and CO_2) which filled the electron beam with unwanted C and O ions.

At this point we decide to replace the 16.5 cm warm bore, 0.96m long, 3 T, iron shielded superconducting solenoid by a solenoid manufactured by Nicolet Instruments Corporation of Madison Wisconsin. The Nicolet solenoid, made for their Fourier transform mass spectrometer, is unshielded, 0.79 m long, has a 15.2cm diameter warm bore, and at 3 T a field profile very similar to that of our old solenoid. It operates in the persistent current mode and its main advantage over the old solenoid is the liquid helium, LHe, and liquid nitrogen, LN_2, consumption of 50 and 120 liters respectively per month, compared to 35 liters of LHe and 150 liters of LN_2 per day consumed by our old solenoid. The very low cryogen consumption will allow us to operate the source continuously, instead of intermittently in two week runs. Because of the slightly smaller magnet bore and shorter length, the main vacuum chamber had to be rebuilt, and we took the opportunity to redesign the support structure of the collector-ion extractor to drastically increase pumping in the collector region. At the same time, we have incorporated a number of other modifications and improvements, such as shortening the gun side positioners, replaced the vacuum chamber positioners by ones using linear bearings and ball lead screws, reduced the overall mass of the cryopump, improved the thermal isolation of the cryopump shields and a host of other minor modifications to make source operation easier. This newest version of the source is shown in fig. 2 and is currently undergoing tests.

CONCLUSIONS

The flexible design of the Cornell cryogenic, superconducting EBIS has allowed us to investigate the effects of vacuum, the degree of electron beam alignment with respect to the magnetic field and/or drift tube structure, correct and incorrect launch of the electron beam into the magnetic field, etc. on the behavior of the source. Individually, the effects of each of the above on electron beam transmission may not be that critical,

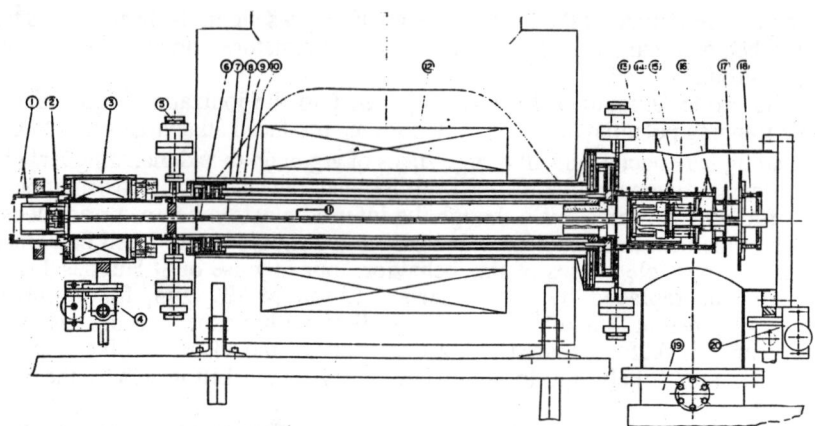

Figure 2. Latest version of CEBIS II utilizing a Nicolect Instrument Corporation 3 Tesla, superconducting solenoid.

but often the combination of two or more improper conditions has a drastic and deleterious effect on source behavior.

The outstanding features of our source are the manner whereby the electron beam is launched into the magnetic field and the provision for beam alignment. Minor mechanical modifications of the entrance magnetic shim and the cathode to shim distance for an individual gun allow us to use whatever gun we choose. We have successfully launched electron beams from three different guns, and in the near future plan to install a 60 kV, 0.2 microperv gun from a Hughes model 605 H travelling wave tube. Our method of beam launch combined with the built in adjustment capabilities have led to almost ideal beam behavior in the magnetic field.

However, 100% electron beam transmission and good beam alignment seem to be only a necessary but not sufficient condition for the EBIS to operate as an ion source for very highly charged ions. Clearly the presence of ions in the electron beam modifies its properties, something well known from early work on travelling wave tubes and more recently rediscovered in work on EBIS sources. Whether the effect of ions is due to slight misalignments of the electron beam containing ions with respect to the magnetic field axis or due to plasma instabilities, or both, is to our mind still not clearly resolved at this point. In a similar vein, the question of ion heating by the electron beam and cooling of heavier ions by lighter ions, the mechanism proposed by Levine et al.[18] to explain the very high charge states and long trapping times obtained with the EBIT requires further investigation. Clearly there is a difference between a 2.5 cm long ion filled beam in an EBIT and a 50-100 cm long beam in an EBIS. In CEBIS I we have observed the replacement of lighter, residual gas ions from the electron beam by the heavier argon feed gas ions, and the replacement of argon ions by xenon ions. We plan to investigate this point further by injecting a beam of singly charged lead ions from a sputter PIG source[19] into the EBIS ionization region and mix the ion beam containing lead with lighter gas ions.

ACKNOWLEDGEMENT

The contributions of my colleagues, B. Amini, E. N. Beebe and J. Perotti are gratefully acknowledged. The work carried out was supported in part by the U.S. Department of

Energy, Office of Basic Energy Sciences, Division of Chemical Sciences.

REFERENCES

1. R. Geller and B. Jacquot, Physica Scripta T3, 19 (1983).
2. E.D. Donets, Physica Scripta T3, 11 (1983).
3. V.O. Kostroun, E. Ghanbari, E.N. Beebe and S.W. Janson, Physica Scripta T3,47,(1983).
4. E.D. Donets, Fiz. Elem. Chastits At. Yadra 13, 941 (1982), in Sov. J. Part. Nucl. 13, 387 (1982).
5. L. Brillouin, Phys. Rev. 67, 260 (1945).
6. J. Arianer, A. Cabrespine and C. Goldstein, Nucl. Instr. and Meth. 193, 401 (1982).
7. C.R. Moster and J.P. Molnar, unpublished notes, Bell Laboratories, 1951.
8. J.F. Gittins, Power Travelling Wave Tubes, (American Elsevier, New York, 1965).
9. V.O.Kostroun, Nucl. Inst. and Meth. in Phys. Research, B10/11, 771 (1985).
10. P.T. Kirstein, G.S. Kino and W.E. Waters, Space Charge Flow, (Mc Graw-Hill, New York 1967). p. 163.
11. J.L. Powell and B. Crasemann, Quantum Mechanics, (Addison-Wesley, Reading MA 1961), p.140.
12. S.W. Janson, Proceedings of the Third EBIS Workshop, Cornell University, May 1985, V.O. Kostroun and R. W. Schmieder Eds.
13. Ren Yao, Ph .D. Thesis, Cornell University, 1986. Unpublished
14. R.W. Hamm, L.M. Choate and R.A. Kenefick, IEEE Trans. Nucl. Sci. NS-23, 1723 (1976).
15. E.N. Beebe, these proceedings.
16. W. Thompson and S. Hanrahan, J. Vac. Sci. Technol. 14, 643 (1977).
17. J.V. Frank, A.A. Arthur, L.A. Brusse and W. Low, Lawrence Berkeley Laboratory Preprint LBL 6382, (1977).
18. R.E. Marrs, M.A. Levine, D.A. Knapp and J.R. Henderson, Phys. Rev. Lett. 60, 1715 (1988).
19. P/N 2-21 ion source for metals and other solids by Physicon Corp., 221 Mt. Auburn St., Boston, MA 02138, USA.

DISCUSSION

Becker: Did you also make these profile measurements on the Litton gun, and was it more rectangular?

Kostroun: We have not done this on the Litton gun. However, from the way the transmission and everything else went, clearly the Litton gun is more superior to our homemade gun. This is not surprising since Litton has more experience. From my own experience, we should not waste time trying to build guns, but should buy them from Raytheon, Litton, Hughes or whomever. I particularly like the Litton gun. It gave some very good results.

Marrs: I would just like to remind people that there is one very important effect which is not in your beam envelope equations and which prevents you from ever getting electron beam collapse in an EBIS, even if there is zero magnetic field on the cathode and even if you have no scalloping or misalignment. The effect is that the Liouville theorem will not let you compress the transverse phase space of the beam infinitely small.

Kostroun: Sure, but that is because you can never have a perfect beam from any cathode.

Marrs: Take a typical cathode where you have, say, 2 A/cm^2 emission and 0.1 eV temperature, the compression....

Kostroun: Do not forget that in Brillouin flow the beam is launched from a zero temperature cathode in zero magnetic field.

Marrs: But the real world does not.

Kostroun: But that was precisely my point.

Marrs: That is in the Herrmann theory, for example.

Antaya: Has anyone observed an electron beam collapse since Arianer?

Kostroun: No, but there is some very early work by Senise, who was working in travelling wave tubes in 1958 or thereabouts, who observed collapse in electron beams.

Schmieder: Val, you said that invariance of beam transmission to geometry and other parameter variations was necessary. Do you also feel it is sufficient?

Kostroun: No, it is necessary. Clearly there are other effects. The vacuum plays an important role as does the way the ions are injected, the problem of heating, and other things. Although I had not mentioned it, one of the things we hope to do is to use tube 11, our gas injection tube, to inject coolant gas into the source. Our sputter PIG source will be used to inject various ions and we plan to cool these ions by injecting neon, argon, xenon, and krypton. The problem of ion heating has been known for some time in the EBIS game, although it has not been advertised. Donets alluded to this back in 1982 at the Stockholm meeting. At that time it just went over my head; I did not realize what he was talking about. But he knew about it then. He uses ions to cool ions. The LBL people use atoms to cool the ions. That is the scheme we plan to use too.

Becker: I just want to comment on Ross' question concerning Liouville's theorem. In that formulation $\phi \neq 0$ at the cathode, which means you include transverse temperature as an equivalent to cathode flux. And then you get a term $1/r^3$.

Kostroun: I think that, strictly speaking, one should not talk about Brillouin but equilibrium flow. Brillouin flow is an ideal situation.

Becker: It is just a matched beam for the magnetic focussing technique and you can control it completely equivalent to what you get in the envelope equation for any rf accelerator.

Kostroun: I might add that, so far, we have not seen effects due to any instabilities in our source. Instabilities may be there, but in the past, whatever was observed was due to bad misalignment, not knowing what the magnetic field was in critical places or some very obvious condition, that once corrected, helped the situation. By the way, the trap length in our new source, is about 35 cm long. It is getting shorter and shorter but purely by accident.

Becker: What is the helium consumption of the solenoid?

Kostroun: Very important. The helium consumption of the solenoid is about 50 liters per month. Nicolet makes the solenoid for Fourier transform mass spectrometers that they sell to laboratories all over the world. They do not want the chemists to worry about filling them often. It is an instrument that is made commercially and so it has very good consumption.

Stockli: Did you try to map the field and did you transmit an electron beam?

Kostroun: No, not yet. From the field measurement that we have from Nicolet I think that the field is as good as, probably even better than, what we had in our previous solenoid. I think it is more important to be able to find the magnetic axis, and we have the means of doing that. My feeling is that all the effects and defects in the cathode positioning and so forth are what kills you ultimately. Certainly a nice straight magnetic field is a great thing to have but that is only one of the prerequisites. There are lots of gremlins hiding in there that make it difficult to propagate the beam.

Stockli: How did you measure the beam current inside the solenoid?

Kostroun: On a Faraday cup which operates with the beam scanner. It is just a long tube. Of course there is a problem using a Faraday cup in a magnetic field.

Stockli: You mentioned that for alignment you follow through from low field to high field. Could you elaborate on this statement, please?

Kostroun: Well, we can tilt our structure. So we start out with the drift tubes centered in the chamber. Then we move the outside positioners and we can follow the beam where it is hitting tube 1, etc. until it reaches the collector. That is, you have maximum transmission to the collector.

Stockli: So you measure the losses on the drift tube?

Kostroun: Not quite at this point. You could have half an ampere on some drift tube. We use a short pulse, 10 microseconds long, with a low duty cycle. As the source is gradually aligned, the length of the pulse is increased. Gradually you have to do less and less alignment, until you get to the point where you are getting the beam through. One interesting thing about this is you can get to the point, at say 2 Tesla, where you get 100%, or in the case of the in-house built Hughes gun, 98% transmission. If you move the drift tubes around, the beam gets noisy, but you still get most of it through. To get a quiet beam, the beam has to be aligned with the magnetic axis and then the drift tubes have to be aligned with respect to the beam.

Stockli: Could you tell me how you align the entrance solenoid?

Kostroun: The entrance solenoid defines an axis on which the beam is. That axis is defined by the hole in the entrance shim and by the mechanical construction of the solenoid.

Becker: If you say you calculate the beam behavior with the envelope equation, are you putting in the cathode flux?

Kostroun: We always put in about 3 G on the cathode which includes the temperature of the cathode.

Marrs: You might be interested in the way we did the alignment of our device in Livermore. It is a little easier because we only have one magnet plus a bucking coil around the gun. We used a Hall probe to map the field before assembly so we knew where the magnetic axis was and where it pointed within 10^{-4} radians. Then we just put it together so that the central magnetic field line went through the center of the cathode and the same field line went through the center of the collector. We never had any crooked beams and the transmission is 99.99+%. I wonder why you did not map ......

Kostroun: Well we did map. But, remember, that while in your device you have to really worry about a 2 cm distance, we have to worry about a distance of 1.25 meters. We tried mapping field lines over this distance and found that maps and attempts to find the magnetic axis were not good enough when we reinserted the EBIS structure.

Amboss: On one of the slides you showed beam profiles. If you look at #1184, you see the main beam and you see a ring of charge around it. I think that one of the problems that you have is the cathode is in the wrong place, with respect to the focus electrode.

Kostroun: Well, it could very well be because this was, as I said, a home-built Hughes gun.

Amboss: That will teach you!

Kostroun: I could not afford the price you guys wanted for it.

Amboss: Well, when you do not charge a lot of money you do not get people like me coming to conferences! Anyway, it could be that the cathode is in the wrong place, so that the focus electrode hole is too large so you are getting side emission. Has it ever occurred to you to put two or three sets of deflection coils along the side of the beam tunnel?

Kostroun: No, I have not. But it is a good point.

Amboss: And then do your beam steering by just putting on small transverse magnetic fields.

Kostroun: That is certainly possible to do.

Amboss: Something I have had some success with is putting thermocouples along the circuit; they can tell you where things are getting hot.

Kostroun: By the way, when the beam is aligned, our total current falling on the drift tube is something like 20 µA. That is to all of the drift tubes. One of the things that we found with the Hughes gun, and again let me reiterate, this gun is a homemade version of a gun Texas A&M bought from you about 10 years ago, is that it had all sorts of problems. However, when we installed the Litton gun all of the problems went away. However, there is some evidence that the Litton gun, which is a very good gun, has a small, non-laminar component which does get reflected as the electron beam enters the 2 or 3 Tesla magnetic field. This is a potential problem in our EBIS because, even though the reflected component may only be a few microamps, this hits the drift tubes and causes desorption of gas which then causes other problems. The difference between a homemade gun and a commercial gun is like night and day.

Levine: I agree that you really need the gun aligned well. In EBIT, the alignment that we find critical is not in terms of millimeters, but in 1/40th of a millimeter. We find .025 mm will affect the beam quality. I think a good way to define the beam quality is to use the Herrmann equation for beam size to calculate an effective cathode temperature. Our effective cathode temperature calculated from beam size is 0.1 eV. As Reinard Becker has pointed out, it is critical to use the correct magnetic field gradient at the gun cathode in the launching region.

Kostroun: Well, you missed the point of how we launch and compress the electron beam. In our source we do what the people in travelling wave tubes do to launch a beam. We do not have this long, sloping magnetic field where we have to match the field lines at the cathode. That is precisely what we want to avoid. Basically what we do is take the beam, follow it out of the magnetic field, and then match the free space expansion with the beam coming from the cathode. In other words, we match beam profiles. Once the beam is established, we then compress it adiabatically by the main field.

Levine: And what effective cathode temperature do you get?

Kostroun: We use a cathode temperature of about 1100° K and we use the magnetic field that we measure at the cathode position with a flux gate.

Levine: But what I mean is, after you have compressed your beam to get the highest current density possible, and if you considered just thermal compression of the electron beam, what effective temperature would you have to suppose for your beam?

Kostroun: I do not know, whatever 3 G plus 1100° add up to.

Levine: Does that check your beam density?

Kostroun: To measure a current density inside this device is not trivial. The current density we infer from the time evolution of the charge state distribution of ions measured by time-of-flight model dependent. Until I can get a hard number that I can measure in some way, I am not going to make any claims about the temperature or anything. I am just saying that the current density is sort of consistent with what these envelope codes calculate; self-consistent meaning agreement within, say, 50%. I do not know anybody who can do better than that with any confidence.

Faure: Is your measured profile the one you intend to have?

Kostroun: Well, no. These are lousy profiles. The only thing that I meant to say is that when we take these measured profiles and compare them with the calculated envelope profiles using the magnetic field that we measure on axis and some reasonable cathode temperature, the current densities sort of agree. In other words, the envelope code may give 300 A/cm^2 and we measure 200 A/cm^2. For me, an ex-nuclear physicist, that is excellent agreement.

Stockli: I would like to comment on the importance and difficulty in aligning a long horizontal solenoid, such as CEBIS. The length of the structure and its orientation cause everything to sag, which has to be considered in the design of the probe or in interpreting its measurements. Vertical solenoids, such as the Sandia Super EBIS, or horizontal and short solenoids, such as EBIT, are much easier to align.

EBIT: Electron Beam Ion Trap

M. A. Levine*, R. E. Marrs, C. L. Bennett, J. R. Henderson, D. A. Knapp and M. B. Schneider

Lawrence Lawrence Livermore National Laboratory, Ca 94550

ABSTRACT

An Electron Beam Ion Trap (EBIT) has been built as an instrument for *in situ* studies of atomic physics. Based on the EBIS concept, EBIT incorporates several novel features including ion cooling using light ions and plasma instability control using a short trap length. To understand the operation of EBIT, measurements have been made of the electron beam behavior. The radius of the beam is observed to follow Herrmann Theory during compression. The electron beam displays an energy dispersion that is larger than theory. However, this energy dispersion is only about 15% of the electron temperature in the trap due to the adiabatic compression of the beam.

*Permanent address Lawrence Berkeley Laboratory, Ca., 94720

INTRODUCTION: The Electron Beam Ion Trap (EBIT)[1] at the Lawrence Livermore National Laboratory was built for the study of atomic physics in Highly Charged Ions (HCI). This instrument is based on the EBIS[2] concept. However, instead of extracting ions as in EBIS, the electrostatic ion trap is used for *in situ* measurement of ion-electron collisions by observing the emitted X Rays. Thus in EBIT, the electron beam is used not only to trap and ionize, as in an EBIS, but also to probe and excite the HCI.

Dielectronic Recombination is one of the more interesting measurements that can be made in EBIT. DR is of interest because it is a dominant mechanism in determining the equilibrium in low density, coronal plasmas such as found in some gas lasers, controlled fusion devices and astrophysics. Previously, DR had only been measured in atoms with a charge, q, of 6+ or less. The problem has been that inelastic cross sections, such as ionization[3], vary as $1/q^2$ and HCI are difficult to make. Nevertheless, high densities are needed for measurements. To solve this problem, EBIT is designed both to create and trap HCI at densities $\approx 10^{10}/q$ per cc. It should be added that because of these unique characteristics, EBIT has turned out to be a very versatile instrument which can perform a host of atomic physics measurements, and may even be useful in some nuclear physics studies which require fully stripped atoms.

Prior to the construction of EBIT an attempt was made to perform DR measurements in the Berkeley EBIS[4]. The measurement planned were similar to those made since in Cryebis[5]. However, an examination of the Berkeley EBIS indicated that the desired HCI could not be produced[6]. In an EBIS, ions are electrostatically trapped, radially by the electron beam and axially by a set of biased drift tubes or electrodes. Ions in the electron beam are sequentially stripped of electrons by the electron beam and, incidentally, heated by collisions with the electron beam. The electron beam radial potential is usually less than 20 eV and the total radial potential to the inside of the drift tube electrode is only a few hundred volts. As ions are heated they first escape the electron beam so that they are not ionized and then escape the trap and are lost.

Ions are heated by collisions with the high energy electron beam. Elastic collisions with electrons were predicted[7] to give a charge limit q<50+. In EBIT "evaporative cooling"[8,9], discussed elsewhere in this conference, is used to carry away the heat generated by collisional heating of the HCl by electrons. However, if the electron beam is unstable, so that the beam is bunched, then the ion heating rate is too large for cooling.

An examination[10] of the Berkeley EBIS indicated that it was subject to many plasma instabilities, so that the ions were being heated and lost before reaching a high charge state. A catalogue of the instabilities found in the Berkeley EBIS indicated that most were convective-like in nature in that they had a growth rate that increased with cavity length and decreased with electron velocity.

Theoretically, many of the instabilities observed in EBIS can be stabilized by judicious shaping of the cavity or the use of absorber material. However, the "rotational-two-stream"[11] (RTS) instability is confined to the ion trap and can only be stabilized by varying conditions in the trap. There is some controversy[12] on the condition for stabilizing the RTS instability. In the EBIT design, the more pessimistic criteria is used. In EBIT, the trap length is shorter than the wavelength of a plasma oscillation as seen by the ions in the moving beam-electrons. This implies a two centimeter trap length. To help control other instabilities, the overall length of the electron beam was kept to ≈ 40 cm.

To facilitate *in situ* study of trapped HCI, EBIT is constructed with a vertical axis and four radial ports. This provides good access to the ion trap for spectroscopic observation of ion-electron collisions. As in the Berkeley EBIS, the radial ports have also been very helpful for beam diagnostics, an important component for planning and understanding the spectroscopic measurements. See fig. 1.

<u>Electron Beam Design</u>: The EBIT electron beam is produced in a Pierce gun[13] with a 3mm cathode diameter, a perveance, p=0.5 μperv, and a focal spot radius, r_x=0.036cm. The electron beam is compressed to a smaller radius and a higher current density with a 3 T superconducting coil. Herrmann theory[14] is used to calculate the electron beam radius, r_o, (80% current) as given by,

$$r_o = r_B \left[\frac{1}{2} + \frac{1}{2}\left\{ 1 + 4\left(\frac{8kTr_c^2}{m\eta^2 r_B^4 B^2} + \frac{B_c^4 r_c^4}{B^2 r_B^4} \right) \right\}^{\frac{1}{2}} \right]^{\frac{1}{2}}, \qquad (1)$$

where r_c is the cathode radius, r_B is the Brillouin radius, T the electron temperature at the cathode, m the electron mass, η the ratio of electron charge to mass, and B the magnetic field intensity. If one assumes that the magnetic field at the cathode, B_c, is zero, then a measurement of r_o at a known magnetic field can be used to obtain an effective cathode temperature which can serve as a figure of merit for electron gun performance.

As can be seen in eq. 1, to obtain a minimum beam radius and a maximum current density, the magnetic field, B_c must be near zero. In EBIT, the gun cathode is mounted in a ferromagnetic shield with a bucking coil so that the magnetic field goes to zero at the cathode and the magnetic vector potential contour lies along the spherical cathode contour.

The transition region in which the electron beam enters the magnetic field requires careful design because the electron beam diameter tends to oscillate in space or scallop due to space charge effects. An empirical relation for the magnetic field gradient, near the cathode, which minimizes the electron beam scallop[15] or spatial oscillation, as it enters the magnetic field, is given by:

$$\frac{dB(T)}{dz(cm)} = 7.9 \frac{p(V^{3/2}/A)V^{1/2}}{r_x^2} \qquad (2)$$

Where, V is the electron-gun anode voltage. The Herrmannsfeldt[16] code was used to design the electron beam dynamics in the gun region.

A major problem in the operation of EBIT is the presence of secondary electrons. These electrons are trapped in a Penning mode on magnetic field lines in regions of positive potential. To minimize secondary electron production, a negative suppresser electrode is placed between the collector and the drift tube assembly and a positive ion extractor inside the collector is used to help spread the electrons radially.

The electron-gun cathode is used as the electrical reference and maintained at ground potential. The electron beam current, 0 to 130 mA, is controlled by the gun anode. The electron beam energy, in the trap region, is controlled by the potential on the drift tubes consisting of a set of three electrodes. The central electrode, in which ions are trapped, is about 2 cm long and is the high voltage potential reference. The high voltage reference is controlled by a precision amplifier which has a settling time of 2 ms (0 to 30 kV). The up-stream and down-stream drift tubes are individually driven by 400 Volt amplifiers relative to the central drift tube. The rise time of the amplifier on the down-stream drift tube is effectively 1 microseconds so that the trap can be gated open during injection.

Ion Injection: Ions are introduced into the trap either by injection from an ion source along magnetic field lines, or by the ionization of background gas by the electron beam. The ion source used is a MEtal Vapor Vacuum Arc source (MEVVA)[17]. A MEVVA was chosen because it is simple, easily pulsed, produces almost any metallic ions, gives an ion current of 1 A/cm^2, and has a reasonable center-of-mass temperature of about 15 eV. Its principle disadvantage is that ion production tends to vary from pulse to pulse.

In EBIT, the MEVVA source is placed on axis, about 75cm above the extractor and the ion beam expands to a 6 cm diameter before it reaches the collector. The aperture in the collector is only 3 mm in diameter so that only a very small percentage of the ions, with a transverse temperature of ≈0.015 eV, is accepted. After entering the collector, the ions are guided and compressed by the electron beam on the path to the trap. After compression, the ions are estimated to have a temperature of about 15 eV so that they can be contained within the electron beam.

Machine Alignment: In EBIT, the electron gun, the center of the drift tube or trap assembly and the collector is designed to lie on the same magnetic axis. The trap magnetic field is produced by a superconducting, cold-bore magnet with a Helmholtz configuration. The magnetic field uniformity is 0.02 percent on axis for the 2 cm trap length and the measured deviation of the axial magnetic field line from the mechanical axis is <.003 radians from gun to trap. The drift tube assembly is supported on precision machined insulators in the center of the magnet. The collector and magnet are mounted and mechanically aligned before assembly. The axis of magnet, collector and drift tube assembly is vertical and is pivoted on a ball joint at the top of EBIT. The electron gun assembly is rigidly mounted at the bottom of EBIT. Four stainless steel wires attached to the magnet housing at the lower end, near the electron gun are used to tilt the magnet until the system is aligned optically. The support system is such that the alignment is observed to remain constant to within ≈ .005 cm during cooling of the superconducting magnet. The final adjustment is made after cooling when the electron beam can be turned on. At this time, the current to the "snout", an electrode just above the anode on the gun assembly, is measured. The snout current is a measure of the number of secondary electrons created when the electron beam "scrapes" various apertures. Final alignment is made by minimizing the "snout" current which can be done by adjusting the tension on the stainless steel wires and then varying the magnetic field in two external steering magnets. The steering magnets produce a weak field of a few Gauss for the beam length, orthogonal to each other and the axis of EBIT.

Spectroscopic Measurements: (For a more complete description of EBIT spectroscopy see Marrs et al in this conference). Ions have been and can be extracted from EBIT, however, only crude time-of-flight measurements can be made at this time and these measurements are inferior to spectroscopic measurements. With the radial ports and berylium windows, spectroscopic measurements with solid-state germanium and lithium-drifted-silicon detectors are simple and fast. Spectroscopic measurements give information on species, charge state, beam density and cross sections for ion-electron collisions. However, solid-state detectors are limited to an energy resolution ≈3 percent. High precision spectroscopy is performed using diffraction instruments[18]. In these measurements it is convenient to use the electron beam of EBIT as the "slit" for the spectrometer. Important to the interpretation of spectrographic results is knowledge of the characteristics of the electron beam current density and energy resolution.

Measurement of the Electron Beam Diameter: The average electron beam current density can be found by measuring the electron beam diameter. A cartoon of the measuring scheme is shown in fig. 2. The region of overlap of the electron beam and the trapped-ion cloud is an x-ray source. A slit, aligned parallel to the EBIT axis, is placed 1 cm from the beam and an x-ray image of the beam is cast on a position sensitive proportional counter about 50 cm away.

The analysis of the proportional counter image assumes the electron beam density has a Gaussian-density distribution in radius and the ions have a uniform distribution throughout the region. Actually, the ion distribution is given by $\rho_i = \rho_0 \exp(-qV(r)/kT_i)$ where q it the ion charge, V(r) the potential and T_i the ion temperature. In order that the ion distribution be uniform, the ion temperature must be several times larger than the potential at the characteristic beam radius i.e. $kT_i > \alpha qV(r_0)$, where α is large.

If α is large, it can be shown that the the distribution at the detector is given by,

$$A(y') = A_o \left\{ \text{erf} \left[\frac{1}{\sqrt{2}\ \sigma\mu} \right] y' + \frac{w}{2}(1 + \mu) \right] - \text{erf} \left[\frac{1}{\sqrt{2}\ \sigma\mu} \right] y' - \frac{w}{2}(1 + \mu) \right] \right\}$$

where μ is the slit magnification, w is the slit width, y´ the distance on the image plane and σ is the variance of the beam radius.

Computer calculations of the ion temperature for equilibrium conditions[19], indicate that the ion temperature is approximately proportional to the axial trap potential. Thus, a large trap potential tends to produce a high ion temperature and a uniform ion density in the region of interest.

The intensity of the x-ray radiation, A(y´), from a mixture of neon-like, sodium-like and magnesium-like gold, cooled with titanium was examined as the trap potential was varied from 40 V to 340 V. The observed variance was almost constant from 140 V to 340 V so that it is assumed that at 340 V, α is large. These preliminary measurements indicated a beam radius ≈ 25 micron to 30 microns for 80% of the beam current. In Fig. 3 the measured beam radius is shown on a plot with curves for the Brillouin theory and the Herrmann theory.

From eq.1, a 25 micron beam radius implies an effective cathode temperature, $T_c \approx 0.1$ eV. The radial compression of the electron beam from cathode to trap implies an electron temperature in the trap, $T_t \approx 300$ eV. However, if the electron beam flow is laminar from the cathode to the trap, the variation of electron energy across the beam, is the difference in potential due to the beam space charge, ≈ 15 V.

The beam-electron energy, or more precisely, the energy in the center-of-mass for ion-electron collisions can be measured by looking at electron-ion recombination. DR is sensitive to collisional energy[20] and the measurement can be extracted with a knowledge of the several resonances and their relative strength. Alternatively, precision spectroscopy of x-rays from radiative recombination (RR) can be used. The advantage of the RR measurement is that recombination to single states in bare and hydrogen-like ions can be used to give an unambiguous measurement. The disadvantage of RR is that the cross section is relatively small and precision spectroscopy using refracting crystals has a low counting rate. To improve the counting rate and make the measurement practical for diagnostic purposes, a K-edge filter can be used to make the measurement.

Fig. 4 shows a diagram of the apparatus for making a K-edge absorption measurement through a krypton gas cell. Krypton is used because it has a relatively sharp and smooth K-edge[21] as shown in fig. 5. RR to the hydrogen-like state is the most convenient to use for the measurement. In a measurement using iron, the electron beam energy is adjusted to maximize the number of in the hydrogen-like state. The electron beam energy is then lowered periodically to map out a K-edge absorption curve. Fig. 6 is an iron spectrum as seen by a germanium solid state detector with the beam energy low enough so that RR to both the bare and hydrogen-like is visible. In fig. 7 the beam energy is increased so that only the line from the hydrogen-like iron is seen. The apparent absorption curve of the gas cell shown in fig. 8 is obtained by varying the electron beam energy. The ion-electron collisional energy is the deconvolution of the K-edge curve from this apparent absorption curve.

To abstract the intrinsic beam-energy variance from the ion-electron collisional energy, the influence of power supply ripple, and ion temperature must be subtracted. It is also possible that the relation of the trap voltage to the drift tube voltage can change due to a variation in ion density and/or a variation of the number of Penning electrons surrounding the beam. Preliminary measurements have given a larger than expected electron beam variance of 25 eV to 50 eV depending on gun and magnetic field conditions. Measurements are in progress in an attempt to better understand the electron beam energy variance in EBIT.

Conclusions: Measurements of dielectronic recombination, impact excitation, radiation polarization and radiative recombination in EBIT have now been made in ion with charge less than 70+. This limit on charge state is due to the limit in attainable beam voltage in EBIT. Plans are now underway to modify EBIT for higher voltage operation and turn it into a SuperEBIT. SuperEBIT would increase the beam energy to 300 kV and make it possible to strip any atom in the atomic table. Recent experiments have shown that ions can be contained in the EBIT for hours. In SuperEBIT this capability will make it possible to study nuclear decay in atoms, such as dysprosium-163, which are predicted to β-decay when fully stripped of electrons.

The performance of EBIT has shown that it is possible to build an electron beam electrostatic trap that is stable and able to contain HCI for hours. Radial access for observation of the beam is an essential part of this instrument for understanding the trap and for atomic physics measurements. SuperEBIT will add the study nuclear physics to the range of experiments in and EBIT.

Acknowledgement This work was performed under the auspices of the U.S. Department of Energy by the Lawrence Livermore National Laboratory under contract W-7405-ENG-48.

References

(1) Levine, M. A., Marrs, R. E., Henderson, J. R., Knapp, D. A. Schneider, M. B., Physica Scripta, T22, 157 (1988)
(2) Donets, E. D., and Ovsyannickov, V. P., JETP 53, 466 (1981)
(3) Lotz, W., Z. Physik 216, 241 (1968)
(4) Brown, I. G., Feinberg, B., Mucl. Instr. and Meth. 220, 251 (1984)
(5) Briand, J. P., et al Phys. Rev. Lett. 52, 617 (1984)
(6) Levine, M. A., Marrs, R. E., Schmeider,R. W. Nucl. Inst. Meth. A237, 429 (1985)
(7) Becker, R., Proc of the 2nd EBIS workshop, Saclay-Orsay, (edited by J. Arianer and M. Olivier) (1981)
(8) Penetrante et al. (this conference)
(9) Schneider et al. (this conference)
(10) ibid 6
(11) Litwin, C., Vella, M., Sessler, A., Nucl. Inst Meth.,198, 189 (1982)
(12) Jacquet, L., Tagger, M., Laboratoire National Saturne, LNS/87/106 (1987)
(13) Raytheon Co., Waltham Ma. USA
(14) Herrmann, G. J., Appl. Phys. 29, 127 (1958)
(15) ibid Becker
(16) Herrmannsfeldt, W. B., Report No. SLAC-226, Stanford Linear Accelerator Center
(17) Brown, I. G., Galvin, J. E., MacGill, R. A., Wright, R. T., Appl. Phys. Lett., 49, 1019 (1986)
(18) Marrs, R. E. et al. this conference
(19) ibid Penetrante et al
(20) Knapp D. et al., Phys. Rev. Lett. to be published
(21) Soules, J. A. and Shaw, C. H. Phys. Rev. 113, 470 (1959)

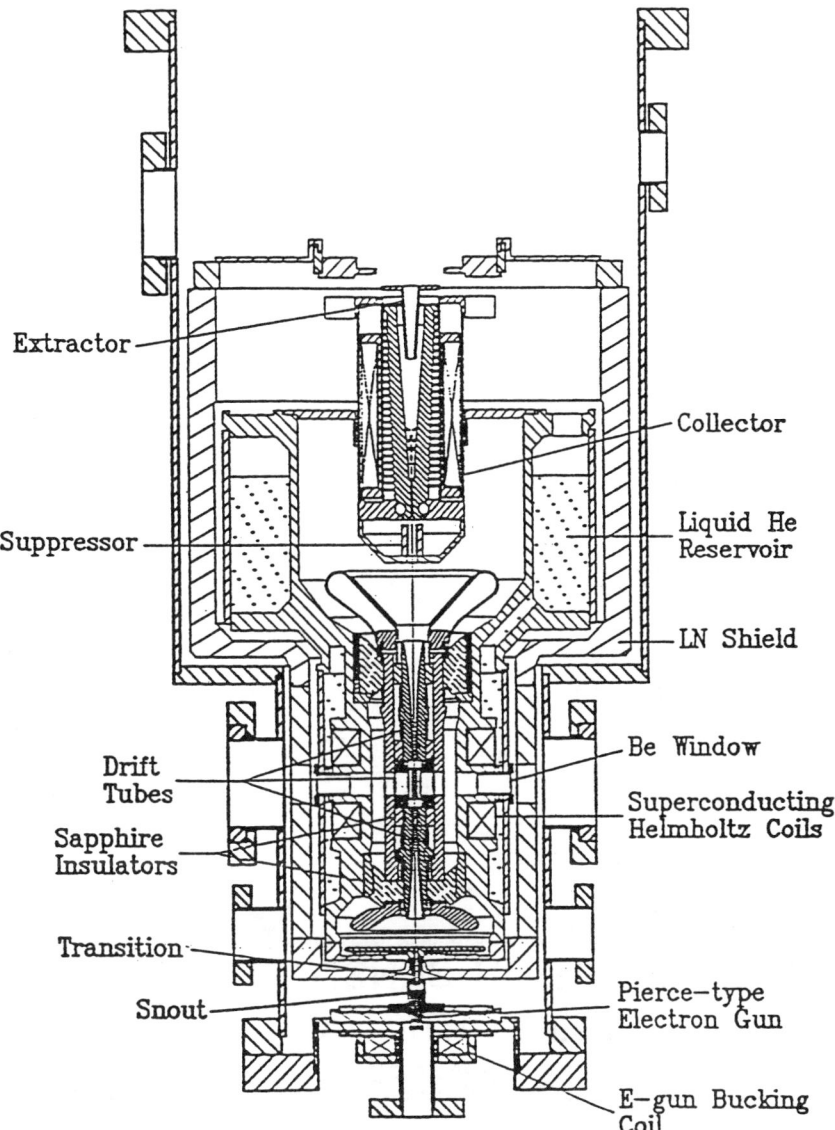

Fig. 1 Cross section drawing of the EBIT. The upper suspension system mounted on the top flange is not shown.

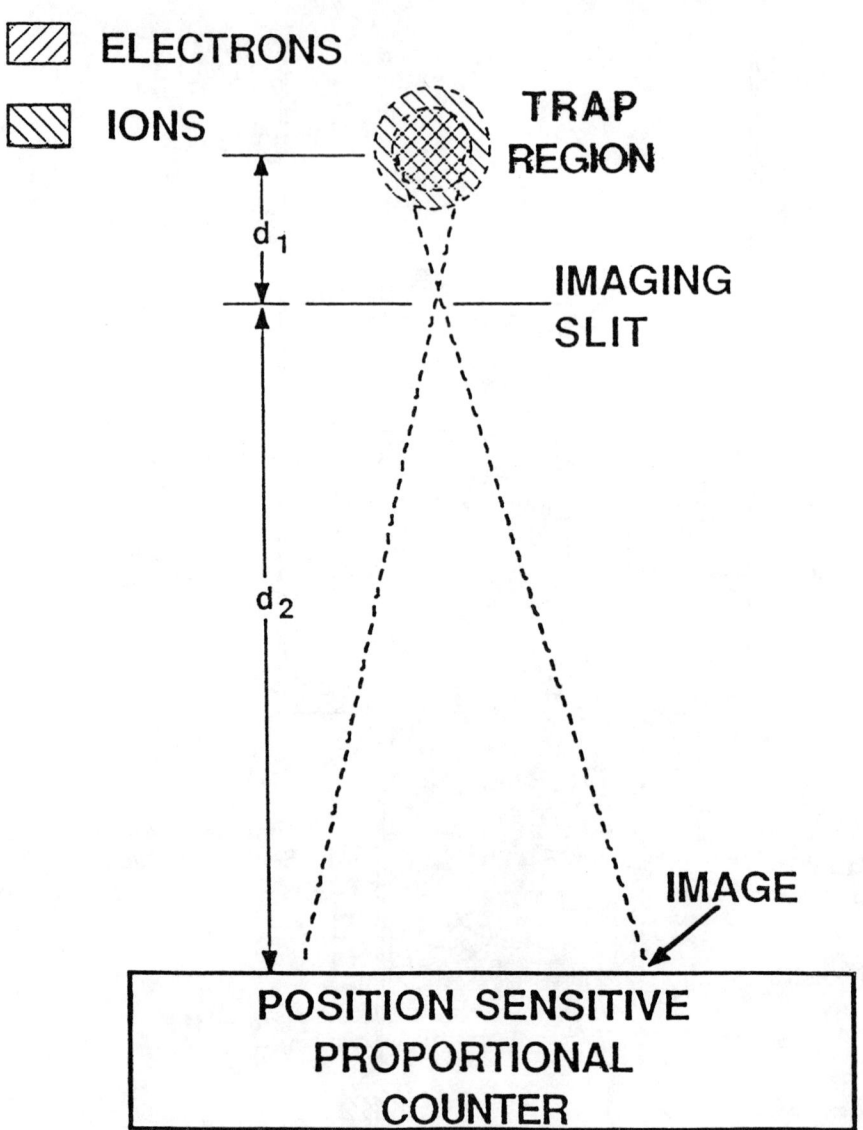

Fig. 2 Diagram of the experimental arrangement for measuring the electron beam radius.

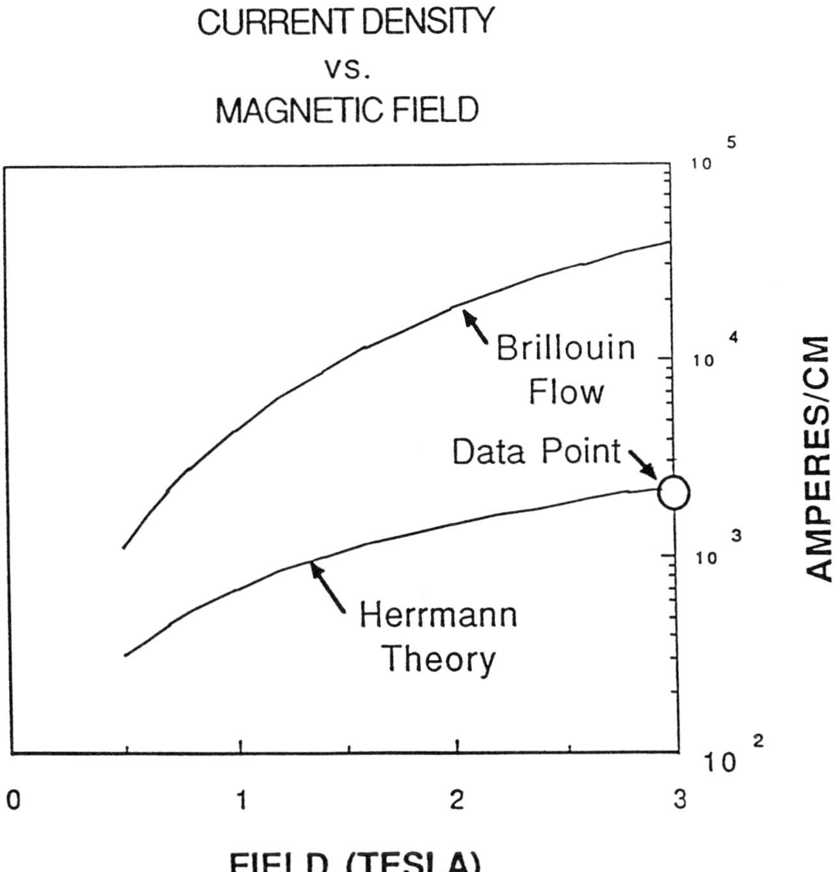

CURRENT DENSITY vs. MAGNETIC FIELD

Fig. 3 Plot showing the current density for a measured electron beam radius in EBIT relative to that predicted by Brillouin Flow and the Herrmann Theory. The theoretical curves are plotted assuming zero magnetic field and an electron temperature of 0.1 eV at the cathode

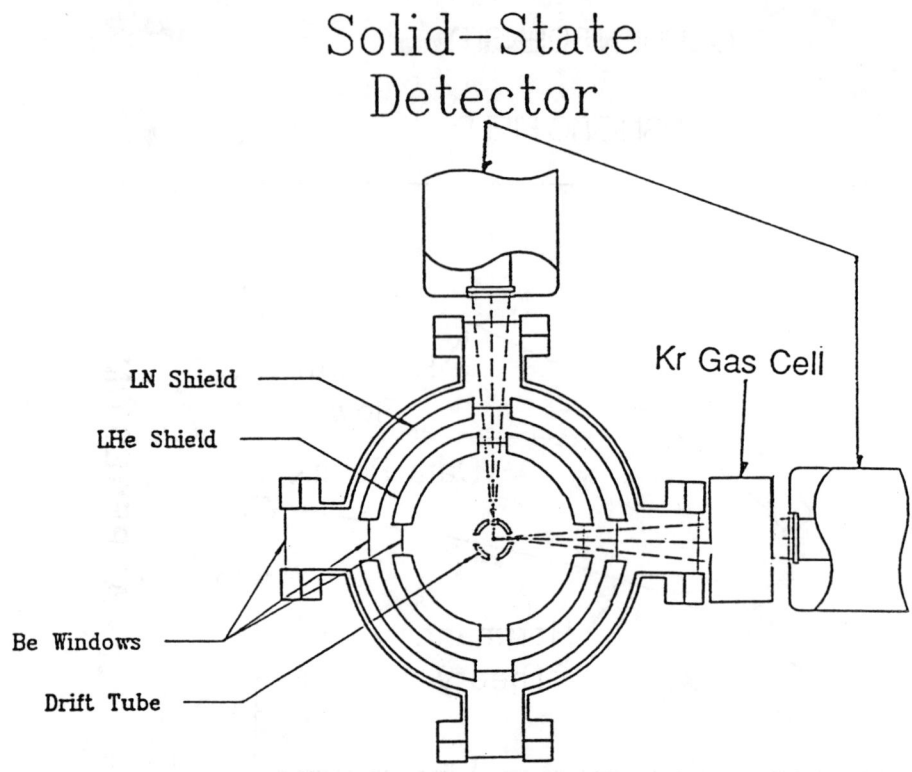

Fig. 4 Experimental arrangement for K-edge absorption measurements in EBIT. The krypton gas cell has berylium windows and is filled to a pressure of 5 atmospheres.

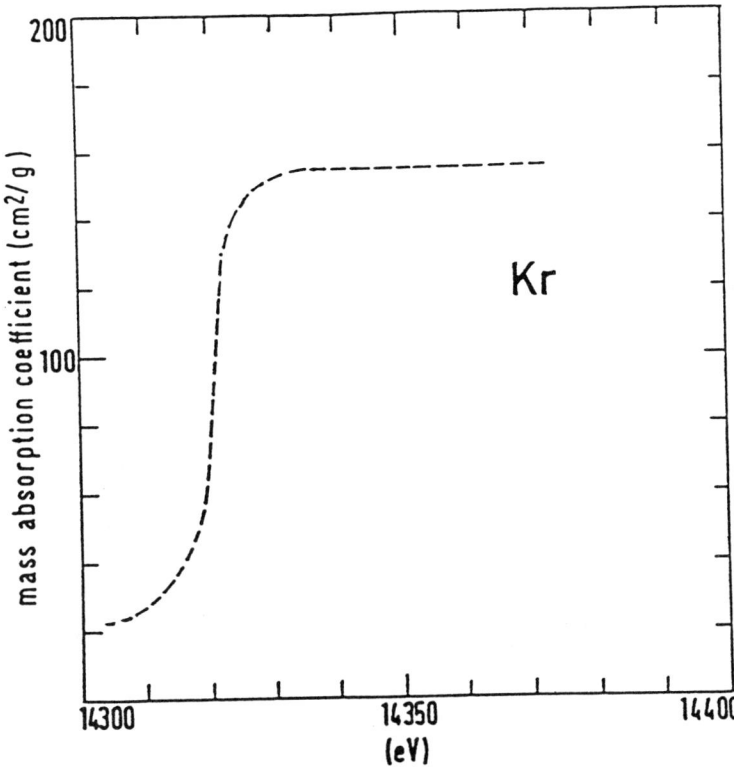

Fig. 5 Mass absorption coefficient of gaseous Kr near the onset of the 1s transitions.

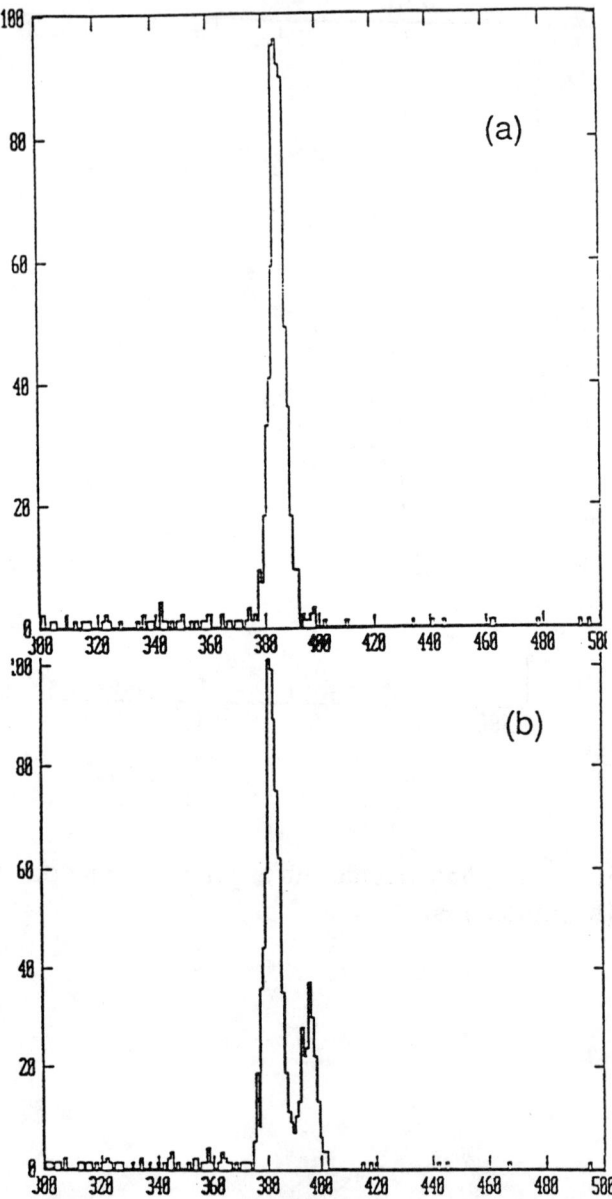

Fig. 6 Iron x-ray spectra as seen through a krypton gas cell with (a) the drift tube voltage at 3.71 kV showing radiative recombination to hydrogen-like iron and (b) the drift tube voltage at 3.61 kV showing recombination to both bare and hydrogen-like iron.

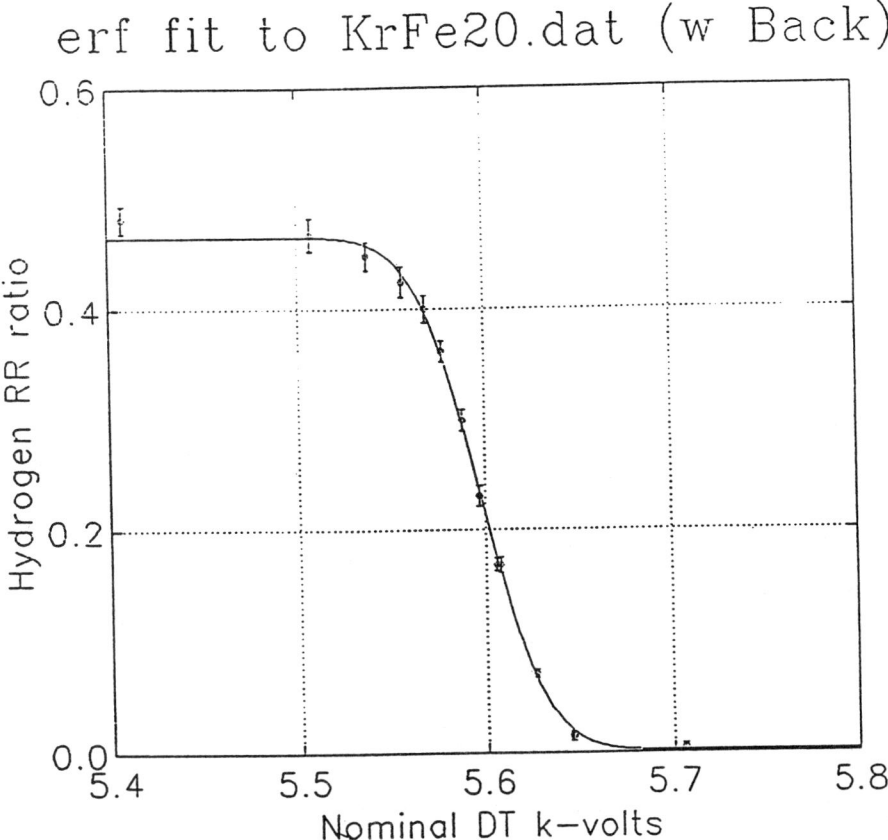

Fig. 7 An erf fit to x-ray intensity through a krypton gas cell as a function of applied drift tube voltage. The uncorrected data has a full width half maximum of 56 ± 3 eV with a chi square of .995.

DISCUSSION

Kostroun: You show that you get Au^{68+} and Au^{69+}. However, you have a 20-30 kV beam, and you could go into the 2p shell. In principle, you could get a higher charge state. What is stopping you?

Levine: Nothing, except we are doing atomic physics. It is very easy to work with a closed shell. Neon-like is a closed shell so that Au^{69+} is preferred.

Kostroun: That is why it does not matter. It is nice to have a 2 or 3 electron system. It is even simpler to understand.

Levine: Well, that is why we do a lot with the helium-like ions.

Kostroun: It is done for titanium and others. I do not quite understand what is stopping you here.

Levine: Yes, but see, we do not exactly know our charge state. When you come to a closed shell you can get very high percentages of a single charge state. It does not mean we are not going to do these things. It also does not mean that we are not looking at other atoms at the time that are not quite neon-like. This is not the full range of things we have looked at. But we have looked at gold because we are interested in the neon-like system and in dielectronic recombination. It is difficult to do and know what percentage you have of ions other than neon-like.

Kostroun: If you could analyze the radiation, there is no problem.

Levine: Well it is more difficult.

Kostroun: There are shifts of a few hundred eV.

Levine: I will let Marilyn answer if there is no problem.

Kostroun: You could certainly separate those x-ray lines. That is not a problem.

Levine: It is possible but it is an added complication.

Schneider: You usually look at things with a solid state detector and there you do not have very good resolution.

Kostroun: The different charge states are certainly separated by more than 300 eV.

Marrs: The real problem is that the electron energy is marginal for opening up the L-shell. The gun was run at typically 18 or 20 kV which is set just below the L-shell ionization potential. We cannot go much farther than that.

Kostroun: OK, I thought you said 30 kV.

Marrs: So even if we reach, say, helium-like gold, there would be a very, very small percentage of it.

Kostroun: That is the answer I was looking for.

Beebe: You flashed a viewgraph up very quickly and I did not catch it. Was that the amount of titanium that you were injecting?

Levine: Yes, the amount of titanium that you put in is a direct measure of the total amount you get.

Beebe: Do you inject titanium continuously or is it introduced in infrequent bursts?

Levine: It is a continuous stream. The titanium is heated up, the shutter is left open so we have a continuous flow of titanium.

Schmieder: I have two questions about heating and cooling. One is: In your code, it was not clear to me that you were accounting in a central way for the flow of energy from the electron beam ionization of ions and the consequent increase of potential energy converting into kinetic energy, and then collisionally heating the light ions. Is that in there?

Levine: Yes.

Schmieder: Is the evolution of charge states an intrinsic and essential part of the code?

Levine: Yes. It is in the code, and one of the reasons why the code runs very slowly is because we do have to use it. It was not in the slide I showed. What we assumed was that they just went up to a particular charge state. But the code does have the ability to calculate the charge state and Bernie Penetrante, who is working on the code, can probably give you much more detail on that.

Schmieder: Can you describe, very briefly and simply, how you flop between the evolution of charge states and the evolution of plasma parameters? That is, energy, velocity, space distributions and so on. Charge states are normally evolved by a rate equation which has a coefficient which is a global average of everything. But in order to model the advance of charge states ....

Levine: What you have to do is iterate around the temperature. To do that you guess the charge state and then you iterate on the ionization time and the heating to solve for the charge state, the energy transfer and the final temperature. The requirement is that the average time of an ion in the beam, the ionization time, and the rate of energy transfer be consistent.

Schmieder: So then you are spatially resolving the densities according to the Boltzmann potential distribution?

Levine: Right, according to the Boltzmann potential.

Schmieder: Did you stop the calculation at some point in each element of phase-space and evolve the charge states a little bit?

Levine: No. We do not take each element of phase-space. The LCI's were assumed to be in equilibrium with each other over the entire volume and they are part-time out of the beam and part-time in the beam. So one has to consider what percentage of time they are in the beam to get their average charge state.

Schmieder: So a rate of evolution of the charge states is just a fraction of the time inside the beam?

Levine: That is right.

Schmieder: So you scale time? The residence time?

Levine: You scale time according to where ions are and how hot they are. And the code looks at the temperature of each individual charge. You know, the q = 5 is cooler than the q = 7 because it has not had time to make so many collisions, etc. So that is in there.

Schmieder: Very, very light ions like hydrogen are inefficient as coolers because they have such small mass. The ideal cooler would be, for gold...

Levine: Yes, essentially because they have such small q, not small mass. Because what they take out is q times the voltage over which they go. What they take away is qV. Well it is not quite qV, but almost.

Schmieder: You showed that hydrogen was a very poor cooler, nitrogen was better, and titanium was even better. It seems to me there is going to be an optimum cooler; one that is heavy enough, massive enough, or large enough, or has a high enough charge state to do efficient cooling. But if you go any farther, it has too many electrons and starts to spoil the trap. Do you have any idea what it is?

Levine: We use titanium to cool gold, chosen with a hand waving guess. One limit on the amount of coolant is charge exchange. If the neutral atom density of the coolant is too large, charge exchange from the neutrals will degrade the charge state of the desired ion. Charge exchange between ions is neglected because at low ion temperatures the coulomb barrier between ions is too large for charge transfer.

Schmieder: So, is your answer - titanium seems to be best for gold?

Levine: Titanium looks about best for gold, but this is only a guess.

Becker: You found that the cold Brillouin theory is wrong for beams as you have it. A real surprise is that the Herrmann theory, which is considered for low perveance beams only, works fine with very high perveance beams. Bernd Fogen, who did his Ph.D. thesis about 10 years ago, found that doing true Pierce-Walker calculations, defining a beam radius by 67% of the contained beam, gave exact agreement with Herrmann's theory. This is something which is probably surprising in this field, but shows that Pierce-Walker's theory can be well-approximated with the much simpler formula from Herrmann's theory, predicting the increase of focussing field.

Levine: Let me add one caveat. We would like an opportunity to vary the parameters on our gun to insure that we have truly minimized beam diameter, that we truly have the optimum magnetic field gradient near the gun, and that our gun is really a good diode, a good Pierce gun. Let me also say that this 57 eV should be compared with the radial energy that explains this particular beam of 300 eV as the electron temperature in the radial direction.

True: I think that it is unwise to be left with the impression that Herrmann's optical theory works for high perveance beams. The theory appears to work for low perveance beams and your beam is low perveance (100 mA at 20 kV). When you get to beams of 1 or 2 µperv, it has been my experience that the theory tends to break down. I am just trying to inject a note of caution. There are regions where the theory works, but in the high perveance region, it does not seem to.

Levine: Is it worse at high perveance?

True: It just does not give the right answer.

Levine: Does it underestimate or overestimate the current density?

True: I do not know off the top of my head.

Becker: Bernd Fogen has shown that it worked at 2 µperv. It was important to define the beam radius, and he did it by saying it contained 67% of the beam.

True: But then that is not Herrmann's theory.

Becker: No, it is Pierce-Walker's theory. You need a certain excess magnetic field to focus the beam, and the result is exactly the same as from Herrmann's theory.

DIONE STATUS REPORT

J.Faure ,P.Antoine J.C.Ciret,L.Degueurce ,P.Gros
A.Courtois,B.Gastineau,R.Gobin,P.Leaux,P.A.Leroy,J.P.Penicaud
Laboratoire National SATURNE
91191 Gif/Yvette France

INTRODUCTION

DIONE has been built and tested from 1985 to 1986 and set up on its high voltage plateform in 1987. The first accelerated beam was served to physicists in 1987 fall.

Since that time, the source has been running 1500 hours as a preinjector and nuclear physics experiments ,distributed in four runs of two weeks , have been performed according to the SATURNE schedule.

In the meantime we have proceeded to several tests and developments.

The next table gives the results obtained recently.
The emittance is $2\ 10^{-7}$ m.rd (normalised value).
The ions beam charges quantity range is $1.5\ 10^{10}$ to $2\ 10^{10}$ all species added up.

Table 1

Ion species	Source exit (charges)	Number of injected pulses	Intensity in SATURNE
N^{7+}	$7\ 10^9$		
C^{6+}	$4\ 10^9$	4	$4\ 10^9$
Ne^{10+}	10^9	4	10^9
Ar^{16+} Ar^{17+} Ar^{18+}	10^9 $6\ 10^8$ $2\ 10^7$	3	10^9
Kr^{30+}	10^8	1	$3\ 10^7$ (MIMAS only)

In this paper , we give a description of DIONÉ and we point out the theoretical and experimental remarks leading to an improvement program.

DESCRIPTION
1. Magnetic field

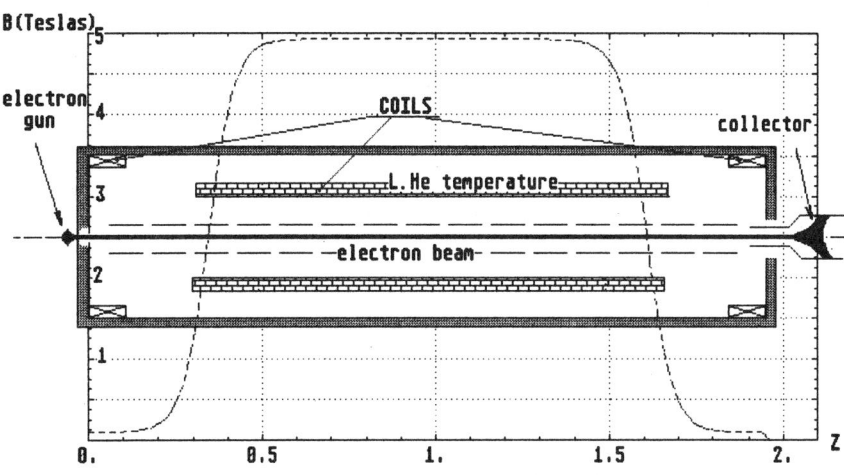

Fig 1 Magnetic configuration

The magnetic field is provided by a cryogenic coil. The maximum value is almost 6 Teslas but we usualy run 5 Teslas. Two room temperature coils at both ends are used to adjust the magnetic field value. They ensure electron Brillouin flow tuning at the gun side and ion beams focusing at the collector side.

2. Electron gun

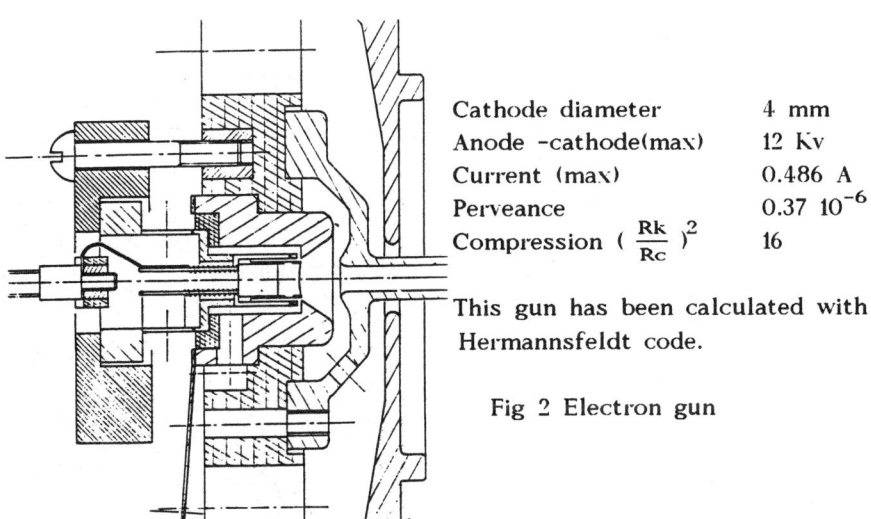

Cathode diameter	4 mm
Anode -cathode(max)	12 Kv
Current (max)	0.486 A
Perveance	$0.37\ 10^{-6}$
Compression ($\frac{Rk}{Rc}$)^2	16

This gun has been calculated with Hermannsfeldt code.

Fig 2 Electron gun

3. Drift tubes polarisation

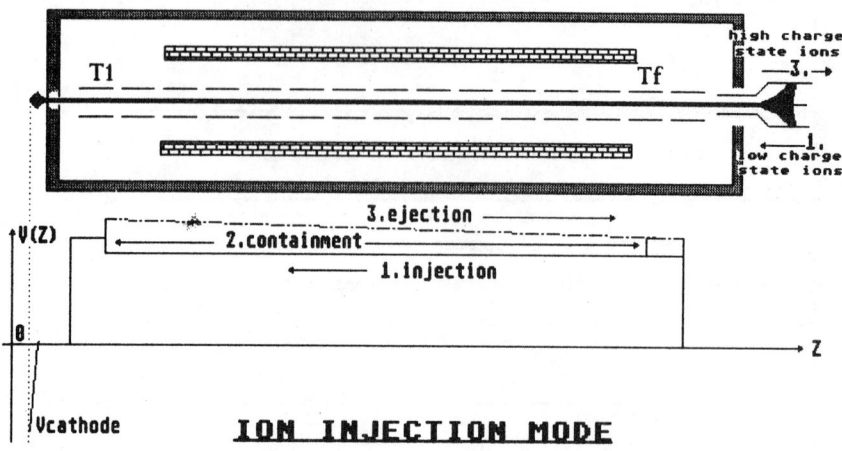

Fig 3 Successive potential distributions

DIONE is working in ion injection mode since this method has been successfully tested on CRYEBIS 1 at SACLAY. The advantages can be summarised in the following way :
- largest range of elements
- containment region pollution neglectable
- perfect stability during many hours

The ionization process is made up of three steps:

1. ***Injection*** . The drift tubes voltage is equal to the accelerating voltage of the external source beam so that the injected ions are slowed down. They are reflected by the tube T1 biased to a slightly upper voltage. When the first injected ions come back to the collector region the Tf tube voltage is rised up, trapping the ions in the containment section. Theoreticaly the injection time could be very long acccording to the ions energy along the axis. Actually 100 μs is the usual figure corresponding to 10-20 eV injected ions remaining energy.

The injection energy is 10 kV.

2. ***Confinement***. During the stripping process the electrons energy can be optimised according to the ion species. Decreasing this energy increases also the space charge capacity of the source.

3.Extraction. The potential of the containment region is rised up so that the ions are able to escape over the traps. For synchrotron injection the process has to be as fast as possible because the required ions impulsion length is 50 µs. The ejected beam energy corresponds to a 10 kV acceleration potential.

4.Cryogeny and vacuum

The Fig.4 shows the cryogenic and vacuum system arrangement. The options are:
- The liquid helium cryostat is common for the super-conducting coil and the vacuum pumping surfaces
- The containment volume is isolated from the cryostat insulating vacuum vessel
- Two 2 K charcoal coated pannels located at both ends provide efficient pumping speed for hydrogen.

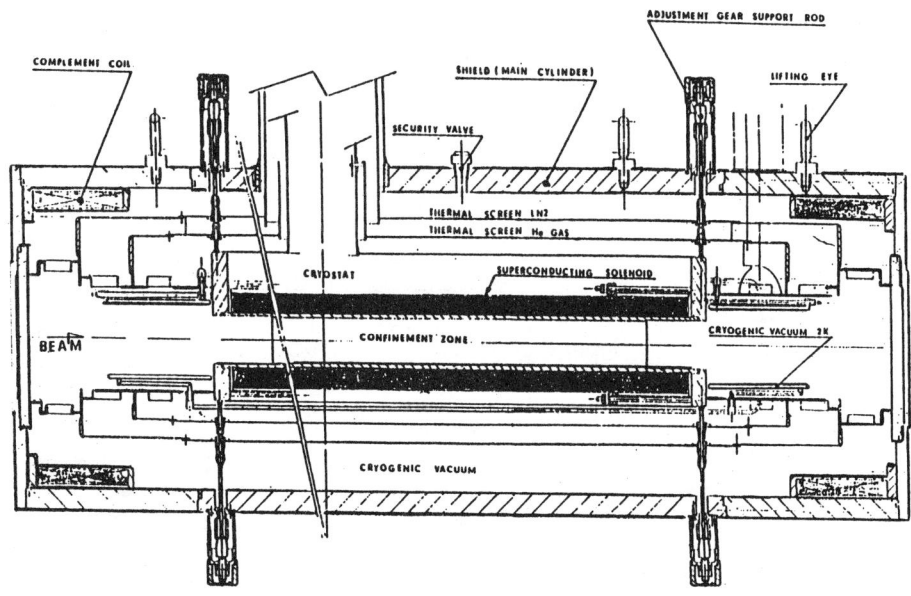

Fig. 4 DIONE Cryogenic details-Vacuum configuration

Unfortunatly the cryogenic system does not work satisfactorily:

-The surrounding liquid nitrogen thermal screen temperature is 110 K instead of 77 K. Therefore the connected drift tube temperature is too high degrading the vacuum performances of the source.
-The consumption of helium is much higher than we expected: 5 liters per hours instead of less than 1 liter.
-Due to the consumption of liquid helium ,cold helium vapors escape ,freezzing the liquid nitrogen circuit.A warming circuit has been installed for the reliability of the source.

5. External source installation

The injection principle has been already described [1] .In brief,the injected beam, provided by a conventional ion source biased at 10 kV, is focused into the collector so that the electronic space charge captures low charge state ions .Then these ions are decelerated ,as we have seen, in the injection phase.

A pulsed electrostic deviator switches from injected ions path to ejected ions path. Details are shown on Fig. 5. The extension for ^{6}Li polarized source is also represented.

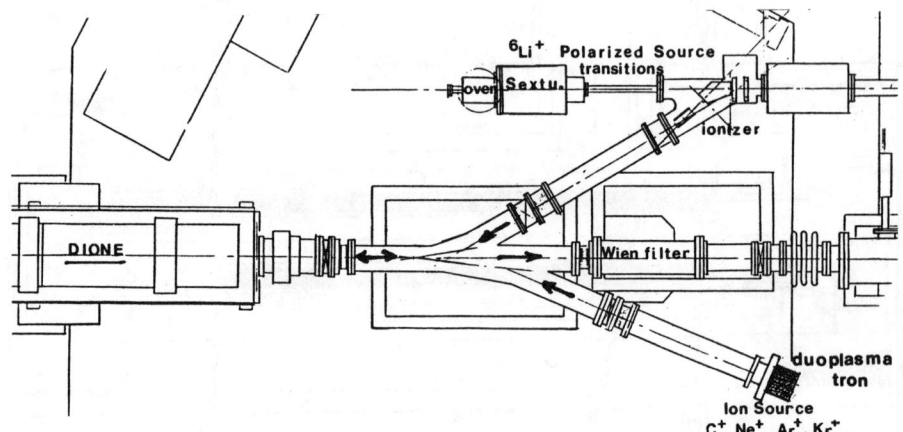

Fig. 5 DIONE and External Sources Assembly

6. Electronic devices

All the parameters are computer controlled. Optical fibers are used for the transmission to the high voltage platform. The order of magnitude is 80 different adjustable voltages or magnetic parameters .30 adjustable delay times and 20 physical acquisitions like pressures or temperatures.

This feature is essential for a good operation when DIONE is connected to the synchrotron and to an easier optimisation of the parameters.

7. Diagnostics [2]

Faraday cup: they are used to measure the ion beams intensity. Their sensitivity is 1 nA.

Current transformers: They are located in the vacuum vessels. Therefore their size is small and the chamber itself is a part of the shielding device. Their outgassing flow is very low so that they are compatible with a 10^{-9} torr pressure section. Their sensitivity is $< 1\mu A$.

Beam profile monitors: They are alumina plates .1 mm thick upon wich 32 or 64 strips of gold are printed by a silk-screen process; the spacing of the strips ranges from 250µm to 1 mm and the thickness of the layer of gold is 20µm. Their use is essentialy devoted to emmitance measurements.

 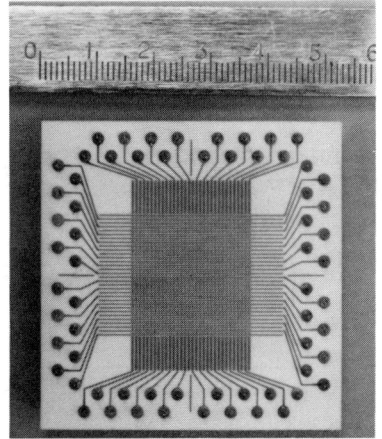

Fig.6 Beam current transformer   Fig.7 Double alumina beam profiler

RESULTS AND DISCUSSIONS

The results are given in Table 1 and they must be completed by the following experimental remarks:

-The extracted intensity decreases if the containment time increases

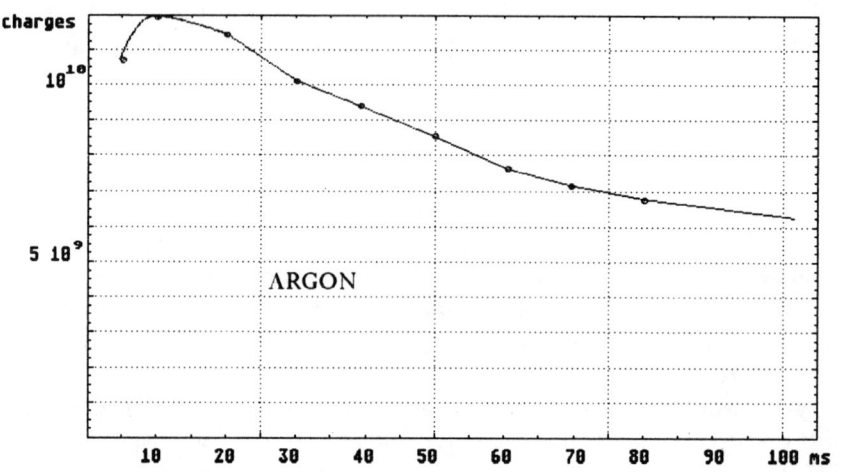

Fig. 8 Ions intensity variation versus confinement time

-The apparent electron density decreases if the space charge neutralisation increases

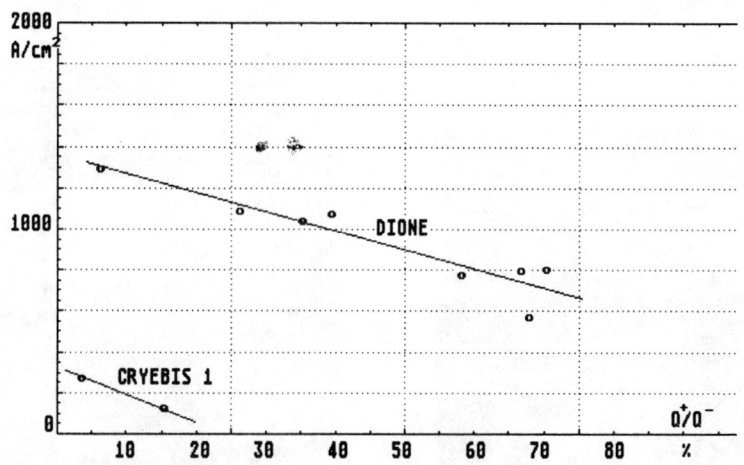

Fig. 9 Density variation versus space charge neutralisation

-The maximum density is lower than one can expect from theoretical magnetic compression

So far we have made calculations in order to investigate the last point. The motion of ions and electrons has been simulated by a multi-particles code taking into account electromagnetic forces ,space charge and electron beam distribution [3]. The Fig. 10 shows the results using uniform or realistic gaussian density. The parameters correspond to actual DIONE adjusted parameters.

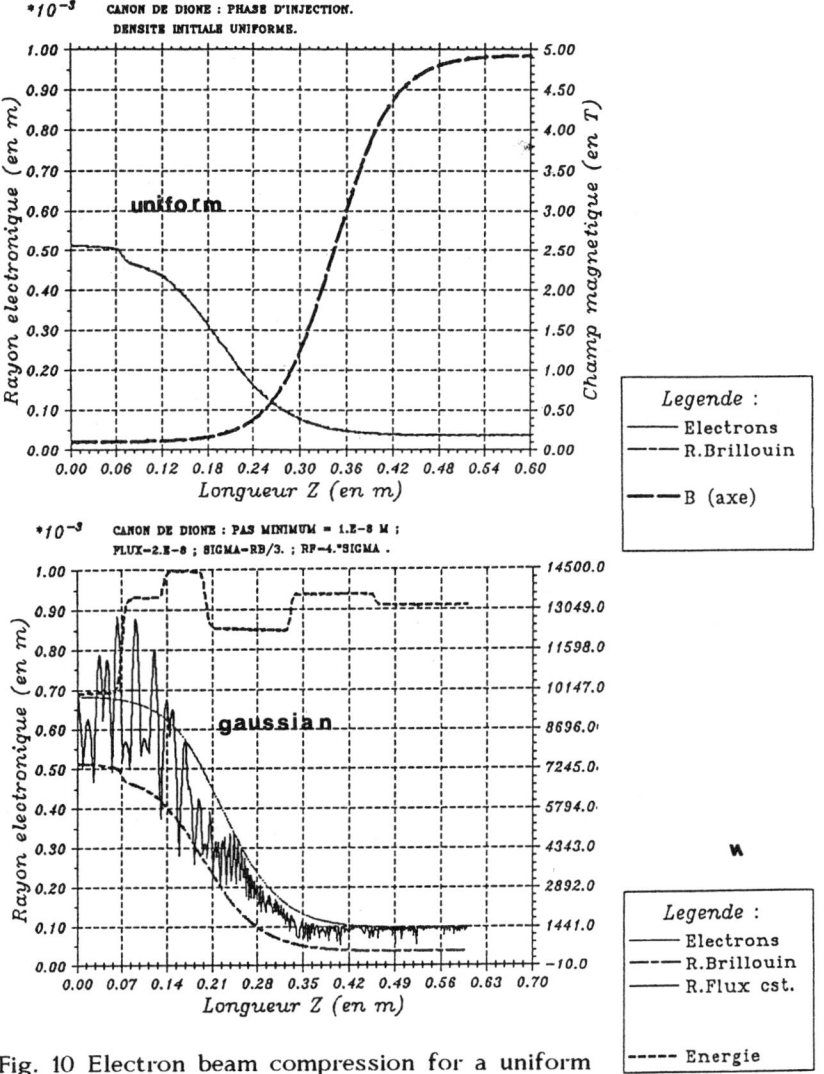

Fig. 10 Electron beam compression for a uniform and a gaussian distribution

These results show that for a uniform distribution the compression is adiabatic according to the Brillouin law.In the case of gaussian distribution the compressed electron beam radius is within Brillouin and constant magnetic flux radius.The density corresponds to the measurements.

Another confirmation of the electrons density distribution influence is given by the emittance measurement of the ejected ion beam.The results are shown Fig. 11 and one can see that experimental result agrees with theory.

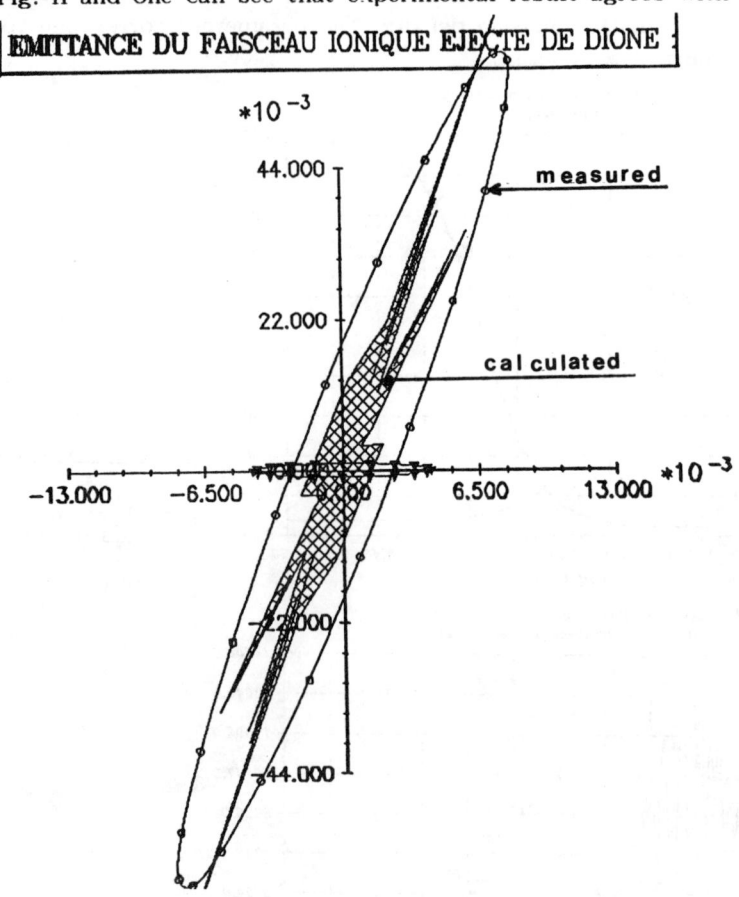

Fig. 11 Comparaison between measured and calculated emittance

Another electron distribution influence has been proved by the following experiment. After optimisation , using Hermannsfeldt code ,a modification of the whenelt electrode was made —see Fig. 12— in order to decrease the electrostatic compression and improve the density distribution.The result is given on the Fig. 13.The shift in relative quantity of Ar^{16+} shows that the electron beam density has been improved.

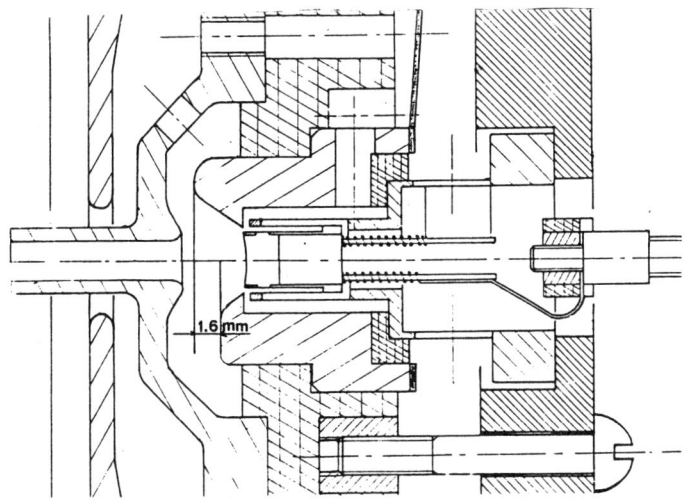

Fig. 12 Whenelt modification

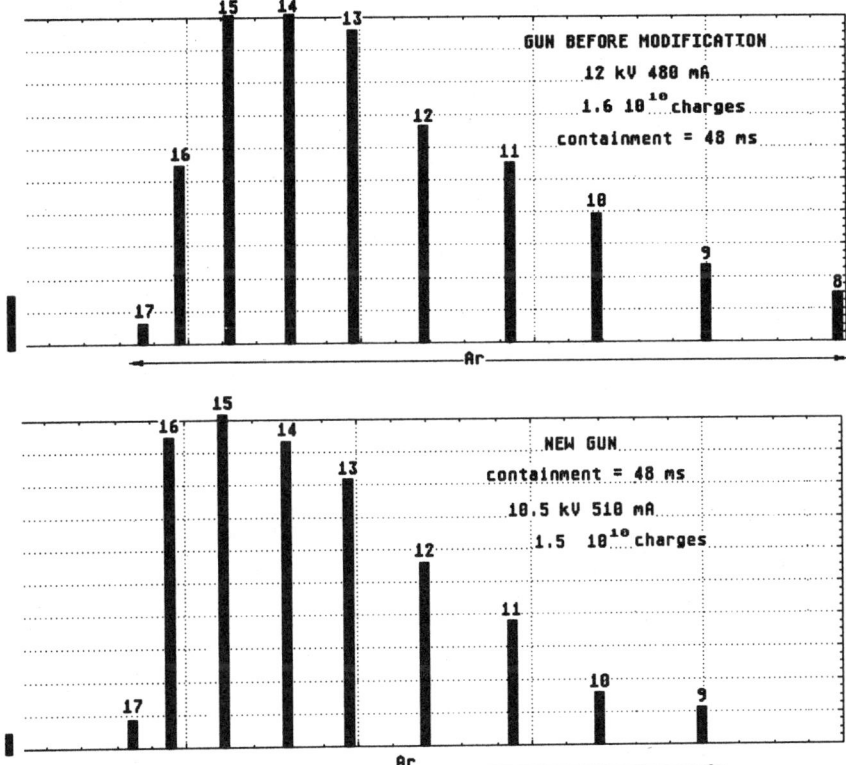

Fig. 13 Argon charges state for two whenelt geometries

M.Tagger [4] has investigated other electrons-electrons or electrons-ions instabilities looking at waves propagation if the beam is moved from equilibrium state. The conclusion seems that we cannot adjust the electron beam parameters to the ideal Brillouin flow values. The results can be caracterised by $\eta/\eta_b \leq 0.85$, (η_b Brillouin density , with zero flux on the cathode and η the attainable density).

CONCLUSION

The experiments show that we can improve the DIONE performances if we gain on the electron beam density distribution. Our program consists in the design of a test bench to measure electron beam density profile and the elaboration of a new code for electron gun calculation.

REFERENCES

1) J.Faure & al.,External ion injection into CRYEBIS, NIM 219 (1984) 449-455
2) L.Degueurce ,Beam diagnostics for the EBIS ....NIM A260(1987) 538-542
3) O.Delferriere ,Thesis .To be published.
4) M.Tagger Electron Beam Dynamics in EBIS Source .This symposium

DISCUSSION

Becker: Did you always observe this charge distribution which has considerable amounts of lower charge states, or is this dependent on injection? Does it vary from time to time?

Faure: You mean for argon?

Becker: Yes. For instance, you showed this improvement with a new gun or by variation of the distance between tubes, but in pure containment mode you should really have no low charge states.

Faure: That is right, we are always observing this type of time-of-flight. But if we look inside the structure and if we isolate the box, we can see that all these spectra are not the same. The average density decreases from the gun to the end of the source when we add all the spaces. Density is not constant through the beam. On the one meter beam the electron beam density is larger at the beginning than the exit. Maybe this is the reason that we have too many species.

Becker: If you run the source with an overall spectrum, you seem to have one main ionization volume; you do not divide it up.

Faure: Yes, we can assume that. If we assume that the ions do not move inside, they are waiting for the extraction. But we have heating, so it is more complicated.

Hershcovitch: Why, is the density of the electron beam not uniform?

Faure: It comes from some aberrations during the compression or something like that. I do not know exactly. From the electrostatic compression of the gun, we are obliged to have aberration on the outside.

Levine: I did not quite understand how you measured the current density.

Faure: We use carbon or oxygen because the spectrum is simple. We say that we have relative amounts of O^{8+}, O^{6+}, O^{7+}, etc. Then we try to fit on the curves.

Levine: But that assumes that the ion is inside the beam.

Faure: That is right.

Stockli: One of your last slides showed a wire strip detector. Was this detector used to measure the emittance?

Faure: Yes.

Stockli: In DIONE, the 77° K drift tubes and the warm drift tubes are all mounted on different supports. How do you know that they are aligned when cold?

Faure: Because when we built DIONE we provided windows on both sides and we had references inside that we looked at. Things moved during the cooling, but after that they went back to their previous place.

THE KSU-CRYEBIS PROGRAM

Martin P. Stöckli, C. L. Cocke and P. Richard
J.R. Macdonald Laboratory, Department of Physics
Kansas State University, Manhattan, KS 66506, USA

ABSTRACT

We are starting up a CRYogenic Electron Beam Ion Source which was designed to produce bare, and almost bare, slow heavy ions and which will allow us to explore the physics of slow collisions involving highly charged ions. The collision energy can be varied over two orders of magnitude. This paper covers the design and layout of our CRYEBIS.

INTRODUCTION

The J. R. Macdonald Laboratory at Kansas State University is dedicated to the studies of atomic interactions of ion-atom collisions. Presently, we are upgrading our facility with a stand alone CRYogenic Electron Beam Ion Source located on a high voltage platform and a superconducting LINAC to boost the ions emerging from the existing 6 MV EN tandem Van de Graaff.[1] These additions will substantially extend the energy- and nuclear-charge-range over which we can produce beams of highly charged, including bare, ions and will allow us to explore the physics of collisions of highly charged ions with solids, molecules, atoms, other ions, electrons, or radiation.[2] For example, we should be able to produce bare chlorine from 140 MeV down to a few keV using the LINAC in the upper and the CRYEBIS in the lower energy range. This report will address only the CRYEBIS ion source which was designed similar to CRYEBIS II in Orsay,[3] but includes significant modifications which make it also similar to DIONE.[4]

THE HIGH VOLTAGE STAND

As mentioned above, the CRYEBIS is located on a high voltage platform, which allows us to increase the energy/charge of the ions from a few keV up to 200 keV. Figure 1 shows the layout. The surrounding fence is assembled from off-the-shelf modular panels, each of which can be removed to increase accessibility. That also allows us to mount the main platform on a 2.7 ton swivel dolly[5] and to roll it outside the cage to perform heavy lifting jobs that cannot be performed in place because of the overhead restrictions. However, the normal access is through the door, which is safety interlocked with the high voltage power supply and the electrically operated grounding rod. Inside the cage are two high voltage platforms, the main platform supporting the actual CRYEBIS and the smaller one supporting the associated electronics. Both are mounted on standard strength station post insulators, which are rated at 235 kV low frequency dry flashover.[6] The space above both platforms is enclosed by a corona guard frame which allows for sidepanels to be

fully enclosed where needed. The high voltage power supply is
installed underneath the main platform. The unit provides 1 ma of
current at a voltage up to 200 kV with a ripple of less than 5 V
peak-to-peak.[7] The low ripple is achieved with the combination of a
driver stack and a filter stack connected with a line of resistors.
Secondary arching between the driver stack and the high voltage
stand is inhibited by a 12 mm polypropylene sheet. The unit is
safety interlocked and remote-controlled from a control panel
outside the access door. Underneath the electronic platform are
three isolation transformers.[8] Each transformer delivers one phase
of electric power, with a total of 38 kW; 28 kW at 480 V are
required for our 2.5 ampere 10 kV power supply,[8] which would be
needed to operate a high power electron gun, such as the ones
developed in Orsay.[9] The remaining 10 kW at 120/208 V operate the
rest of the equipment. The secondary windings are grounded to the
high voltage platform over resistors and potentiometers which can be
adjusted to balance out induced ripple. Most of the thermal power
produced on the high voltage stand is removed with a loop of
deionized water, which is cooled to 14 degrees in a secondary heat
exchanger.[10] The ion concentration in the deionized water is kept
below 1 ppb by bypassing a fraction of the water constantly through
two deionizer cartridges.[11] Even under the maximum thermal load of
30 kW the system will maintain a total resistance of 3 Gohms. It
can be pressurized up to 3.4 bars to reduce steam production under
high thermal loads. The electronic platform carries two 19"
electronic racks in addition to the 2 amp 10 kV power supply. Some

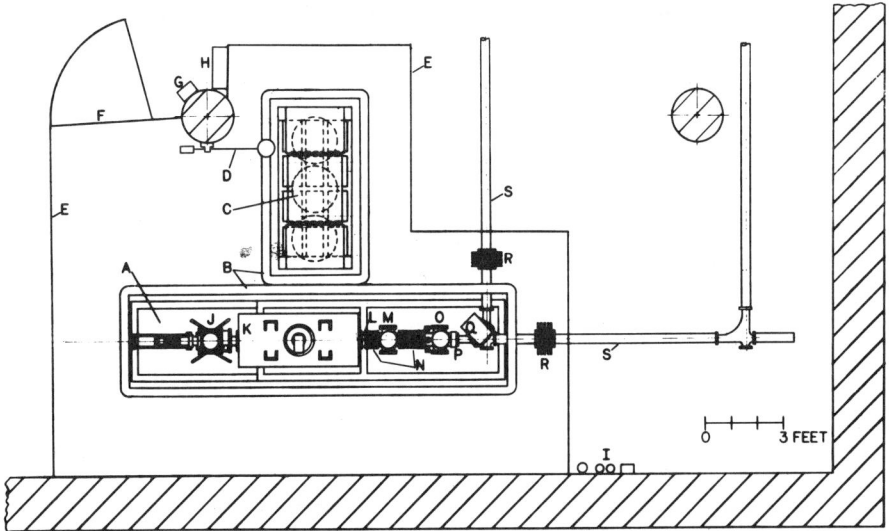

Figure 1: Layout of KSU-CRYEBIS:(A) high voltage platform for ion source with high voltage power supply underneath, (B) corona guards, (C) high voltage platform for electronics with 3 isolation transformers underneath, (D) automatic grounding rod, (E) modular fence with (F) access door, (G) main powerbreaker, (H) high voltage control unit, (I) heat exchanger, deionizer, centrifugal pump and control system for cooling water circuit, (J) electron gun housing, (K) shielded superconducting solenoid, (L) 6" gate valve, (M) collector housing, (N) welded bellows sections, (O) diagnostics- and injection-housing, (P) fast pneumatic 4" gate valve, (Q) 90 degree analyzing magnet, (R) accelerator columns, (S) beam lines on ground potential, (shaded area) building walls and columns.

of these units are insulated from the platform to simplify future
polarization of the gun and/or drift tube electronics to allow post-
and/or internal acceleration.[12] These units are enclosed in a lucite
cage and controlled with tenite rods, which couple easily to the
individual control knobs. All the units are presently controlled
from inside the cage, but also can be controlled from outside the
cage using longer tenite rods.

THE SOLENOID

The superconducting solenoid is the main component of our
CRYEBIS. We took great care to obtain a magnet which is straight
and has a clean bore vacuum.[13] The 5 Tesla solenoid was produced in
England[14] according to our specifications. The surrounding vacuum
vessel, made from 3 to 6 cm thick nickel-plated low carbon steel,
was fabricated in Germany and is primarily responsible for the
weight of over one ton. Even under the maximum field this shield
reduces the outside magnetic field to less than one gauss at
adequate distance from the axial penetrations. Figure 2 shows the
outside of the solenoid vessel which has a diameter of 60 cm and a
length of 141.4 cm. The indicated six large knobs are used to align
the solenoid within the shield. We added dials, which allow us to
position the solenoid reproducibly. We also developed a magnetic
alignment procedure and a magnetic mapping method which yield a
precision of approximately 0.02 mm, but that has to be discussed
elsewhere.[15] Figure 3 shows a cross section through a corner of the
solenoid. One can see how the solenoid itself is suspended from the

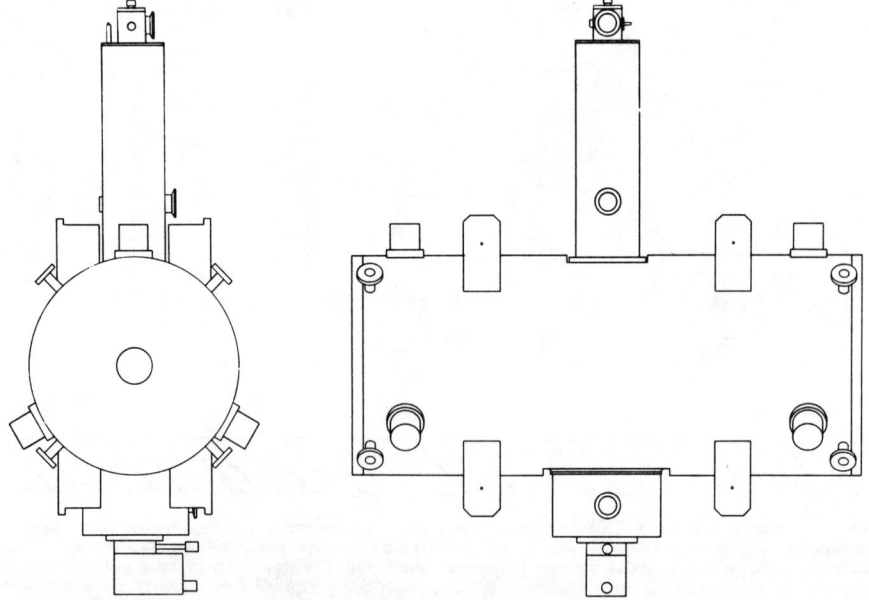

Figure 2: End and side view of the superconducting solenoid with the cryo generator at the bottom and cryogenic-, current- and instrumentation ports on top.

shield with six axial and six radial titanium rods. It also shows
how stainless steel bellows separate the cryogenic vacuum from the
cold bore, which is 110 cm long and has an inner diameter of 8 cm.
Wound onto the cold bore is the main solenoid, which is 100 cm long
and yields a maximum of 5 tesla with a current of 129 amps. At both
ends, 10 cm long end coils are wound over the main coil which yield
a maximum of 1.25 Tesla at 120 amps, and which allow us to alter the
fringing fields on the axis close to the shields by 40%. The three
coils each have an individual persistent switch and are connected
with four current leads so that they can be individually energized
to create a variety of field profiles. When in the persistent mode,
the liquid He consumption is approximately 8 liters per day, which
requires batch filling of the 48 liter reservoir at least every 5
days. A cryogenerator[16] keeps the two radiation shields at 80 and 25
degrees Kelvin. For precooling we blow approximately 100 liters of
liquid nitrogen through a separate heat exchanger loop which is
wrapped around the liquid helium reservoir. Subsequently, 40 liters
of liquid helium are used to cool the structure to 4.2 degrees
Kelvin. The two 20 W heaters mounted on the reservoir require 3
days for warm up, a time span we are trying to reduce to 1 day. All
the cryogenic numbers above exclude additional thermal loads and
masses added inside the bore. The instrumentation consists of 8
thermometers, 2 teslameters and one level gauge. We added safety
valves, pressure gauges, flowmeter, pressurization port and cold gas
vent to the helium exhaust to simplify liquid helium transfer and to
reduce the risks associated with condensation of air. The cold gas
**vent serves also as the pump port for initial evacuation or to pump
on the liquid helium for limited testing at higher fields.**

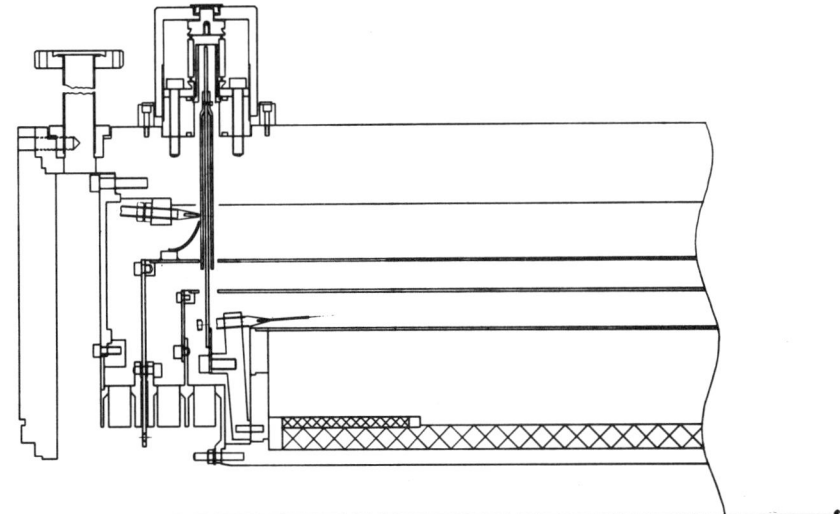

Figure 3: Cross section through a corner of the solenoid above the central axis which is indicated with the dashed line.

ELECTRON GUNS

We did not proceed with the initial plan to copy one of Orsay's high perveance guns,[9] but implemented and tested two smaller electron guns instead. First, we copied and tested the gun which is used on Cornell's CEBIS I,[17] with 2.2 µP. Figure 1 shows the housing in which this gun is mounted. Its housing allows us to retract and isolate the gun from the solenoid bore vacuum for servicing and conditioning, which is very useful because of the time it can take to pump down the solenoid bore. The gun housing itself can also slide back to provide physical access to the gun or can be completely removed to load the solenoid bore. The gun housing allows for 15 kV "postacceleration",[12] has 5 degrees of freedom for the gun adjustments, a thermal sink for the gun, and external bucking coils to vary the magnetic field in the gun region. A combination of a turbo pump, an ion pump, and a titanium sublimation pump allows the pressure to reach the 10^{-10} Torr range after a 150 degree C bake out and gun conditioning. Although there were some heavy discharges caused by inadequate temporary electrical connections, all tests gave adequate results.

We felt that the start up should be easier with the commercial gun which is used on the LLNL EBIT.[18] This 0.5 µP gun[19] is produced for traveling wave tubes. LLNL helped us to essentially copy their setup. The gun is surrounded with a bucking coil which reduces the magnetic field from the main solenoid in the cathode region. Figure 4 shows the axial field component in the gun region for several bucking coil currents. Figure 5 gives the mounting details. The electron gun is pressed into a low carbon steel diaphragm, and hence has no alignment provisions. The diaphragm and the attached skimmer, which has a 2.5 mm inner diameter, are mounted on a ceramic insulator, but are grounded internally. The temperature of the cathode is measured through a rear viewing port with a pyrometer. The rear end of the gun is pumped down through a right angle valve with a 50 liter turbo pump, which is normally removed after bakeout, since it is also needed to evacuate the cryogenic vacuum in the solenoid, and to operate our mobile RGA station. The vacuum is then maintained with an 8 ℓ/s ion pump which yields a pressure below 10^{-8} Torr which is the sensitivity limit of the ion pump. This gun was tested on our test solenoid with up to 4 kV and 100% duty cycle and showed very small losses to ground.

THE DRIFT TUBE STRUCTURE

To minimize the risk of transverse field components we avoided or symmetrized magnetizable materials in our solenoid[13] as well as in our drift tube structure, which was designed and built in-house. The drift tubes are made from OFE copper, insulations (where necessary) are made from Teflon and 2024T4 aluminum was chosen for bolts and nuts which could not be arranged in a symmetric fashion. Figure 5 shows the upstream section of the drift tubes. After the skimmer and a short drift tube, follows a gas cell, all of which are mounted and centered from the opening in the entrance shield and so are near ambient temperature. The gas cell was put in this location to avoid condensation of the seed gas, but the low magnetic field level in this area may be inadequate. However, we have tested a setup with a heater wire, which would allows us to inject in the cryogenic area which is at full field. No drift tube is attached to the liquid nitrogen shield, which presently serves only as a radiation shield. The cryogenic drift tubes are all mounted on a common drift tube bed which is centered inside the solenoid bore. Figure 6 shows how copper rings fitted with teflon O-rings provide a zero-clearance, well-centering fit between the bore and the drift tubes. The difference in thermal contraction of the aluminum bore and the copper rings neutralizes the rather large thermal contraction of teflon. A solid 8 mm thick copper plate is bolted to the drift tube bed and to the bore end, as shown in Figure 5, which connects the drift tube bed thermally and electrically to the bore. Presently the 22 cold drift tubes are connected to only 8 electrical leads, because we have only a few radial penetrations for the feedthroughs and we want to put high voltage on each lead and drift tube. The electrical connections are made from 304 stainless steel thin wall tubing which has very low thermal conductivity considering the large 1.8 mm outer diameter, which reduces high voltage discharges. These tubes were drilled to facilitate pumping and can be very easily connected and disconnected by inserting smaller size

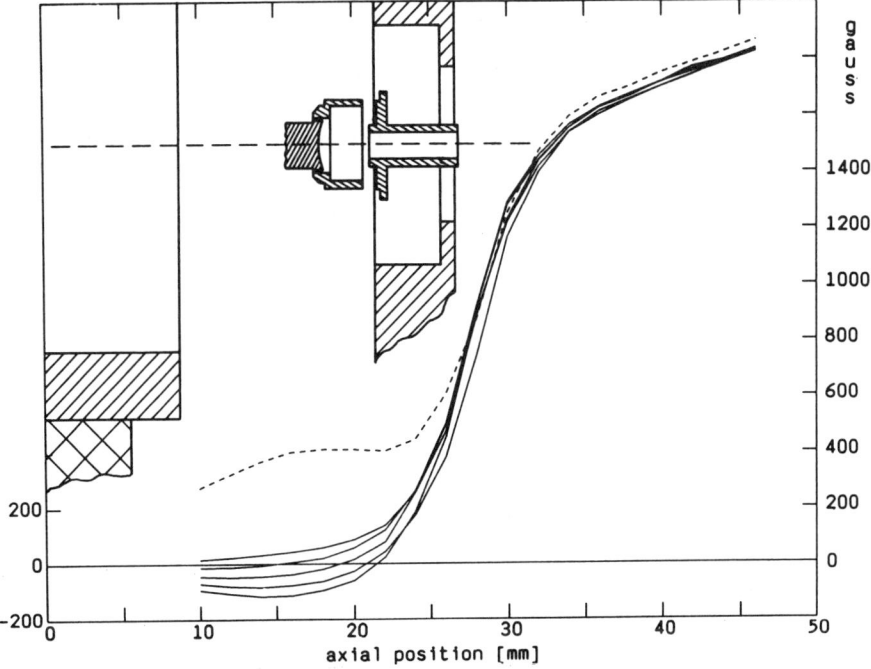

Figure 4: Measured field profiles in the electron gun region. The dashed curve shows the axial field from the main coil only, set at 5 Tesla. The five solid curves show the axial field component for different bucking coil currents between 2.25 and 3.25 amperes which reduce the field in the area of the cathode. The insert in the upper left shows the location of the electron gun (fine shaded), parts of the low carbon steel shims (coarse shaded) and a part of the bucking coil (cross hatched) all of which are circular symmetric around the dashed axis.

tubing or rods. These rather stiff electrical connections also have the advantage of remaining in place even under thermal and mechanical stress and thus are fail-safe. Since they are carefully placed a suitable distance from ground and other leads, we can test the connections to the drift tube by measuring the capacity of the connector versus ground, since the capacitance of one drift tube is approximately half of the capacitance of the electrical lead itself. Cryogenic temperature sensors[20] monitor the temperature of the drift tube bed, which presently indicate a temperature of 12 degrees Kelvin. The last three drift tubes in the downstream fringing field are supported and centered from the end shield and hence near ambient temperature. The last drift tube penetrates through and **shields the magnetic diaphragm. The outside cone matches the collector shape so it can be used to repel backstreaming secondary electrons.**

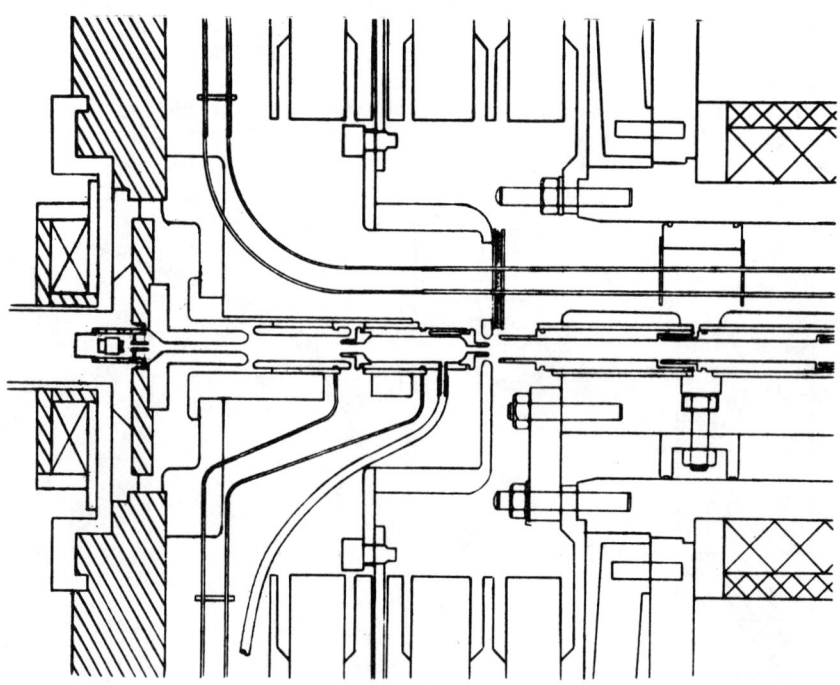

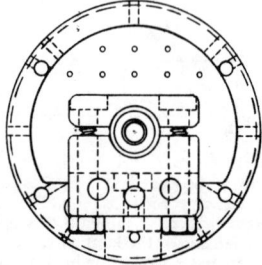

Figure 5: Cross section through electron gun surrounded by its bucking coil, the warm drift tubes, the liquid nitrogen shield and the cold drift tubes. The windings of the various coils are cross hatched and the low carbon steel shield and shims are shaded.

Figure 6: Front view on the cold drift tubes mounted on their bed with eight electrical leads.

THE COLLECTOR AND EXTRACTION REGION

Figure 7 shows our electron collector which is a simplified version of the collector used on CRYEBIS II in Orsay.[9] It can absorb 25 kilowatts, the maximum power of our electron beam power supply. The collector was machined from OFE copper, cleaned, pickled, and assembled in-house with a heater[21] and brazing wire.[22] Then it was sent to a contractor[23] to be degassed at 700 degrees C and brazed at 838 degrees C in a vacuum furnace. Figure 1 shows how the collector housing is mounted between two welded bellows segments, which allows us to retract the collector from its operational position and to close the gate valve. Then, without affecting the bore vacuum, we can degas the collector by activating the internal heater wire in absence of the cooling water. All collector feedthroughs are mounted on a ceramic standoff which allows postacceleration (or a depressed collector)[12] up to 15 kilovolts.

A 110 mm long tube mounted behind the collector acts as electron repeller and ion extractor. At the end of this tube the ions are focussed with a grided einzel lens which ends with a tube grounded to the potential of the vacuum vessel. A deflector is incorporated in the second bellows section. The ions form a waist at the location of the diagnostics cross, which is equipped with

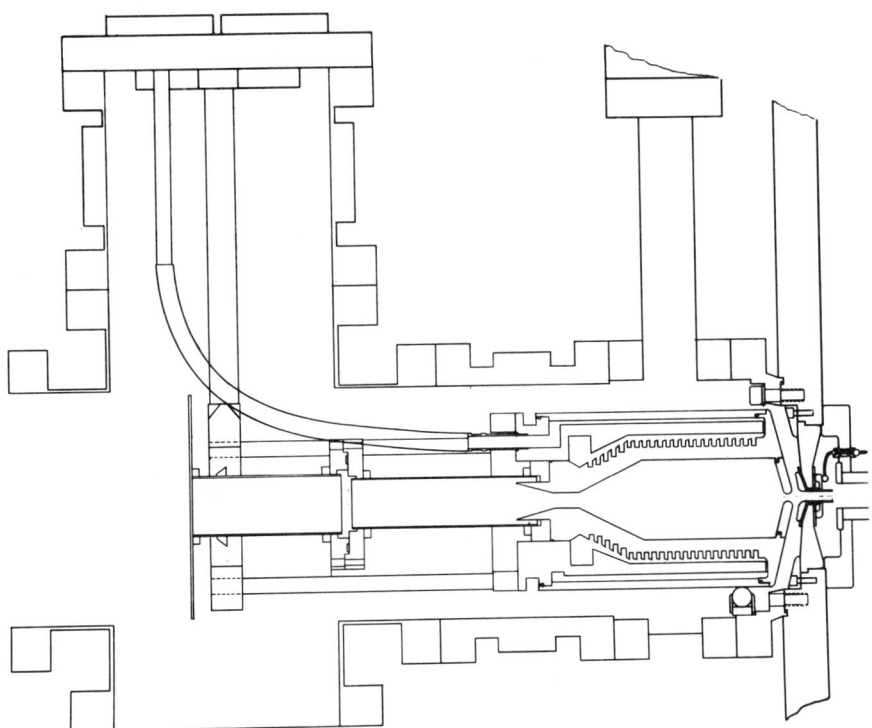

Figure 7: Cross section through the collector and ion extraction region.

four jaw slits and some simple diagnostic. This custom cross also can be used for ion injection from an external ion source to seed the CRYEBIS with low charged ions, as demonstrated by the Saclay group.[24] Then follows a 100 mm gate valve with a nominal closing time of 1 second.[25] We made some minor modifications and added a fast pneumatic circuit which enables the valve to be closed in 0.05 seconds in case of catastrophic vacuum failures. The ions can be analyzed with a double focussing analyzing magnet with a mass energy product of 58 keV/amu which we copied[26] and was commercially produced.[27] We redesigned the vacuum chamber of the magnet to reduce the costs of the commercial production.[25] Whether or not the ions are analyzed, they are then injected into one of the accelerator columns[28] where they can be accelerated, depending on the voltage applied to the high voltage platform. From there on, the ions are guided in beam lines to various experimental stations where people are anxiously waiting for the beam to come.

VACUUM SYSTEM

The vacuum in the electron gun mounts were discussed above. The solenoid bore and the extraction region are presently pumped with turbo molecular pumps[29] which are backed by oil-sealed rotary vane pumps.[16] In a UHV system one has to be concerned about hydrocarbon contamination, because it can form insulating layers on surfaces. Such layers can be electrically charged and can cause instabilities, especially in a UHV system where charged particles are involved. Therefore, we designed a protection system against hydrocarbon contamination. The vessel is roughed through the accelerating turbo pump, which is, when running, a very effective barrier against backstreaming hydrocarbons. The turbo pump controller can be interlocked to protect against various failures. If the controller loses power, or the pump is switched off, the valve between the pump and the UHV vessel closes, the fore pump stops and its intake valve closes, and the turbo pump together with the foreline is vented to prevent hydrocarbon migration to the hydrocarbon-free high vacuum side of the turbo pump. Unfortunately, the system does not protect against breaking of the electrical connection between the pump and its controller. Figure 8 shows a recent RGA spectrum after the cable was disconnected by accident. The evacuation of the bore vacuum is difficult because we minimized penetrations through the shield to assure high field quality. Two of the radial ports shown in Figure 3 were redesigned, which improved the conductance by 50%. Now we are pumping with one single turbo pump at both ends of the solenoid with a total of 19 ℓ/sec. An additional limitation is the thermal shield in the bore vacuum, which was designed to minimize thermal radiation losses, and therefore acts like a big virtual leak, because it has no pumping holes. It takes presently approximately one day to reach $1 \cdot 10^{-5}$ Torr and more than a week to reach $1 \cdot 10^{-6}$ Torr, with water vapor being dominant. The planned bakeout is severely limited, because the field quality of the solenoid depends on the precise position of the windings inside an epoxy matrix. Naturally the pressure drops rapidly, as soon as the solenoid is cooled to liquid nitrogen. The cool down to liquid

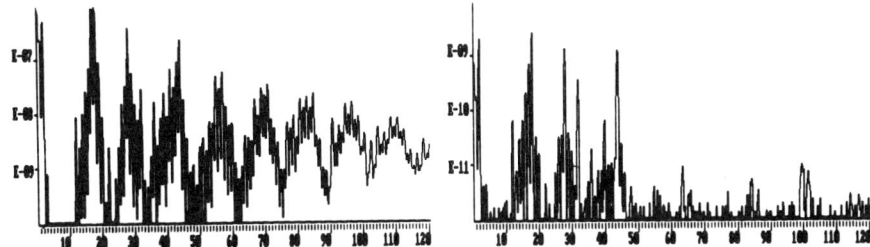

Figure 8: RGA spectra from the bore vacuum: Left: The residual gas during a pump down following an accidental disconnection of the turbo pump from its controller. The total pressure is $1.4 \cdot 10^{-5}$ Torr, which is mostly water vapor which has an overflow in this spectrum. The upper mass range shows clearly the hydrocarbons which were deposited during the mishap on the surfaces including the RGA. Right: After 7 days of cryopumping a pressure below $1 \cdot 10^{-8}$ was reached, and the RGA showed a fairly clean vacuum considering the vessel was never baked. However, most of the hydrocarbons were still on the walls and returned when the system was warmed up.

helium does not help much, because of the thermal shields with their small conductance. $2 \cdot 10^{-8}$ Torr measured at one of the unbaked ports is reached within a week. However, it is certain that the cryopumping reduces the pressure inside the drift tubes substantially below this value. A few simple changes and added equipment will speed up the pump down in the near future.

ASSOCIATED EQUIPMENT

It is important to minimize the time consuming cycling with a system as complex as a CRYEBIS. Hence we test many components on a simple test stand before they are installed on our ion source, so we will discuss briefly the test equipment which turns out to be very beneficial. The first item is a test stand with a classical 2 kG solenoid with identical mounting and shielding arrangement as on the main solenoid to allow preliminary and off-line testing of electron guns, collectors and associated electronics. The beam from the tested electron guns are absorbed by a test collector with a 12 cm deep conical cavity. The magnetic field inside the cavity can be substantially reduced with one or two bucking coils to improve the collection efficiency. An electrical insulated large entrance aperture can be electrically charged to repel secondary electrons. Our tests showed a very high collection efficiency and only a weak dependence on the bucking field and repeller voltage. If one mounts a fine wire on the entrance aperture or replaces it with a pinhole aperture, one could also measure the beam profile. A spherical rear surface allows a highly efficient heat transfer to the circulating cooling water and hence allows for a substantial heat load. The collector can be degassed with a heater installed in the cooling cavity in the absence of water. During some tests the collector was cooled with liquid nitrogen, but it did not indicate a substantial improvement of our operating conditions.

We also built a vacuum test stand which essentially consists of a turbo pump, a vacuum gauge, a conflat cross and a tee, all UHV compatible, which are mounted on a mobile frame and equipped with various kinds of adapters. The unit can reach UHV condition in a

rather short time scale, because it can be degassed with a
commercial vacuum bakeout unit[30] which uses a 600 W quartz lamp[31]
mounted inside the vacuum chamber. This unit turns out to be very
helpful to test the operational limits, reliability and/or UHV
compatibility of individual components.

The RGA equipment,[32] which is used to operate the RGA sensors
installed on the CRYEBIS was installed on a cart. An additional
sensor, hardware and a small turbo station, which is protected
against most failures, was added to the equipment. The turbo pump
is used to analyze gases at higher pressures or to install the
sensor on a system under vacuum, which does not have to be vented if
there is an adequate valved port available. Electric power is the
only required utility, which allows us to troubleshoot easily in any
required location.

SUMMARY

We described the layout and the design of our ion source,
located on a 200 kV platform, which was built to study slow
collisions involving highly charged ions. We had to minimize
equipment and production costs, to stay within our budget and man
power limits. We had to expend considerable effort on safety and
reliability, because we have no experienced or trained staff. We
sought a modular, versatile design and used the most economical
off-the-shelf items, where possible. We designed the system in a
way that simplifies future upgrades and the diagnostics of many
potential problems. We are in the start up phase and ion production
is expected in the very near future.

Such a project would not have been possible without our friends
from the CRYEBIS community, who shared their knowledge and expertise
with us, or gave us their designs for copying, many of them
mentioned above. J. Arianer contributed substantially in the design
through a collaboration between Kansas State University and the
Institut de Physique Nucleaire in Orsay.

This project is funded by the Division of Chemical Sciences, U. S.
Department of Energy.

REFERENCES

1. Martin P. Stöckli, K. Carnes, C.L. Cocke, B. Curnutte, T.J.
   Gray, S. Hagmann, J.C. Legg and P. Richard, Proceedings of the
   XI National Conference on Particle Accelerators, Dubna, USSR
   (1988).
2. C.L. Cocke, P. Richard, J.S. Eck and R. Pardo, Nucl. Instr.
   Meth. B10/11, 838 (1985).
3. J. Arianer, M. Brient, C. Collart, C. Goldstein, J. MacFarlane,
   M. Malard, P. Nicol, A. Serafini, A. Steinegger, Proceedings of
   the 2nd EBIS Workshop, Saclay-Orsay, France, p240, (1981).
4. A. Curtois, J. Faure, Proceedings of the 3rd International EBIS
   Workshop, Ithaca, NY, USA, p27 (1985).
5. Magline Inc., Pinconning, MI 48650, USA: 42"x48" swivel
   magliner, 6000 lb capacity.

6. Lapp Insulators, LeRoy, NY 14482, USA: 9521A/T.R.216.
7. Glassman Inc., Whitehouse Station, NJ 08889, USA: PG200P1-LR.
8. Hipotronics Inc., Brewster, NY 10509, USA: IT 200-38/3, PS 810-2.5A.
9. J. Arianer, C. Goldstein, H. Laurent and M. Malard, IEEE Trans. on Nucl. Sci. NS-30, 2737 (1983).
10. ITT Standard, Buffalo, NY 14240, USA: P/N 5-160-03-036-001.
11. Ion Exchange Inc., Chicago, IL 60640, USA: LAB-FLOW#1702; our measurements indicated that the Flow (GPH) = $6.16 \cdot$ Pressure $(PSI)^{0.42}$ or Pressure (PSI) = $0.013 \cdot$ Flow $(GPH)^{2.4}$, which is useful for designing a system.
12. Martin P. Stöckli, Proceedings of the International Conference on ECR Ion Sources and their Applications, East Lansing, MI, USA, 1987, NSCL report #MSUCP-47, Ed. J. Parker, p219 (1987).
13. Martin P. Stöckli, K. Carnes, C.L. Cocke, B. Curnutte, J.S. Eck, T.J. Gray, J.C. Legg and P. Richard, Nucl. Instr. and Meth. B10/11, 763 (1985).
14. Cryogenic Consultants Limited, The Vale, London W3 7QS, England.
15. Martin P. Stöckli, C.L. Cocke, J.A. Good and P.Wilkins, Proceedings of 4th International Symposium on EBIS and their Applications, Upton, NY, USA (1988).
16. Leybold-Heraeus GMBH, D-6450 Hanau, FRG: RGD-330 with RW-3, D4A.
17. V.O. Kostroun, E. Ghanbary, E.N. Beebe and S.W. Janson, Physica Scripta T3, 47 (1983).
18. M.A. Levine, R.E. Marrs, J.R. Henderson, D.A. Knapp, M.B. Schneider, Physica Scripta T22, 157 (1988).
19. Raytheon Co., Waltham, MA 02254, USA: QKX1966.
20. CRYO-CAL Inc., St.Paul, MN 55114, USA: cryo-carbons (2%).
21. Thermocoax & Cie, F 92150 Suresnes, France: SEI10/200.
22. Lucas-Milhaupt Inc., Cudahy, WI 53110, USA: 68Ag/27Cu/5Pd VTG.
23. Bennett, Ivyland, PA 18974, USA.
24. J. Faure, B. Feinberg, A. Courtois and B. Gobin, Nucl. Instr. and Meth. 219, 449 (1984).
25. High Vacuum Apparatus, Hayward, CA 94545, USA: 125-0400, magnet box custom made.
26. R.W. Schmieder, Sandia Nat. Laboratory, Livermore, CA 94550, USA.
27. Jem City, Dayton, OH 45401, USA.
28. National Electrostatics, Middleton, WI 53562, USA: 2JA004152 with 8" CF and 550 Mohm resistors.
29. Balzers AG, FL-9496, Balzers, Lichtenstein, TPU170 with TPC300.
30. Vacuum Research Corp., Pittsburgh, Penn 15222, USA.
31. General Electric, Cleveland, OH 44112, USA: FCB.
32. Dycor, now Ametek, Pittsburgh, PA 15238, USA: M100M.

DISCUSSION

<u>Hershcovitch</u>: I noticed you have heaters for quick warmup. In my cold atomic beam, my experience has been that quick warmups made the system more susceptible to leaks. Maybe it is my system that is not very good, but are quick warmups necessary?

<u>Stockli</u>: During startup it is likely that one has to open the system several times, which causes serious delays if the warmup requires several days. I am aiming for a warmup period of approximately one day, which is risk free, because I routinely cool down the system in the same period of time without observing any leaks. In addition, I should point out that our system was designed to minimize such risks.

The Reliable Operation of CRYEBIS 1

J.Faure
Laboratoire National SATURNE
91191 Gif/Yvette France

Introduction

CRYEBIS has been built in ORSAY UNIVERSITY by ARIANER group as a heavy ions source for SATURNE. It was foreseen to use it to ionise polarised hydrogen and deuterium atoms too.

In spite of very encouraging results at the very beginning of the experiments ,in ORSAY, CRYEBIS 1 delivered very modest heavy ions beams without any reliability. It was decided ,in 1980, to set it up ,in SACLAY, near to SATURNE , to try to improve the performances and to adapt it with the usual reliability necessary in the vicinity of a 24 hours a day running accelerator.

When CRYEBIS 1 was intalled on the high voltage platform in Dec. 1983 many changes has been done. This source will deliver beams during 2 years with a very good reliability .

Cryogenic system

The system has been improved on two points:
 1) Some Helium leaks occured at the place where the system could be dismantled in two parts using special Brown type joints .
 This possibility was suppressed and the two parts welded together.
 2) The cryogenic insulating vacuum was separated from confinement volume. Therefore the residual gas became independent of cryostat liquid Helium level.

Injection system

The injection from external source has been described[1] many times and succesfully tested . It was the more important improvement leading to an operational source.

Results and conclusion

ION	I Total (charges)	I Ions (charges)	τ (ms)	Q accel. (charges)
Nitrogen	$6\ 10^9$	$3.5\ 10^9\ N^{7+}$	150	$4\ 10^8$
Carbon	$4\ 10^9$	$2.3\ 10^9\ C^{6+}$	150	$3\ 10^8$
Neon	$5\ 10^9$	$1.2\ 10^9\ Ne^{10+}$	150	10^8

In conclusion CRYEBIS 1 has been a good injector for SATURNE but the time of containment was too long. For that reason we have built DIONÈ.

Reference

1) J.Faure & al.,External ion injection into CRYEBIS,NIM 219(1984) 449-455

EBIS PHYSICS

ELECTRON BEAM DYNAMICS IN EBIS SOURCES

Michel TAGGER
C.E.N. SACLAY, 91191 Gif sur Yvette, FRANCE

ABSTRACT

I consider possible sources of departure of the electron beam from the ideal Brillouin regime. They would result in imperfect magnetic compression of the beam, as observed in present EBIS sources. I discuss the sensitivity of the compression to imperfections of the beam quality at the field buildup, and then to collective effects (emittance growth by initial relaxation or by beam instability).

INTRODUCTION

Optimal operation of EBIS sources relies on the possibility to magnetically compress the electron beam while keeping it close to the Brillouin regime where the density increases as B^2, whereas any other regime results in "flux–conserving" compression, $i.e.$ $Br^2 = Const.$ so that the density increases only as B. In practice it has been difficult to obtain much better than flux–conserving compression, although no reason has been found to strictly forbid it. In a series of recent reports[1,2,3] we have checked that any small departure from the Brillouin regime is very strongly amplified during the magnetic field rise, thus preventing a good compression. However the effects considered this far (e.g. the cathode magnetic field and temperature, the oscillations due to a bad adaptation of the beam) remain too weak to explain the experimental results. We will see here that the initial beam density profile can provide an answer.

Indeed the adaptation of the beam links its density to the magnetic field which must confine it. Thus if the density is not radially constant the beam cannot be everywhere in equilibrium between its space charge and the force exerted by the magnetic field. In another context (that of linear accelerators) this problem has been studied long ago by LAPOSTOLLE[4] and by SACHERER[5], and it has been meeting a renewed interest linked to the development of Heavy Ion Linacs and to the availability of large computers, making possible realistic numerical experiments. These works have shown that the initial inhomogeneity of a charged paricles beam is converted into radial emittance by a fast relaxation at the magnetic field buildup, while the density profile becomes perfectly flat.

In a first step I will first repeat these calculations, generalizing them to the case of a beam with cylindrical symmetry and non–zero azimuthal momentum (linked to the cathode magnetic field). I will conclude that an extreme inhomogeneity would be needed to strongly affect the Brillouin flow.

However other works[6] have shown that this emittance growth is very brutal, taking place immediatly after the magnetic buildup. This is indeed observed in the numerical simulations by WANGLER et al^7. The emittance grows very suddenly and then oscillates about its new equilibrium value. The beam profile also oscillates (becoming in turn peqked qnd hollow) and finally settles at a perfectly flat equilibrium. These oscillations of the profile are of course the equivalent in a high space–charge beam of the alternance of cathode images and crossovers, found in the optical theory[12,13] at low space charge.

Thus I will consider in a second step what happens if these oscillations of the beam profile survive until the phase of magnetic compression – a problem which was not encountered in Linacs. I will show that these oscillations can resonate with the harmonics of the frequency of the radial motion (Betatron frequency) of the electrons. This causes a new growth of the transverse emittance, as these resonances can very efficiently convert longitudinal energy into transverse energy, diluting the beam away from the Brillouin regime.

I will conclude by showing that, according to the works of I. HOFFMAN[8], these fluctuations might even be unstable, *i.e.* they might develop spontaneously, whatever the initial beam quality. Only particle simulations, such as those developped for plasma physics, could give a final answer on this point.

EMITTANCE GROWTH IN THE SPACE CHARGE REGIME

I repeat here the calculations of SACHERER[5], which I adapt to the case of a beam in cylindrical symetry in a constant magnetic field B_z. The equations of the radial motion of an electron are:

$$\dot{r} = P_r \tag{1}$$

$$\dot{P_r} = \frac{P_\theta^2}{r^3} - q^2 \frac{B_z^2}{4} r + F_s \tag{2}$$

where P_θ is the azimuthal momentum (fixed by the cathode magnetic field), F_s is the space-charge force, and $q = -e$ is the electron charge (in order to simplify the expressions I take its mass equal to unity). I compute the first two moments of the electron distribution, *i.e.* the averages (noted by a bar):

$$\dot{\bar{r}} = \bar{P_r} \tag{3}$$

$$\dot{\bar{P_r}} = \overline{\left(\frac{P_\theta^2}{r^3}\right)} - q^2 \frac{B_z^2}{4} \bar{r} \tag{4}$$

as $\overline{F_s} = 0$ by the law of action and reaction, then:

$$\dot{\overline{r^2}} = \overline{\dot{r^2}} = 2\overline{rP_r} \tag{5}$$

$$\dot{\overline{rP_r}} = \overline{\dot{r}P_r} + \overline{r\dot{P_r}}$$

$$= \overline{P_r^2} + \overline{\left(\frac{P_\theta^2}{r^2}\right)} - \bar{r}^2 \frac{\Omega_c^2}{4} + \overline{rF_s} \tag{6}$$

$$\dot{\overline{P_r^2}} = -\frac{\Omega_c^2}{2}\overline{rP_r} + 2\overline{\left(\frac{P_r P_\theta^2}{r^3}\right)} + 2\overline{P_r F_s} \tag{7}$$

where $\Omega_c = qB_z/m$ is the cyclotron frequency.

Then one defines the RMS radial emittance of the beam as

$$\epsilon_r^2 = \overline{r^2 P_r^2} - \overline{rP_r}^2 \tag{8}$$

and Equations (5)—(7) give its evolution:

$$\epsilon\dot\epsilon = [\overline{r^2\ P_r F_s} - \overline{rP_r}\ \overline{rF_s}] + [(\overline{\frac{P_r P_\theta^2}{r^3}})r^2 - \overline{rP_r}(\overline{\frac{P_\theta^2}{r^2}})] \tag{9}$$

The first bracket vanishes for a beam of constant density ($F_s = qE_r$ proportionnal to r); the second one vanishes if P_θ is proportionnal to r^2, in particular if the beam has remained laminar and homothetic from the cathode.

One then defines the average radius as :

$$\tilde{r} = (\overline{r^2})^{\frac{1}{2}} \tag{10}$$

(thus $\tilde{r} = a/\sqrt{2}$ for a flat beam of radius a). By derivation of Equation (5) and using (6) one easily obtains the enveloppe equation:

$$\ddot{\tilde{r}} + \frac{\Omega_c^2}{4}\tilde{r} - \frac{\epsilon_r^2}{\tilde{r}^3} - \frac{\overline{rF_s}}{\tilde{r}} - \frac{1}{\tilde{r}}(\overline{\frac{P_\theta^2}{r^2}}) = 0 \tag{11}$$

One then obtains the condition of beam RMS adaptation by asking that $\ddot{\tilde{r}}$ should vanish:

$$\frac{\Omega_c^2}{4}\tilde{r} = \frac{\epsilon_r^2}{\tilde{r}^3} + \frac{\overline{rF_s}}{\tilde{r}} + \frac{1}{\tilde{r}}(\overline{\frac{P_\theta^2}{r^2}}) \tag{12}$$

One checks easily that for a flat beam ($F_s = \frac{nq^2 r}{2\epsilon_0}$, where n is the density), at $P_\theta = 0$, one obtains the Brillouin equilibrium condition:

$$\alpha = \frac{n}{n_B} = 1 \tag{13}$$

where $n_B = \epsilon_0 B^2/2m$ is the Brillouin density. For non vanishing emittance and P_θ one obtains for a flat beam:

$$\frac{\Omega_c^2}{4}(1-\alpha)\tilde{r}^2 - d^2\tilde{r}^2 - \frac{\epsilon_r^2}{\tilde{r}^2} = 0 \tag{14}$$

where I have assumed $P_\theta = dr^2$ with $d = qB_c r_c^2/\tilde{r}^2$, B_c and r_c being the values of B_z and $\tilde{r}$ at the cathode, and thus that the beam has remained laminar since its production : this hypothesis, together with that of a flat profile, gives the essential limit of such calculations with envelop equations; in fact they give the necessary conditions to "cut" the hierarchy of moment equations, without which (14) would be expressed in terms of third order moments etc...

Equations (9) and (11) couple the evolutions of ϵ_r and of $\tilde{r}$ with z. Let us first check that, during the adiabatic compression, the RMS emittance of a flat beam in equilibrium is conserved. Rather than returning to Equations (1—9), and including the variations of B_z, one can directly use the existence of adiabatic invariants of particle motion[2]. This motion is described by :

$$r^2 = R^2 + \rho^2 + 2R\rho \cos\varsigma \qquad (15)$$
$$\dot\varsigma = \Omega_G = \Omega_c\sqrt{1-\alpha} \qquad (16)$$
$$P_r = -\frac{R\rho}{r}\Omega_G \sin\varsigma \qquad (17)$$

where R and ρ are defined by the adiabatic invariants :

$$\mu = \frac{\Omega_G}{2}\rho^2 \qquad (18)$$
$$P_\theta = \frac{\Omega_G}{2}(\rho^2 - R^2) \qquad (19)$$

At equilibrium the distribution function depends only on the invariants of the motion :

$$f = f(\mu, P_\theta) \qquad (20)$$

hence

$$\overline{P_r^2} = \frac{\int d\mu dP_\theta f(\mu,P_\theta) R^2 \rho^2 \Omega_G^2 \int_0^{2\pi} d\varsigma \frac{\sin^2\varsigma}{r^2}}{\int d\mu dP_\theta f(\mu,P_\theta) \int_0^{2\pi} d\varsigma}$$

One verifies easily by using (15), and the fact that Ω_G depends only on z in a flat beam, that $\overline{P_r^2}$ remains proportionnal to $\Omega_G(z)$ during the compression. In the same manner one would show that $\overline{\Omega_G r^2}$ remains constant, and thus that $\overline{P_r^2.r^2}$ is conserved. Since at equilibrium $\overline{rP_r}$ vanishes (by parity) equation (8) shows that ϵ_r is conserved during the compression.

One must remeber that this is a property of the equilibrium, due to the fact that f does not depend on ς, and that the compression is adiabatic. Indeed it is because these properties are not verified at the field buildup and that the compression is not adiabatic that it causes the emittance growth we are discussing.

Let us now return to Equation (14) which can thus be written :

$$\epsilon_r^2 = \frac{\Omega_G^2}{4}(1-\alpha)\tilde{r}^4 - d^2\tilde{r}^4 = C^{te} \qquad (21)$$

and compare two successive states, noted by subscripts 0 and 1, of the beam, between which the field has been increased by a factor b. One obtains :

$$\Omega_{c1}^2 = b^2\, \Omega_{c0}^2$$
$$d_1^2\, \tilde{r}_1^4 = d_0^2\, \tilde{r}_0^4$$
$$\alpha_1\, \tilde{r}_1^2 = b^{-2}\, \alpha_0\, \tilde{r}_0^2$$

where the last equality expresses the conservation of the total number of particles. Equation (21) then gives :

$$\frac{1-\alpha_1}{\alpha_1^2} = b^2 \frac{1-\alpha_0}{\alpha_0^2} \qquad (22)$$

identical to the expression obtained at zero emittance. This means that **the compression law is not modified by inclusion of the transverse RMS emittance** . However the emittance determines the initial departure of the beam from the Brillouin regime, since from (21) one gets :

$$\frac{\Omega_{c0}^2}{4}(1-\alpha_0) = \frac{\epsilon_r^2 + d_0^2 \tilde{r}_0^4}{\tilde{r}_0^4} = \frac{\epsilon_r^2 + \overline{P_\theta^2}}{\tilde{r}_0^4} \qquad (23)$$

This is identical to the result of AZAN[9] concerning the emittance due to the cathode temperature. Since this result has been judged too weak to explain the results of the Saclay sources CRYEBIS and DIONE we look here for another contribution, due to the emittance generated by the initial inhomogeneity of the beam. For this I repeat here the calculations of SACHERER[5] and of WANGLER et al[7]. I return to Equation (9), neglecting terms in P_θ, which are easily shown to be similar to terms in F_s, but $(1-\alpha)$ times smaller. This gives :

$$\epsilon_r \dot{\epsilon}_r = \overline{r^2} \; \overline{P_r F_s} - \overline{rP_r} \; \overline{rF_s} \qquad (24)$$

where

$$F_s = qE_r = \frac{q^2}{r\epsilon_0} \int_0^r dr' \; r' \; n(r') \qquad (25)$$

and thus :

$$\overline{rF_s} = \frac{1}{N} \int_0^\infty dr \; 2\pi \; r^2 \; n(r) \; q \; E_r(r)$$

where N is the density per unit length :

$$N = 2\pi \int_0^\infty dr \; r \; n(r)$$

giving

$$\overline{rF_s} = \frac{1}{4\pi} \frac{Ne^2}{\epsilon_0}$$

independently from the density profile. In the same manner :

$$\overline{P_r F_s} = \frac{2\pi}{N} \int_0^\infty dr \; r \; n \; P_r \; F_s$$
$$= -\frac{2\pi}{N} \int_0^\infty dr \frac{\partial}{\partial r}(rnP_r) \int_0^r F_s dr'$$

One can then use the equation of continuity

$$v_z \frac{\partial n}{\partial z} + \frac{1}{r} \frac{\partial (rnP_r)}{\partial r} = 0$$

where v_z is the velocity along $\vec{B}$ to obtain

$$\overline{P_r F_s} = -\frac{2\pi v_z}{N} \int_0^\infty dr\, F_s \frac{\partial}{\partial z} \int_0^r n(r')\, dr'$$

giving finally

$$\overline{P_r F_s} = -\frac{\pi \epsilon_0}{N} v_z \frac{\partial}{\partial z} \int_0^\infty dr\, r\, E_r^2 \qquad (27)$$

Now gathering Equations (5), (24), (26) and (27) one obtains

$$2\epsilon_r\, \dot{\epsilon}_r = v_z \frac{\partial \overline{\epsilon_r^2}}{\partial z} = -\frac{2\pi\epsilon_0}{N} \overline{r^2}\, v_z \frac{\partial}{\partial z} \int_0^r dr'\, r'\, E_r^2 - \frac{e^2}{4\pi\epsilon_0}\, v_z N \frac{\partial}{\partial z} \overline{r^2} \qquad (28)$$

The first term involves the quantity

$$W = 2\pi\epsilon_0 \int_0^\infty dr\, r\, E_r^2$$
$$= 2\pi \int_0^\infty dr\, r\, n(r)\, q\Phi(r) \qquad (29)$$

which is the potential energy of the beam in its own space charge. In fact the integral in (29) is divergent, and it should be bounded at a fixed radius, larger than that of the beam. Anyway we are only concerned here with its derivative where this divergence does not appear.

Let us then refer to an equivalent flat beam, defined as having the same average radius and the same density per unit length. It is thus a beam of total radius $a = \tilde{r}\sqrt{2}$ and density

$$n_0 = \frac{N}{2\pi\tilde{r}^2}$$

generating a field

$$E_0 = \frac{Ner}{4\pi\epsilon_0\tilde{r}^2} \qquad (r < a)$$
$$= \frac{Ne}{2\pi\epsilon_0\, r} \qquad (r > a)$$

and a potential energy W_0. One can then write:

$$U = W - W_0$$

$$\frac{dU}{dz} = 2\pi\epsilon_0 \frac{\partial}{\partial z} \int_0^\infty dr\, r\, [E_r^2 - E_0^2]$$
$$= 2\pi\epsilon_0 \frac{\partial}{\partial z} \int_0^\infty dr\, r\, E_r^2 + \frac{N^2 e^2}{4\pi\epsilon_0 \tilde{r}^2} \frac{\partial \tilde{r}^2}{\partial z} \qquad (30)$$

or
$$\frac{\partial}{\partial z}\epsilon_r^2 = -\frac{\tilde{r}^2}{N}v_z\frac{dU}{dz} \qquad (31)$$

This equation can also be written

$$\frac{\partial}{\partial z}\epsilon_r^2 = -\frac{Ne^2}{8\pi\epsilon_0}\tilde{r}^2\frac{\partial}{\partial z}\left(\frac{W}{W_0}\right) \qquad (32)$$

It relates the variations of the beam emittance and density profile. WANGLER et al[7] have performed numerical particle simulations to study these variations. They observe that in a first, very sudden phase, the emittance increases very sharply while the beam relaxes to a nearly flat profile. Then for a long time the emittance oscillates about its new equilibrium value. The density profile, and the corresponding quantity W/W_0, also oscillate following exactly relation (32).

For an RMS adapted beam (verifying Equation (14)), $\tilde{r}$ still changes if the emittance increases, but as a first approximation one can neglect this variation; then Equation (32) can be integrated exactly. On the other hand HOFFMANN and STRUCKMEIER[10] have shown that the equivalent flat beam ($U = 0$) minimizes energy, at fixed average radius. One can also easily show that the additional constraint of the conservation of angular momentum does not affect this result. Let us thus assume that the final result of the radial oscillations is to bring the beam to this state of minimum energy. This gives :

$$\epsilon_r^2 = (\epsilon_r^i)^2 + \frac{Ne^2}{8\pi\epsilon_0}\tilde{r}^2\left(\frac{W}{W_0}\right)_i \qquad (33)$$

where the subscript i applies to the values at the magnetic field buildup and ϵ_r is the asymptotic value reached when U has vanished. One can calculate this value for various density profiles; let us consider a truncated parabolic profile :

$$\begin{aligned} n &= n_0\left(1 - \lambda\frac{r^2}{r_0^2}\right) \quad (r < r_0) \\ n &= 0 \quad (r > r_0) \\ \lambda &= 1 - \frac{n(r_0)}{n(0)} \end{aligned} \qquad (34)$$

The RMS radius is
$$\tilde{r}^2 = \frac{r_0^2}{2}\frac{1 - \frac{2}{3}\lambda}{\frac{1-\lambda}{2}}$$

The density per unit length
$$N_0 = \pi r_0^2 n_0\left(1 - \frac{\lambda}{2}\right)$$

and the average density

$$\bar{n} = n_0 \frac{(\frac{1-\lambda}{2})^2}{1 - \frac{2}{3}\lambda}$$

The beam is hollow for $\lambda < 0$, peaked for $\lambda > 0$. The equivalent beam has a radius $a = \tilde{r}\sqrt{2}$. Its potential energy is

$$\frac{U}{W_0} = g(\lambda) = \frac{\frac{\lambda}{3} - \frac{\lambda^2}{8}}{(1 - \frac{\lambda}{2})^2} + 2\log\frac{1 - \frac{2}{3}\lambda}{\frac{1-\lambda}{2}} \qquad (35)$$

As mentionned above the flat beam has the minimum energy, thus $g(\lambda)$ has a minimum $g(\lambda) = 0$ for $\lambda = 0$.

Now gathering Equations (22), (23), (31)' and (35) one can study the compression of a beam which at the field buildup has an RMS adapted radius and a vanishing emittance, but a density profile given by Equation (34). I note respectively by subscripts 0 and 1 the values after the phase of rapid relaxation (just after the field buildup) and after the magnetic compression by a factor $b = B_1/B_0 = 60$. Table 1 shows as a function of λ the values of g, α_0, α_1 and $\beta = b\,\alpha_1/\alpha_0$ (β is thus the factor of improvement relative to "constant-flux" compression which would give $\alpha_1 = \alpha_0/b$).

λ	g	α_0	α_1	β
$-\infty$	.075	.9640	.081	5.
-1.	.0070	.9965	.244	14.7
-.8	.0051	.9975	.280	16.9
-.6	.0032	.9984	.337	20.2
-.4	.0017	.9992	.435	26.1
-.2	.0005	.9998	.644	38.6
-.1	.0001	.9999	.838	50.3
0.	0.	1.	1.	60.
.1	.0002	.9999	.819	49.1
.2	.0007	.9997	.591	35.5
.4	.0031	.9985	.345	20.7
.6	.0080	.9960	.230	13.9
.8	.0159	.9921	.170	10.3
1.	.0224	.9889	.145	8.8

TABLE 1

One sees that even a totally hollow beam ($\lambda = -\infty$, $n(0)/n(r_0) = 0$) can be initially very close to the Brillouin regime and that, even though the compression amplifies very strongly the departure from this regime β remains large for realistic values of λ. This also applies to different profile shapes, though they tend to give larger departures from the Brillouin flow.

One can thus conclude this part by counting the rapid growth of emittance, due to the initial beam density profile, among effects which can degrade the magnetic compression but should not (except maybe by adding a sufficient number of these effects) forbid a significant improvement over constant-flux compression.

EQUILIBRIUM, STABILITY AND OSCILLATIONS OF THE BEAM

We would like now to define the equilibrium which the beam can effectively reach after this initial relaxation. The ideal equilibrium we would like to obtain is a flat beam at Brillouin density, but it is likely that this equilibrium is unstable. A partial indication is given by GLUCKSTERN[11] and HOFFMANN[8] who studied the stability of a beam with a Kapchinskij - Vladimirsky (K-V) distribution function :

$$f = f_0 \, \delta(H - E_0) \qquad (36)$$

where f_0 and E_0 are constants. They showed that a flat beam with this distribution is unstable for

$$\alpha = \frac{n}{n_B} > .85 \qquad (37)$$

($1/R^2 > 11.5$ in the notations of HOFFMANN). This instability appears as axisymmetric oscillations of the density profile, which should spread the beam and decrease its density. Its growth rate is large (of the order of Ω_G) and it might thus develop very fast after the magnetic field buildup. It belongs to the family (well known in plasma physics) of the loss-cone instabilities, due to the population inversion introduced by the distribution (36).

If we assume that tyhis instability would bring the beam density down to $n/n_B \simeq .85$ right after the field buildup, this would imply $n/n_B \simeq .036$ after magnetic compression by a factor of 60, *i.e.* only 2.6 times better than constant-flux compression : this instability thus constitutes an excellent candidate to explain that this limit has this far proved very difficult to overcome experimentally. However one should remember that this calculation is limited to the stability of the K-V distribution, which is very singular, and that it cannot at this point be generalised to all beams close to the Brillouin flow. Going further is difficult, since we do not know analytically a distibution, other than K-V, giving a flat beam in the space charge regime, and we don't even know whether there exists a different distribution giving such a beam arbitrarily close to the Brillouin limit. HOFFMANN[8] has studied widened distributions (gaussian or "water-bag") and showed that they are more stable than K-V; in fact they can even be stable up to the Brillouin limit. But his result is not fully consistent and can precisely not be taken very firmly close to the Brillouin limit since the distribution functions he uses correspond to non flat density profiles (which he does not take into account in the calculation of the radial electric field), and thus cannot represent properly a beam close to this limit. It can be considered likely that the loss-cone instability strongly limits the density of the beam and thus its compression, even though the limit $\alpha \simeq .85$ might be slightly overcome by a different distribution function.

However another phenomenon would further complicate the approach of the Brillouin regime : Even when the oscillations we just discussed are stable, they still exist as eigenwaves of the system, *i.e.* undamped oscillations which can in particular be excited as the beam is emitted or as it passes at the magnetic field buildup. These are the oscillations observed by WANGLER *et al*[7] after the initial relaxation. They are well known of the designers of electron guns as "scallops" or, in the optical theory of HERRMANN[12], experimentally verified by AMBOSS[13], as fluctuations of density which create the alternance of "crossovers" and cathode images. The former are excited at the creation of an unadapted beam and appear as fluctuations of the radius at constant profile whereas the latter, corresponding to a more subtle difference between the initial distribution function of the beam and its equilibrium distribution, although its average radius is adapted, are oscillations of the profile at constant average radius.

At low density one easily shows that these oscillations are stationnary in the laboratory frame with a wavenumber :

$$k^2 V_z^2 = \omega_H^2 = \Omega_G^2 + \omega_p^2 = \Omega_c^2 - \omega_p^2 \qquad (38)$$

where ω_p is the plasma frequency :

$$\omega_p^2 = \frac{ne^2}{m\epsilon_0} = \alpha \frac{\Omega_c^2}{2}$$

At higher density the results of HOFFMANN show* that the wavenumber for the "fundamental" mode (noted by ω_{11}) is lower ; we can write it as :

$$k^2 V_z^2 = \Omega_G^2 + \lambda \omega_p^2$$

where $0 < \lambda < 1$. These waves can thus resonate with the harmonics of the radial motion of the electrons when $kV_z = l\Omega_G$, $l = 2, 3, ...$ *i.e.* during the magnetic compression when

$$\alpha = \frac{2l^2 - 2}{2l^2 - 2 + \lambda} \qquad (39)$$

It is noteworthy that for $\lambda = 1$ one would find the resonance $l = 2$ for $\alpha \simeq .85$, *i.e.* (by coincidence) the same limit given above as the instability threshold, while the resonances $l = 3, 4, ...$ are accumulated close to $\alpha = 1$.

As it crosses these values of α the oscillations of the beam resonate with the radial motion of the electrons; they are thus absorbed by *Landau damping* and give their energy to the electrons. We will see that on this occasion some electron parallel energy is very efficiently

---

* his figures show $\sigma = kVz/\Omega_G$ as a function of $\frac{1}{R} = \sqrt{\frac{2\alpha}{1-\alpha}}$

converted into transverse energy, thus giving a new growth of the transverse emittance. This effect cannot be described in the numerical simulations of WANGLER et al[7] which do not take into account the perturbed parallel motion.

To understand it we must come back to the electron motion, as I have described it in reference [2] (and repeated briefly above in equations (15—19)): We have written the Hamiltonian, for an electron in a constant magnetic field B_z and in the electrostatic potential of a flat beam :

$$H_0(\varsigma, \mu; \overline{\theta}, J_\theta; z, p_z) = \frac{p_z^2}{2} + \mu \Omega_G + \frac{J_\theta}{2}(\Omega_c - \Omega_G) \qquad (40)$$

and the particle motion is given by

$$\mu = \frac{\Omega_G}{2} \rho^2 \qquad (41)$$

$$J_\theta = \frac{\Omega_G}{2}(\rho^2 - R^2) \qquad (42)$$

$$r^2 = \rho^2 + R^2 + 2R\rho \cos \varsigma \qquad (43)$$

$$\dot{\varsigma} = \frac{\partial H_0}{\partial \mu} = \Omega_G \qquad (44)$$

$$\dot{z} = \frac{\partial H_0}{\partial p_z} = p_z \qquad (45)$$

plus equations not needed here which describe the azimuthal motion. For the sake of simplicity I assume $J_\theta = 0$ (no field at the cathode), and I perturb Equation (40) by a potential*

$$\tilde{\Phi}(r, z) = \Phi_s \left(\frac{r}{r_0}\right)^4 \cos(kz) \qquad (46)$$

i.e. a total Hamiltonian

$$H = H_0 + q\Phi_s \left(\frac{R}{r_0}\right)^4 \left(1 + \frac{\rho^2}{R^2} + 2\frac{\rho}{R}\cos\varsigma\right)^2 \cos kz \qquad (47)$$

In order to study the resonance $l = 2$ I expand the perturbation and retain only the important part (with the relevant dependance on ς) :

$$H = H_0 + q\Phi_s \frac{\rho^2 R^2}{r_0^4} \cos(2\varsigma - kz) \qquad (48)$$

---

* this radial dependance simplifies the calculations but any other functional form, except $\Phi \sim r^2$, would give similar results

(one verifies that the phase of the cosine is stationnary at the resonance $2\Omega_G = kV_z$) or after a new change of variables :

$$H(\xi, P_\xi; z', p_z') = \frac{1}{2}{p_z'}^2 + \frac{k^2}{2}(P_\xi - P_0)^2 - \frac{k^2}{2}P_0^2 + q\Phi_s \left[\frac{16P_\xi^2 - 8P_\xi P_\theta}{\Omega_G^2 r_0^4}\right]^2 \cos\xi \tag{49}$$

where :

$$\xi = 2\varsigma - kz \tag{50}$$
$$z' = z \tag{51}$$
$$\mu = 2P_\xi \tag{52}$$
$$p_z = p_z' - kP_\xi \tag{53}$$
$$P_0 = \frac{p_z'}{k} - \frac{2\Omega_G}{k^2} \tag{54}$$

One sees immediately that in the fluctuating potential, for a resonant particle $(P_\xi - P_0 \sim 0)$ we have

$$[\delta(P_\xi)]^2 \sim \frac{q\Phi_s}{k^2} \tag{55}$$

if $\rho \sim R \sim r_0$. On the other hand we have in order of magnitude

$$q\Phi_s \sim \frac{ne^2}{\epsilon_0}r_0^2\left(\frac{\delta r}{r_0}\right)^2 \tag{56}$$

where $\delta r/r_0$ is the relative amplitude of the scallops. Finally this gives

$$\delta\left(\frac{\rho^2}{r_0^2}\right) \sim \frac{\omega_p^2}{\Omega_G^2}\frac{1}{k^2 r_0^2}\left(\frac{\delta r}{r_0}\right)^2 \tag{57}$$

The term in $1/k^2 r_0^2$ is very large. It shows that on the occasion of its exchange of energy with the wave the particle transforms a much larger quantity of parallel energy into perpendicular energy. We have $\frac{\omega_p^2}{\Omega_G^2} = \frac{\alpha}{2(1-\alpha)}$, thus large, and for DIONE one can calculate :

$$V_z(m.s^{-1}) = 1.9\ 10^7 U^{1/2}(kV)$$

$$r_0^2(m^2) = 2.52\ 10^{-6}\frac{J(A)}{B^2(kG)U^{1/2}(kV)}$$

$$k^2(m^{-2}) = \frac{\Omega_c^2(1 - \alpha/2)}{V_z^2} = 4.3\ 10^5\frac{B^2(kG)}{U(kV)}$$

one thus gets, with $J = .36A$, $U = 10kV$, $B = 2.46kG$,

$$\delta(\rho^2) > 100\delta(r^2)$$

For many reasons this calculation must not be taken at its face value; first because in a non strictly flat beam Ω_G depends on μ, which introduces an additionnal dependance in Equation (49) and in realistic conditions would strongly reduce the estimate (57); the small factor would then not be kr but $dLog\Omega_G / dLogr$, assumed small if we are close to the Brillouin flow. The second reason is that, long before the electrons have acquired such a radial excursion, one will have to take into account its self–consistent effect on the potential and on the particle dynamics (in particular on the resonance condition), which is not done here. However the factors we get are so large that one can safely expect that the residual radial oscillations, albeit of very low amplitude, might cause the beam to expand until resonances are passed, which means approximately (as for the threshold of the loss–cone instability) $\alpha \simeq .85$, with the consequences we have seen on further magnetic compression.

CONCLUSION

I have shown that the emittance growth which can be expected to result from the initial beam profile imperfections should not put a stringent limit on the possibility to compress the electron beam close to the Brillouin regime, *i.e.* much better than constant–flux compression. However two other mechanisms, the loss–cone instability and the longitudinal/ transverse energy transfer by interaction with the radial oscillations of the beam, might represent a limit on the compression ratio one can obtain in EBIS sources. Further analytical and numerical calculations would be necessary to reach a firm conclusion on the efficiency of these mechanisms and on the ultimate limits on the performances of these sources, beyond the present experimental results.

REFERENCES

1. L. JACQUET, M. TAGGER, *Etude de l' instabilité de rotation electrons – ions dans les sources EBIS*, Rapport LNS/87/106 (1987)
2. M. TAGGER, *Compression adiabatique du faisceau d'electrons dans DIONE*, Rapport LNS/87/107 (1987)
3. M. TAGGER, *Mise en rotation et oscillations du faisceau électronique*, Rapport LNS/87/109 (1987)
4. P.M. LAPOSTOLLE, *Energy relationships in continuous beams*, Rapport CERN-ISR-DI/71-6 (1971)

Possible emittance increase through filamentation due to space charge in continuous beams, IEEE Trans. Nucl. Sci. 18 (3),1101 (1971)

5. F.J. SACHERER, *RMS enveloppe equations with space charge*, IEEE Trans. Nucl. Sci. 18 (3),1105 (1971)
6. O.A. ANDERSON, *Proceedings of the 1986 Linear Accelerator conference*, SLAC report 303,p. 65 (1986)
7. T.P. WANGLER et al, *Field energy and RMS emittance in intense particle beams*, Los Alamos Report LA-UR-85-1628 (1985)
8. I. HOFFMANN, Phys. Fluids 23, 296 (1980)
9. J.L. AZAN, *Thèse de 3eme cycle*, Université Paris VI, 1983
10. I. HOFFMANN, J. STRUCKMEIER, Particle accelerators 21, 69 (1987)
11. R.L. GLUCKSTERN, 1970 Proton Linear Accelerator Conference (Batavia, Ill, 1971) p.811
12. G. HERRMANN, J. Appl. Phys. 29, 127 (1958)
13. K. AMBOSS, IEEE-ED 11, 479 (1964)

COMPUTER PREDICTIONS OF "EVAPORATIVE" COOLING OF HIGHLY CHARGED IONS IN EBIT

B. M. Penetrante, M. A. Levine* and J. N. Bardsley
Lawrence Livermore National Laboratory, Livermore, CA 94550
*Lawrence Berkeley Laboratory, Berkeley, CA 94720

ABSTRACT

Evaporative cooling has been used successfully in EBIT to extend the containment time for neon-like gold to several hours. This paper discusses a theoretical basis for evaporative cooling. Also included is an assessment of the processes which affect the temperature and number balance of the trapped and coolant ions in EBIT, and how the basic operating parameters affect these processes. Results of computer calculations using nitrogen as a coolant are presented and compared with an approximate analytic solution.

INTRODUCTION

The electron beam ion trap (EBIT)[1-5] is an instrument designed for the spectroscopic study of highly charged ions (HCIs). In EBIT HCIs are trapped for in situ study. Current studies include the measurement of ion electron collisional processes such as dielectronic recombination and impact ionization. Fig. 1 is a schematic diagram of EBIT and Fig. 2 shows the trap region.

EBIT is in essence a variant of the electron beam ion source (EBIS) differing in emphasis rather than kind. Specifically, EBIT is shorter than most EBIS with a trap length of 2 cm to 10 cm rather than the more usual 100 cm EBIS trap length. This reduced length is made possible because the spectroscopic studies in EBIT require high densities of HCIs rather than the large quantities of HCIs required by a source such as EBIS. The advantage of the short size of EBIT is that it is able to meet the stability criteria for the modified rotational two stream instability predicted for an ion trap[6] and measured in the Berkeley EBIS[7].

© 1989 American Institute of Physics

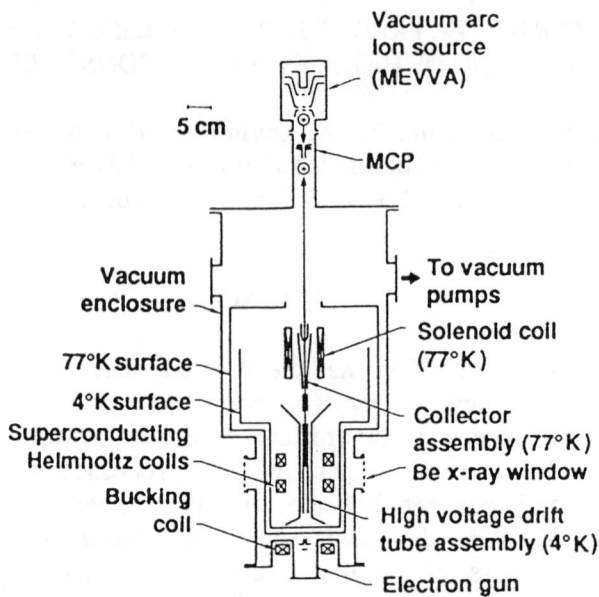

Fig. 1. Schematic diagram of EBIT. The electron beam from the gun is adiabatically compressed by the magnetic field of the superconducting Helmholtz coil which surround the drift tubes. The voltage applied to the drift tubes determine the energy of the electrons. The drift tubes and the Helmholtz coils are operated at a temperature of 4 °K. X-rays are observed at 90° to the electron beam through 2.5 mm slits in the central drift tube. Thin beryllium windows on the liquid nitrogen shield prevent a high background gas load. The ions to be ionized are injected into the trap axially from a MEVVA ion source.

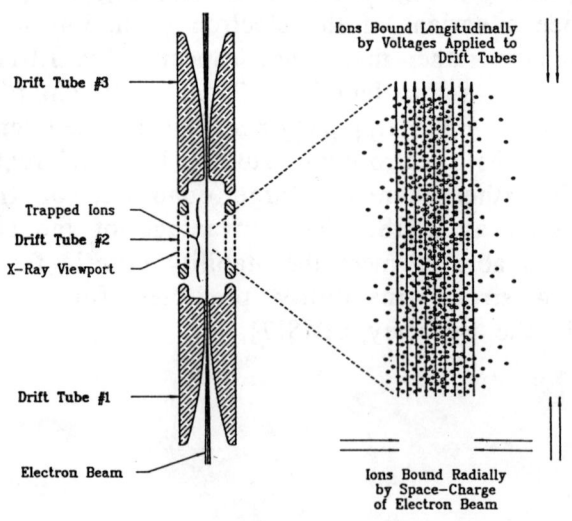

Fig. 2. The EBIT trap. The end drift tubes can be biased positively with respect to the central drift tube to increase the axial trap depth.

In EBIT, as in an EBIS, the electron beam serves the dual purpose of electrostatically trapping the ions and stripping partially ionized atoms to create HCIs. However, in EBIT, the electron beam serves a third purpose. The electron beam is also used as a probe to study HCIs. By adjusting the electron beam energy, electron-ion collisional processes can be explored by measuring the x-rays generated. For this purpose, EBIT has radial ports in the trap region. These ports are used both for spectroscopic studies and for measuring the radial beam profile, ion-electron overlap, etc.

Because of its importance to the operation of EBIT great care went into the electron beam design. The beam is launched in a Pierce gun with a 3 mm barium dispenser cathode, a perviance of 0.5 microperv, and zero magnetic field on the cathode. A 3 T superconducting cold bore magnet is used to guide and compress the electron beam in the trap region. A preliminary measurement of the beam radius inferred from x-ray imaging through the radial port gave a beam diameter of 61 microns for 80% of the electron current. Using the Herrmann[8] theory this would imply 0.1 ev temperature for electrons at the cathode. The electron beam voltage is adjustable to 30 kV and has a current density of the order of 3000 A/cm^2.

As stated above, EBIT is designed to study ion-electron collisions with HCIs trapped in situ. This implies that the HCIs should be within the electron beam. Once the HCIs are outside the electron beam they do not collide with beam electrons, are invisible to the x-ray detectors, are not ionized by the beam, and charge transfer with background gas to decrease the charge state. Schneider et. al.[1], in a discussion of the trap, point out that the radial potential at the edge of the electron beam is only 13 eV. If the ion temperature is > 13 Q_i eV, the ions will be outside of the beam. Unfortunately, there is a constant flow of energy, Γ_i, into the HCIs which tends to raise their temperature. This energy comes from the elastic collisions of the ions with the electron beam (Spitzer heating), given by[9]

$$\Gamma_i = 442.6 \, \lambda \, Q_i^2 \, N_i \, j / (E_b \, A_i) \tag{1}$$

where λ is the coulomb logarithm, N_i the number of ions per unit length within the beam, Q_i the ion charge, A_i its atomic weight, E_b is the electron beam energy in eV, and j the electron beam current density in A/cm^2. This energy transfer from the electron beam to the ions can be expected to heat the HCIs above the 13 Q_i eV limit even before they are fully ionized. To overcome this problem, "evaporative" cooling has been developed.

"EVAPORATIVE" COOLING

Evaporative cooling uses LCIs (low charged ions) to cool the HCIs in the trap. In an electrostatic trap with a well potential, V_w, charged atoms with an energy > $Q_i V_w$ can escape from the trap, where Q_i is the charge of escaping ions. The rate at which ions escape from the trap is given by[10]

$$\frac{dN_i}{dt} = - N_i v_i \left[\frac{e^{-\omega_i}}{\omega_i} - \sqrt{\omega_i}(\text{erf}(\omega_i) - 1)\right] \qquad (2)$$

where

$$\omega_i = \frac{Q_i V_w}{k T_i}$$

and v_i is the coulomb collision rate, which includes both the collisions with the same species and with all other ions.

In a mixture of LCIs and HCIs at near the same temperature, Eq. (2) indicates that the LCIs with a lower charge will escape from the trap at a faster rate than the HCIs. Multiplying by the energy per particle, the rate of energy loss due to escaping ions is given by

$$\frac{d(N_i k T_i)}{dt} = -\left(\frac{2}{3} N_i v_i e^{-\omega_i} - \frac{d N_i}{dt}\right) k T_i \qquad (3)$$

This loss of energy can be used to cool the HCIs.

Before presenting all the elements for a computer calculation of "evaporative" cooling, it is useful to consider an approximate equilibrium solution. The model assumes that LCIs are created by the ionization of a background gas by the electron beam. It also assumes that there is no loss of HCIs from the trap and that all ions are cold enough to remain in the electron beam. The rate of energy lost from the trap/cm is proportional to the rate of loss of LCIs. Since it is assumed that the system is in equilibrium, the rate of loss of LCIs is simply proportional to the rate of ionization of the LCIs from background gas i.e. $I\sigma_0 n_0/e$ where I is the electron beam current, σ_0, the ionization cross section of the neutral species of the coolant and n_0, the background gas density. Each escaping ion of charge Q_i carries away $Q_i V_w$ eV of energy. Equating this energy to the energy generated by elastic collisions with the HCIs gives

$$N_H = \alpha\, n_0 \tag{4}$$

where

$$\alpha = \frac{Q_i V_w \sigma_0 E_b A_H \pi r_b^2}{442.6\, \lambda\, Q_H^2\, e}$$

Here Q_H is the charge of the HCIs and A_H is its atomic weight.

There are several problems in the equilibrium solution. For one thing the rate of heating of the trap ensemble is dependent on the number of LCIs in the trap and it is difficult to evaluate the number of LCIs in the trap. A second problem is that the charge state of the LCIs, when it is lost from the trap, is not known à prioré. In order to answer these questions, a more elaborate computer program has been written which takes into account ion density distribution, ion heating as a function of charge state, ionization balance of the mixture, number of particles which escape as a function of charge state, charge exchange with background gas, energy transfer between ion species, etc.

COMPUTER CALCULATIONS

The calculation involves solving the set of coupled nonlinear differential equations for the energy transfer rates, particle production and escape rates for all ionic species. The processes controlling the energy rate balance are (1) heating of each ion species (HCIs and all charge states of LCIs) by the electron beam, (2) energy transfer from the HCIs to the LCIs, (3) energy transfer between the individual LCIs charge states, and (4) energy escape of each type of ion. The processes controlling the ion number balance are (1) electron-impact ionization and (2) ion escape from the trap. The rates used in both the energy balance and ion number balance take into account the appropriate overlap factors between the different species. For most cases, a significant fraction of the LCIs are outside of the beam region. This factor depends nonlinearly on the ion temperatures, and thus requires a self-consistent solution of the energy balance. The ion temperatures, on the other hand, depend on the densities of the various ions since these would affect the amount of energy transfer between these ions.

Fig. 3 shows the ion densities for a mixture of Neon-like Gold ions and Nitrogen ions for an electron beam energy of 18 keV, a beam current of 100 mA, and a well potential of 200 V. Shown in Fig. 3 are the density distributions for injected neutral coolant densities of 10^4 and 10^5 atoms/cc. In the experiment, the MEVVA injects many more ions than can be trapped. The equilibrium HCIs density is then determined by the amount of cooling available. In the computer calculations, this equilibrium density is obtained by specifying the condition that 100 percent of the energy that the HCIs gained from the electron beam is transfered to the LCIs, and not through the loss of HCIs.. One observes from Fig. 3 that as the coolant density is increased (and hence the Gold density), the proportion of lower charged Nitrogen ions, i.e., N^{1+} to N^{6+}, increases relative to the N^{7+}. This indicates that whereas for low coolant densities the Au^{69+} energy is absorbed mostly by N^{7+}, for high coolant densities a significant fraction of the Au^{69+} energy gets to be absorbed directly by lower charged N ions. This effect is substantiated later in Fig. 6.

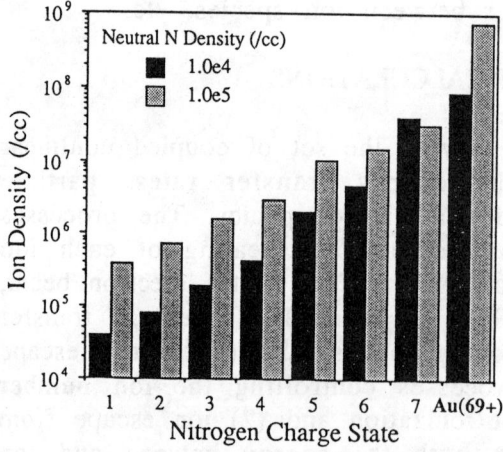

Fig. 3. Ion density distribution in a mixture of Neon-like Gold HCIs and Nitrogen LCIs. (E_b = 18 keV, I = 100 mA, V_w = 200 V)

Fig. 4 shows the characteristic radii of the ions for the same operating conditions. The electron beam radius is 30 microns. Note that the highly charged Gold ions are well confined within the beam. For the Nitrogen ions a large fraction of their time are spent outside the beam. This reduces the rate at which the Nitrogen ions absorb energy from the Gold ions. However, this is partially compensated by the fact that the rate at which these Nitrogen ions get heated by the electron beam is also decreased, thus allowing them to exist at temperatures lower than would be possible otherwise.

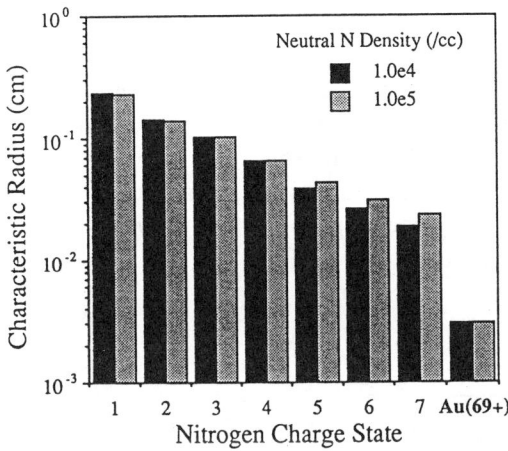

Fig. 4. Characteristic radii of the ion species for the Gold-Nitrogen mixture at E_b = 18 keV, I = 100 mA and V_w = 200 V.

Fig. 5 show the heating rate of the ions by the electron beam under the same operating conditions. Note that as the coolant density is increased, the relative amount of energy absorbed by the lower charged Nitrogen ions from the beam increases. This implies that for high density coolants a significant fraction of the total escape energy will also be carried out by these lower charged Nitrogen ions; whereas for low density coolants the escape energy is carried out predominantly by N^{7+}.

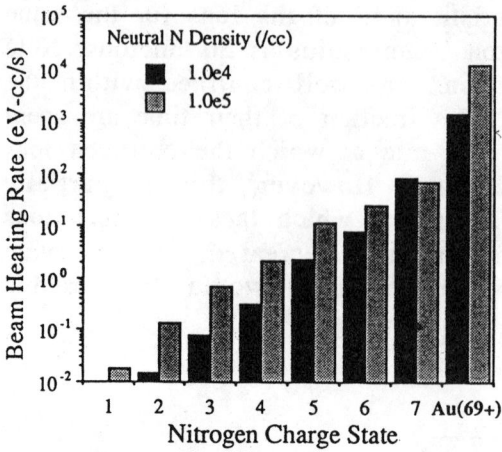

Fig. 5. Heating rate of the ions by the electron beam. (E_b = 18 keV, I = 100 mA, V_w = 200 V)

Fig. 6 shows the fractional energy transfer from the highly charged Gold to the Nitrogen ions. At low coolant densities most of the energy that the Gold absorbed from the electron beam is transfered to N^{7+}. At higher coolant densities, because of the increased densities of N^{6+} and N^{5+} (see Fig. 3), these lower charged ions also pick up energy directly from the Gold.

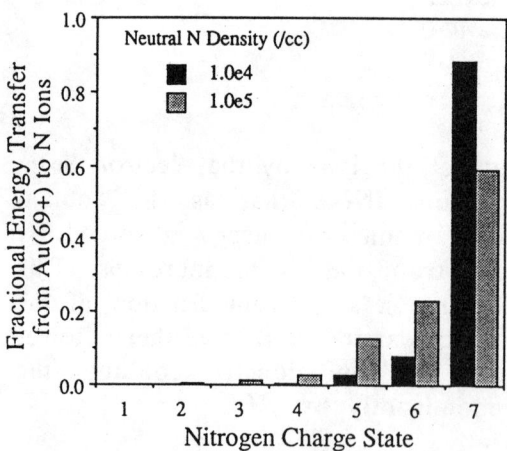

Fig. 6. Fractional energy transfer from Neon-like Gold to the Nitrogen coolant ions. The total energy absorbed by the Nitrogen ions from the Gold ions corresponds to the heat given to the Gold ions by the electron beam, as shown in Fig. 5. (E_b = 18 keV, I = 100 mA, V_w = 200 V)

Fig. 7 shows the energy transfer rates to each Nitrogen ion from all other Nitrogen ions. Note how energy from N^{7+} gets distributed to the lower charged Nitrogen ions. For the 10^4 atoms/cc injected coolant shown in Fig. 7, 18 percent of the energy that N^{7+} absorbed from the Gold ions is transfered to the other Nitrogen ions. At 10^5 atoms/cc injected coolant, 50 percent of the energy that N^{7+} absorbed from the Gold is transfered to the lower charged ions.

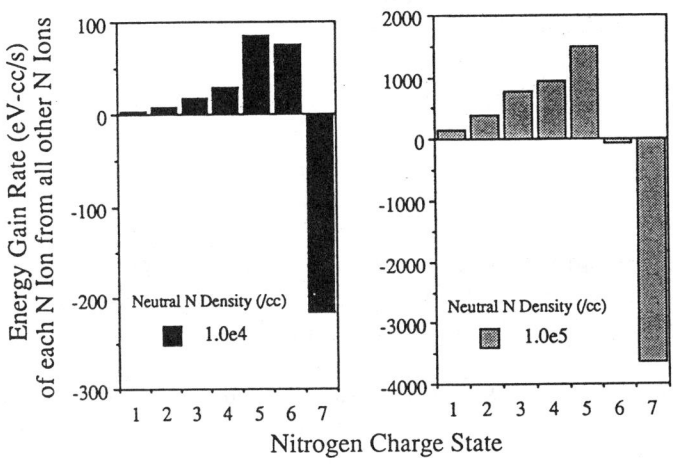

Fig. 7. Energy gain rate of each Nitrogen ion from all the other Nitrogen ions. (Negative gain rate indicates an effective outflow of energy instead of energy absorption.) (E_b = 18 keV, I = 100 mA, V_w = 200 V)

Fig. 8 shows how much of the total energy absorbed from the electron beam is carried off by each of the escaping ions. At low coolant densities, most of the energy evaporates through N^{7+}. At higher coolant densities, almost equal amounts of energy escape out via N^{5+}, N^{6+} and N^{7+}.

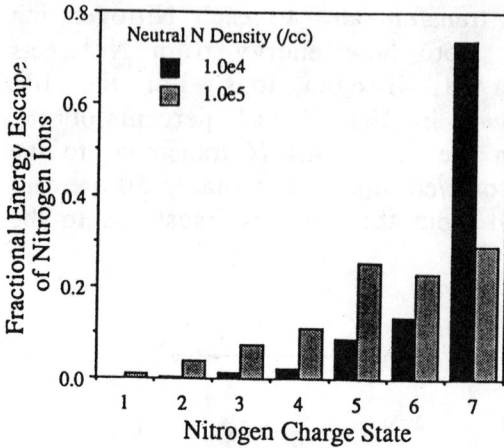

Fig. 8. Fractional energy escape of the Nitrogen ions. The total amount of energy carried off by the Nitrogen ions includes the heat from the electron beam to both the Nitrogen ions and the Gold ions. (E_b = 18 keV, I = 100 mA, V_w = 200 V)

Table 1 shows the calculated mean escape energy for each charge state compared with the quantity $Q_i V_w$. A typical temperature distribution of the ions is shown in Fig. 9.

Charge	Mean Escape Energy (eV)	$Q_i V_w$
1	464	200
2	791	400
3	1035	600
4	1288	800
5	1537	1000
6	1777	1200
7	2011	1400

Table 1. Mean escape energy as a function of charge state for $n_o = 10^4$. (E_b = 18 keV, I = 100 mA, V_w = 200 V)

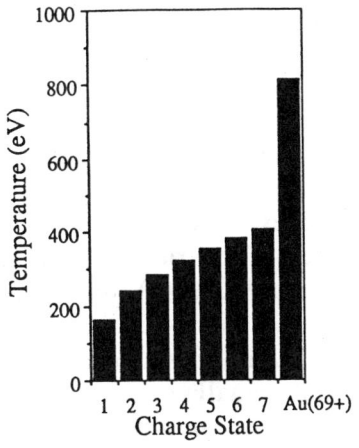

Fig. 9. Temperatures of the Gold HCI and the various Nitrogen LCIs for $E_b = 18$ keV, I = 100 mA, $V_w = 200$ V.

Fig. 10 shows the resulting density of trapped highly charged Gold as a function of the injected Nitrogen coolant density.

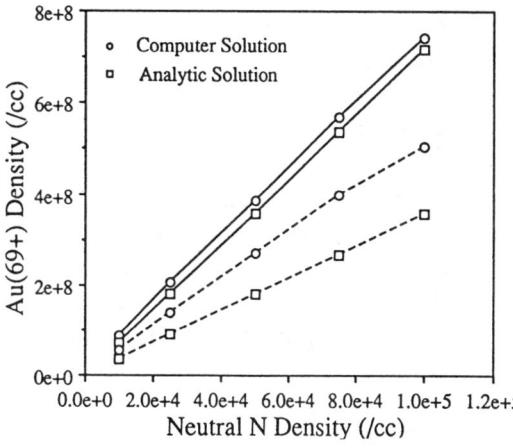

Fig. 10. Density of trapped Gold HCIs as a function of the Nitrogen coolant density. (Full Lines: $V_w=200$ V; Dashed Lines: $V_w=100$ V) ($E_b = 18$ keV, I = 100 mA)

CONCLUDING REMARKS

The heating of the highest charged ions by an electron beam makes it impossible to strip or trap them in the absence of a cooling mechanism. Evaporative cooling has been developed to solve this problem. In EBIT, the objectives of evaporative cooling are threefold: (1) to maximize the number of confined HCIs, (2) to maintain a good and stable charge balance, and (3) to have long containment times. In this paper we have elucidated the mechanisms which underlie evaporative cooling. At low levels of cooling, where the ion space charge is small compared to the electron space charge, we observe that the number of trapped HCIs increases linearly with the coolant density. Higher coolant densities introduce additional nonlinearities which further complicate the program. This is mainly because of beam neutralization. Work is currently in progress to establish the limitations due to this effect.

Although we have not yet completed a similar analysis of the effects of heavier coolants, such as Titanium, it is clear that there are both advantages and disadvantages in using ions which can reach higher charge states. The energy carried out by each escaping ion is approximately proportional to its charge. However, the additional trapping time of more highly charged coolants greatly increases their density within the electron beam. The heating rate of the coolant ions by the electron beam becomes significant and can exceed the heating of the HCIs. This reduces the cooling efficiency of more highly charged coolant ions. Further studies of the relative efficiencies of Nitrogen and Titanium are also in progress.

ACKNOWLEDGEMENT

This work was performed under the auspices of the U. S. Department of Energy by the Lawrence Livermore National Laboratory under Contract No. W-7405-ENG-48.

REFERENCES

1. M. B. Schneider, M. A. Levine, C. L. Bennett, J. R. Henderson, D. A. Knapp and R. E. Marrs, this volume.

2. M. A. Levine, R. E. Marrs, J. R. Henderson, D. A. Knapp and M. B. Schneider, Phys. Scr. T22, 157 (1988).

3. R. E. Marrs, M. A. Levine, D. A. Knapp and J. R. Henderson, Phys. Rev. Lett. 60, 1715 (1988).

4. R. E. Marrs, C. Bennett, M. H, Chen, T. Cowan, D. Dietrich, J. R. Henderson, D. A. Knapp, M. A. Levine, M. B. Scheneider and J. H. Scofield, *Proceedings of the International Conference on the Physics of Multiply Charged Ions, Grenoble, France September 14-16 1988.*

5. D. A. Knapp, R. E. Marrs, M. A. Levine, C. L. Bennett, M. H. Chen, J. R. Henderson, M. B. Schneider and J. H. Scofield, Phys. Rev. Lett. (submitted).

6. C. Litwin, M. C Vella, and A. Sessler, Nucl. Instr. and Meth. 198, 189 (1982).

7. M. A. Levine, R. E. Marrs and R. W. Schmeider, Nucl. Instr. and Meth. A237, 429 (1985).

8. G. J. Herrmann, J. Appl. Phys. 29, 127 (1958).

9. L. Spitzer, Physics of Fully Ionized Gases, J. Wiley & Sons, N.Y. (1962).

10. V. P. Pastukhov, Nucl. Fusion 14, 3, (1974).

EVAPORATIVE COOLING OF HIGHLY CHARGED IONS IN EBIT: AN EXPERIMENTAL REALIZATION **

Marilyn B. Schneider, Morton A. Levine*, Charles L. Bennett,
J. R. Henderson, D. A. Knapp, and R. E. Marrs
Lawrence Livermore National Laboratory, Livermore, CA 94550
* Lawrence Berkeley National Laboratory, Berkeley CA 94720

ABSTRACT

Both the total number and trapping lifetime of near-neon-like gold ions held in an electron beam ion trap have been greatly increased by a process of 'evaporative cooling'. A continuous flow of low-charge-state ions into the trap cools the high-charge-state ions in the trap. Preliminary experimental results using titanium ions as a coolant are presented.

INTRODUCTION

A titanium injector has been built to provide evaporative cooling of highly-charged gold ions in the Electron Beam Ion Trap (EBIT)[1-5] at the Lawrence Livermore National Laboratory. The preliminary experiments reported here demonstrate the success of evaporative cooling in increasing both the total number and trapping lifetime of gold ions in EBIT. We report x-ray intensities as a function of various trap parameters. Experimentally, we define two limits. At low levels of cooling, the number of trapped gold ions in the electron beam increases linearly with the coolant (neutral titanium) density, and the calculated ion space-charge is small. In contrast to this, at higher levels of cooling the number of trapped gold ions in the beam is a weaker function of coolant density and the variation of x-ray intensity with the trap parameters is very different.

We discuss first the nature of the EBIT trap to justify our representation of the cooling process as a one-dimensional problem. Then we present the experimental results, and indicate some preliminary agreement with theory. Finally, we mention some practical considerations of evaporative cooling.

NATURE OF THE EBIT TRAP

EBIT[1-5] consists of a very intense electron beam which passes through and is accelerated by three drift tubes as it follows the field lines of a 3T superconducting magnet. The voltage applied to the drift tubes determines

**Work performed under the auspices of the U.S. Department of Energy by the Lawrence Livermore National Laboratory under contract W-7405-ENG-48.

the electron beam energy. The two end drift tubes can be biased with respect to the central one to provide an axial trap. Slits in the central drift tube allow x-rays emitted perpendicular to the electron beam to be detected.

The ions are trapped in the radial direction by a combination of the space-charge of the electron beam and the magnetic field. Ions and secondary electrons modify the trapping potential. For a Gaussian radial distribution of electrons in the beam, the space-charge potential of the beam, V_b, at the radius of the beam which contains 80% of its charge (r ~ 31 μm), is 0.55 $I/\sqrt{E}$ (Volts) where I is the electron current in mA and E is the electron energy in keV. For 100mA at 18keV, this is about 13V. At the drift tube wall, a radial distance of 0.5 cm from the center of EBIT, the potential is 9.4 V_b, or 122V. Thus while an ion of charge Q needs an energy of 122Q eV to 'escape' the trap radially, it needs only 13Q eV to leave the electron beam. For most experiments, an ion must be in the beam to be useful.

In addition, the magnetic field produces an effective potential[6] that further inhibits radial escape. The orbit of a charged particle in the trap, specified by its energy and generalized angular momentum (or, equivalently, its instantaneous position and momentum) is confined between minimum and maximum radii. In order to move from the electron beam to the drift tube wall an ion must diffuse both in energy and generalized angular momentum space, which is a slow process.

Axially, the trap is purely electrostatic. The 'bare' potential is the sum of the voltage (V_A) applied to the end drift tubes relative to the central one, plus the potential due to the narrowing of the end drift tubes (2.1V_b or ~ 30V at I=100 mA, E=18keV). (The central drift tube is 1 cm in diameter and 2 cm long. The end drift tubes narrow from 0.5 cm to 0.15 cm in diameter 1.7 cm from the center of EBIT.) Finally, the space-charge of the ions contributes to the axial potential.

The minimum ion-ion collison time is expected to be on the order of a few hundred μs [7], but the ion velocities are roughly 10^4 cm/sec, so the ion mean-free path in the axial direction is longer than the 2cm long trap. An ion can escape axially as soon as it has enough axial energy. Because radial escape is much more difficult, the EBIT trap is here considered as a one-dimensional problem with V_{trap} equal to the axial potential.

EVAPORATIVE COOLING

The elastic collisions between the ions and the electron beam primarily increase the transverse momentum of the ions[2]. This kinetic energy is partitioned in all directions and among all the ions by ion-ion elastic collisions. The energy gain of the ions is large enough that some ions will spend a significant amount of time outside the electron beam, and, as their energy grows, eventually escape the trap, often before they reach the maximium ionization state possible for the energy of the electron beam.

In evaporative cooling, a second ionic species (LCIs) with a lower charge, q, is continuously added to the trap containing the desired highly-charged ions (HCIs). The LCIs are heated through ion-ion collisions with the HCIs until both species have approximately the same temperature. The LCIs, having a lower charge, need less energy per ion to escape the trap, so they 'evaporate', removing their kinetic energy from the trap. The continuous evaporation of the

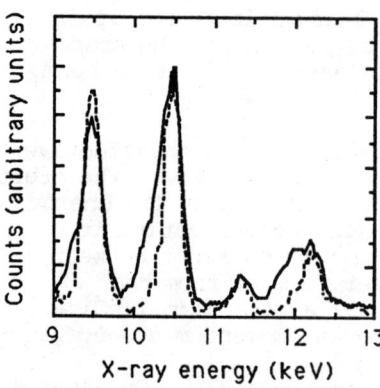

Fig.1. N=3->2 transition lines in near neon-like Au HCIs taken with a solid-state Ge detector. E=18keV, I=100mA. Solid: no evaporative cooling, V_{trap}=80V. Dash: evaporative cooling (Ti ions), V_{trap}=40V. The narrower lines with cooling indicate a narrower charge state distribution.

LCIs cools the HCIs. Fig.1 compares the n=3->2 transition lines of Au HCIs without evaporative cooling (solid) to those with evaporative cooling (dash). The electron beam energy was adjusted to maximize the formation of neon-like gold (Au^{69+}). The narrower lines in the second case indicate a more highly ionized charge state distribution. The 1/e lifetime of uncooled ions in the electron beam was measured to be 18s. The lifetime for the cooled HCIs was 210s, a factor of 10 larger despite a lower trapping potential (40V compared to 80V).

EVAPORATIVE COOLING EXPERIMENTS

To study evaporative cooling, we chose highly-charged (near-neon-like) gold (Au^{69+}, Au^{68+}, Au^{67+}...) for the HCIs and titanium ions (maximum charge = 22+) for the LCIs. Titanium atoms are evaporated from a calibrated Ti wire getter. The hot Ti atoms flow continuously into EBIT through one of the radial x-ray ports and are ionized by the electron beam. The 1/8" diameter Ti wire, 29cm from the center of EBIT, is imaged with a 0.25" hole in a metal mask onto a 1cm length of the electron beam. The mask is attached to a liquid nitrogen shield to minimize the background gas load into EBIT. The Ti flow is monitored by measuring the temperature (~1500 C) of the wire with a pyrometer, and can be adjusted by changing the wire current. The density, n_o(Ti) of titanium atoms in the center of EBIT ranges from 10^5 to 10^7 cm^{-3}. The Ti wire deforms when it is hot and does not

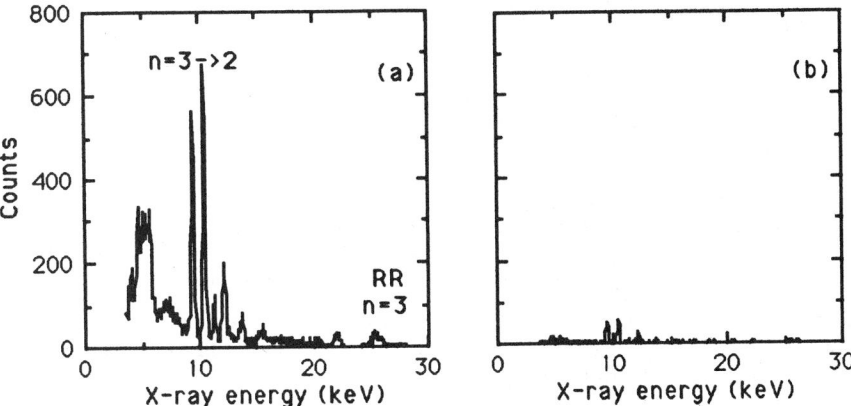

Fig.2. Time-routed Ge spectra of Au HCI (E=18keV, I=100mA, V_{trap}=130V). Ti shutter is kept open during (a) but kept closed during (b). The n=3->2 transition lines and the radiative recomination lines (RR) to n=3 are shown.

always fill the EBIT image, making the calibration of n_o(Ti) inaccurate by the unknown geometrical filling factor. A shutter in the Ti beam path can open or close in 25ms.

The dramatic success of evaporative cooling is shown in Fig.2. The first spectrum was taken for 60s, then the Ti shutter was closed and a second spectrum taken for 60s. (If the shutter is open, the ion lifetime in the electron beam is greater than 10 minutes, Fig.4,6).

Cooled Au HCIs can remain in the electron beam for hours. The longest 1/e trap lifetime observed was about 4 hours (Fig.3).

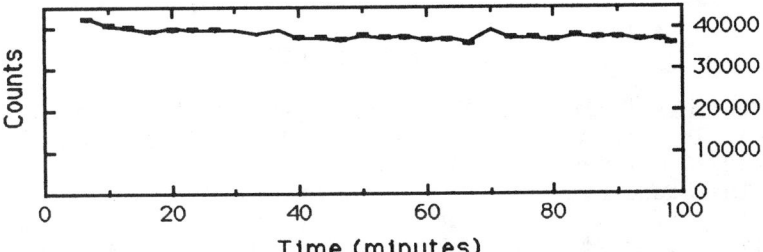

Fig. 3. Count rate in n=3->2 transition lines vs. time. Gold is injected into the trap at 0 minutes. The estimated 1/e trap lifetime is about 4 hrs. Ti shutter is always open.

TRAP LIFETIMES AND ION DENSITIES

Fig.4a shows the count rate in the n=3->2 lines in Au HCIs as a function of time for different value of $n_o(Ti)$. The long lifetime in the case of $n_o(Ti)=0$ results from the

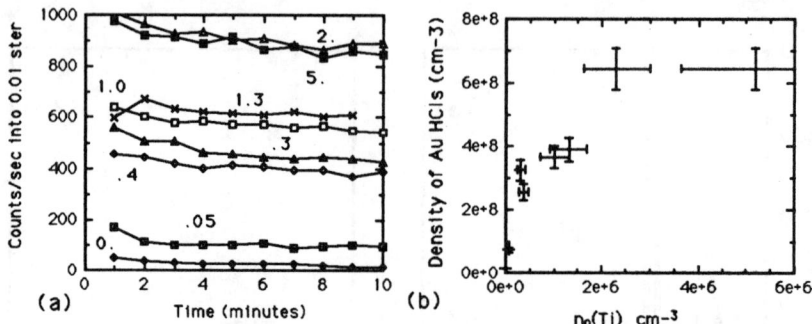

Fig.4. (a) Count rate vs time. The numbers give $n_o(Ti)$ in units of 10^6 cm^{-3}. (b) Calculated density of Au HCIs vs estimated $n_o(Ti)$. Data taken at E=18keV, I=75mA, V_{trap}=320V.

cooling from increased background gases in EBIT compared to the conditions of Fig.1, when there were no open apertures in the liquid nitrogen and helium shields[1-4].)

The density of highly charged Au ions in EBIT is estimated from the intensity of the radiative recombination lines using calculated cross sections[8], and is plotted versus $n_o(Ti)$ in Fig.4b. The charge density of Au HCIs neutralizes 2% of the electron beam at the highest densities shown. There is no direct measurement of the LCIs' (Ti ion) density in the trap, but it is expected[1] to be several times the HCIs' density.

The MEVVA injects many more ions than can be trapped, so the equilibrium HCIs' density is determined by the amount of cooling available. The low level limit of cooling is defined to be the region where the HCIs' density in the electron beam varies linearly with $n_o(Ti)$. From Fig.4b, the low level cooling limit is approximately $n_o(Ti) < 10^6$ cm^{-3}.

LOW LEVELS OF COOLING

In this limit, the HCIs' charge density in the electron beam is less than 1% of the electron charge density and ion screening effects are ignored.

The count rate in the 3->2 lines is proportional to the square of the electron beam current (Table I). This is true for two different values of the axial potential. Note that for a given current, the count rate is higher for the higher axial potential.

Table I   Count Rate vs. Current for Different V_A

Data taken at E = 18keV, $n_o(Ti) = 7 \; 10^5 \; cm^{-3}$

I (mA)	I² (mA²)	V_A = 100V counts/sec	V_A = 300V counts/sec
74	5476	210	350
100	10000	420	600
Ratio:	0.55	0.5	0.6

Table II   Ion Temperature vs. V_{trap} for Different I

Data taken at E = 18keV, $n_o(Ti) = 7 \; 10^5 \; cm^{-3}$.

	I = 74 mA			I = 100 mA		
V_{trap}	Z_{RMS}	T/QV_{trap}	V_{trap}	Z_{RMS}	T/QV_{trap}	
120 V	22.9	0.04	130 V	24.9	0.5	
220 V	21.8	0.03				
320 V	21.3	0.03	330 V	22.2	0.4	

As V_{trap} is increased, the x-ray intensity increases, and then levels off. This observation confirms the one-dimensional trapping model: beyond a certain value of V_{trap}, radial escape occurs more often than axial escape.

In a one-dimensional model, the temperature, T should be proportional to QV_{trap} (the only characteristic energy in the problem). A measurement of how the ions fill the trap should determine the temperature. We have used a slit to axially image the ions in EBIT onto a position sensitive-detector gated to only count n=3->2 Au x-rays. The directly measured ion distribution can be characterized by its root-mean-square width Z_{RMS}, measured in channel numbers on the detector. To deduce a temperature we assume the HCIs' density in the electron beam is proportional to $\exp(-QV(z)/T)$ where V(z), the axial electrostatic potential at the height z neglects the space charge of the ions. Preliminary results, at two different electron beam currents, are that T/QV_{trap} is approximately constant (Table II). Using a similar technique to image the ions in the radial direction, we see a wider distribution for a higher V_{trap} (Fig. 5). This indicates that it is the axial potential that controls the ion temperature, and hence the ion spatial distribution.

HIGH LEVELS OF COOLING

More complicated effects occur at higher ion densities. Fig.6 shows that the x-ray intensity is now

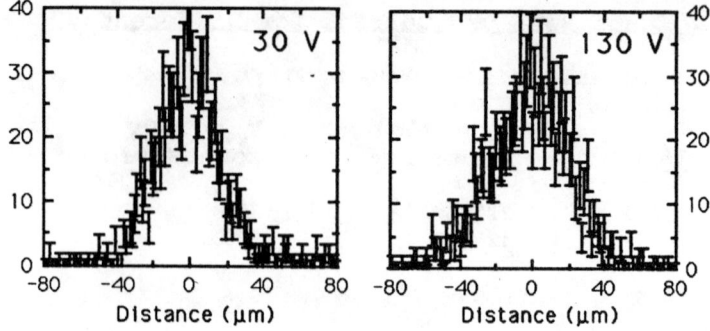

Fig.5 Radial image of ions (counts vs distance in EBIT). E=18keV, I=100mA, $n_o(Ti) = 10^6$ cm^{-3}. (a) V_{trap}=30V (b) V_{trap}=130V.

linearly proportional to the current. A comparison of the count rate for the same current between V_A=100V and 300V in Fig.6 shows <u>fewer</u> ions in the beam for the higher trapping voltage, in contrast to what occurs at lower cooling levels (Table I). This is important because the simple theories

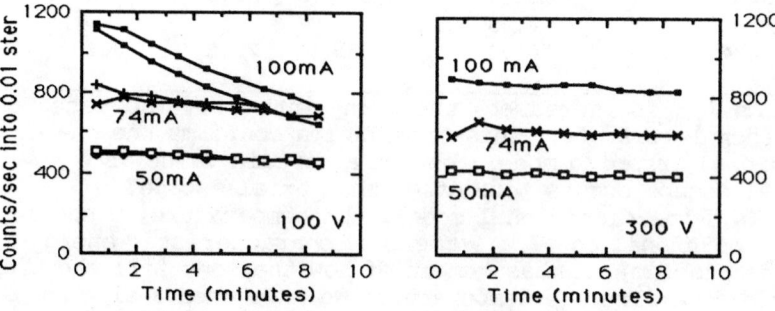

Fig.6. Count rate vs time at high levels of cooling. Data taken at E=18keV, $n_o(Ti)$=1.3 10^6 cm^{-3} for V_A=100V , V_A=300V.

that ignore ion space charge cannot explain it. Also there is a noticeable decay (1/e lifetime = 18 min) at V_A=100V and I=100mA.

The best ratio of Au^{69+} to total Au ions we have obtained in EBIT is about 1:2. This corresponds to the theoretical limit given by the ratio of cross sections of electron impact ionization of Au^{68+} to radiative recombination onto Au^{69+}. This limit is not achieved at the higher levels of cooling. As the ion density increases, the space-charge of the ions begins to screen the electron beam, so the ions spend less time in the beam, and charge-exchange outside the beam destroys the charge state. In addition, charge-exchange is expected to become important even in the beam region at higher $n_o(Ti)$.

ADDITIONAL CONSIDERATIONS

Depending on conditions, Ti excitation and/or recombination lines may be a background to specific desired Au lines in a solid state detector. To overcome this, we have also used nitrogen gas as a coolant. Preliminary results indicate that nitrogen cools as well as titanium.

LCIs will cool anything that is more highly charged than itself, so impurites may become trapped. We have seen a slow buildup of barium and tungsten ions from the electron gun and lead from the titanium wire. The rate of buildup of barium and tungsten is higher with nitrogen cooling.

SUMMARY

Both the number and lifetime of Au HCIs have been increased with the use of Ti LCIs to evaporatively cool the gold ions. The cooling mechanism is consistent with the evaporation of hot ions from a one-dimensional potential well, which is the axial trap. In the limit of low-level cooling, the number of HCIs trapped in the electron beam is proportional to $n_o(Ti)$, and the contribution of the space-charge of the ions to the total potential is small. More complicated effects occur at higher levels of cooling, probably because ion space-charge effects become important and the ions spend much of their time outside the beam. Additional experimental and theoretical[1] work are planned to further our understanding of the cooling mechanism and its effects.

References

1. M.A. Levine, B.M. Penetrante, and J.N. Bardsley, this volume.
2. M.A. Levine, R.E. Marrs, J.R. Henderson, D.A. Knapp, and M.B. Schneider, Phys. Scr. T22, 157 (1988).
3. R.E. Marrs, M.A. Levine, D.A. Knapp, and J.R. Henderson, Phys. Rev. Lett. 60, 1715 (1988).
4. R.E. Marrs, C. Bennett, M.H. Chen, T. Cowan, D. Dietrich, J.R. Henderson, D.A. Knapp, M.A. Levine, M.B. Schneider, and J.H. Scofield, Proceedings of the *International Conference on the Physics of Multiply Charged Ions*, Grenoble, France September 14-16 1988.
5. D.A. Knapp, R.E. Marrs, M.A. Levine, C.L. Bennett, M.H. Chen, J.R. Henderson, M.B. Schneider, and J.H. Scofield, Phys. Rev. Lett. (submitted).
6. Britton Chang, private communication.
7. L. Spitzer, *Physics of Ionized Gases*, J. Wiley & Sons, N.Y. (1962).
8. E.B. Saloman, J.H. Hubbell, and J.H. Scofield, At. Data Nucl. Data Tables 38, 1 (1988).

EVIDENCE FOR ION COOLING AND AN OBSERVATION OF ION HEATING IN CORNELL EBIS I

E. N. Beebe
Department of Nuclear Science and Engineering, Ward Laboratory,
Cornell University, Ithaca, N.Y., USA 14853

ABSTRACT

Preliminary studies made on the Cornell Electron Beam Ion Source I (CEBIS I) indicate that a cooling of high Z ions through energy exchange with lighter, low charge ions may contribute to longer ion confinement times and production of higher charge state ions than would otherwise be possible without this effect. A DC electron beam propagated at energies less than 3.6 keV, at currents of 20 mA or less. Multiply charged ions were produced in a containment mode and charge state evolution was observed for injected nitrogen, argon, and xenon gases as well as residual gases. The extracted C^{5+}, N^{6+}, O^{7+}, Ar^{16+}, and Xe^{28+} ions were unambiguously observed by both time of flight (TOF) spectroscopy and by the use of an analyzing bending magnet. In addition, the time dependence of low to moderate charge state ions escaping from the trap, (presumably due to heating) during the confinement period was observed.

INTRODUCTION

The Cornell electron beam ion source I (CEBIS I) is a small EBIS device whose basic design and details of operation are described elsewhere.[1,2,3] Notable features of CEBIS I are a conventional solenoid typically operated at ~2.8 kGauss and a 140 l/sec distributed sputter ion pump which pumps the ionization region. A continuous electron beam, launched externally to the magnetic field, propagates through a series of stainless steel mesh trap electrodes and is collected by an OFHC copper electron collector. The source is operated in the containment mode, i.e., ions are retained inside the axial trap for a predetermined confinement time. Radial trapping of the ions is by the electron beam space charge depression (see figure 1).[4]

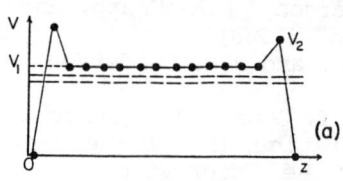
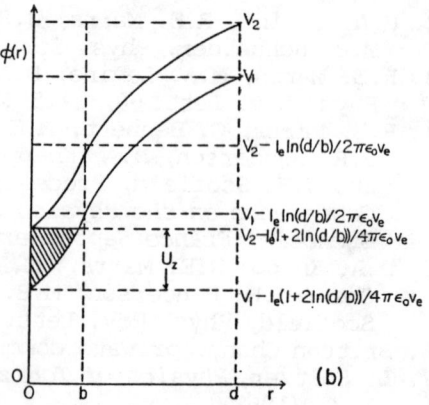

fig. 1 a) Schematic of axial ion trap formed by drift tube potentials.
b) radial ion trapping due to the space charge of electron beam.

© 1989 American Institute of Physics

The ions are extracted by lowering the potential on the axial barrier which allows them to escape the trap in a short burst. The time between pulses is chosen to produce ions with the desired charge state.

The electron gun is an in house built version of a Hughes model 112-2B modified Pierce type convergent gun with a nominal perveance of 2.2 microperv. For the studies reported, CEBIS I was operated with a cathode current of about one tenth the space charge limited current, (typically 20 mA with -2 kV on the cathode). Referring to figure 1b, one can see that there can be a considerable

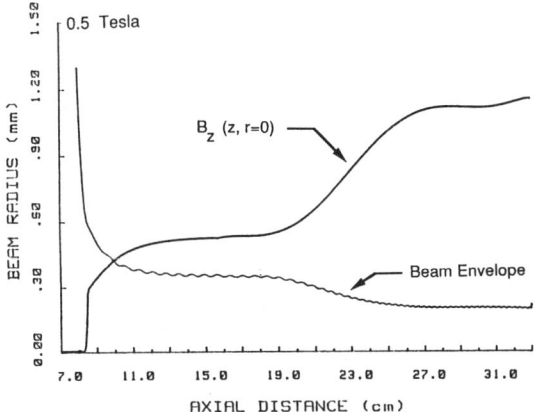

fig. 2 Simulation of an Electron beam launch into entrance an entrance solenoid with further compression by the main solenoid. An electron beam current density of ~150 A/cm2 is predicted at a magnetic field of ~3.8 kGauss.

spread in the kinetic energy of ions extracted from the trap, corresponding to the difference between the minimum effective trap region and barrier potentials. The issue is further complicated by the partial neutralization of the electron beam in the trap region by ions. By reducing the electron beam current, the energy spread of the extracted ions can be reduced, but at the expense of reduced ion trap capacity.

Good results have been obtained on CEBIS II with a two solenoid system.[5,6] Therefore, CEBIS I has also been fitted with an entrance solenoid. Numerical electron beam envelope calculations with the electron beam launched first into an entrance solenoid, followed by compression by the main solenoid, indicate that in CEBIS I, one should expect a current density of ~150 Amps/cm^2 with a space charge limited beam (see figure 2).[7] A space charge limited beam can be propagated through the source with essentially 100% transmission, but faraday cup studies indicate that the ion output current saturates in less than 4 milliseconds, figure 3. The Cornell EBIS Voltage Controller[8] allows flexible programming of drift tube voltages and confinement times, thereby permitting one to observe progressively longer confinement times on a single oscilloscope trace. Initially it was thought that such a rapid compensation of the beam would inhibit ionization to high charge states in CEBIS I.

fig. 3 Oscilloscope trace showing the saturation of trap ions on a faraday cup for successively longer confinement times.

(horiz. scale: 2 mSec/division)

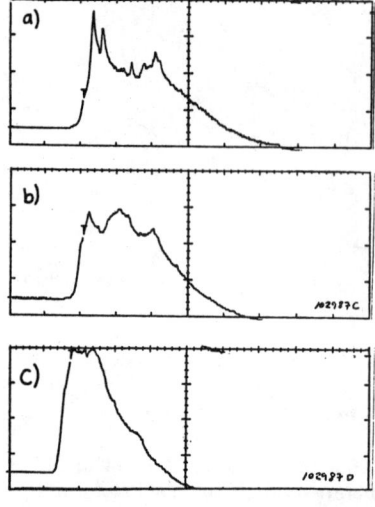
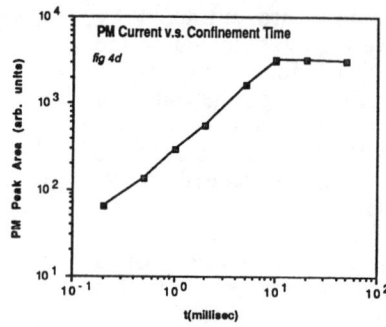

fig. 4 a-c (left) Unchopped TOF signal on modified PM tube in high current mode for 2, 10, and 50 mSec confinement, respectively. Vert. scale: 10, 50, 50 mV/div; horz: 10 μSec/div
d (above) Summary of peak areas indicate saturation; whereas, the charge state is still shifting.

Further time of flight studies of ions extracted from the source, using a photomultiplier dynode chain in a high current mode as an ion detector, were more promising; the change in amplitude and shape of the recorded signal indicated that a shift to higher charge states was occurring for at least up to 50 milliseconds of ion confinement time (see figure 4). The entire extracted ion pulse is allowed to impinge on the first dynode of the chain and the change in pulse shape is due to time of flight effects. At this point, the requirement of 100% electron beam transmission through the source was relaxed, as it was noted that one could obtain good time of flight spectra at lower electron beam currents even with several milliamperes of current loss in the relatively well pumped electron gun region. Very soon thereafter, CEBIS I was producing high charge states of background ions and could confine them for times on the order of one second.

ION COOLING: EVOLUTION OF NITROGEN, ARGON AND XENON CHARGE STATES

Once CEBIS I was producing the highest charge states of the background gases, C^{5+}, N^{6+}, and O^{7+}, gas injection was tested. The first gas to be injected successfully was nitrogen. Nitrogen was selected because its background level in the source remains fairly constant even under poor beam transmission conditions, i.e., when the electron beam grazes a drift tube or other electrodes, the amplitudes of the carbon and oxygen peaks increase, while the nitrogen peaks remain essentially constant. Furthermore, the nitrogen spectrum is rather simple. Gas is injected into the drift region through a radial hole in one of the drift tubes and the leak rate is controlled by a leak valve connected to a reservoir maintained at a pressure of 200 millitorr. By monitoring the sputter ion pump pressure in the interaction region, the leak rate can be balanced by the pumping speed to an equilibrium pressure slightly above the background pressure with no gas injected.

The result of nitrogen injection is summarized in figure 5. One might have expected to see a set of curves in which the charge state q is generated at the expense of the (q-1) charge state and hence a depletion of the lower charge states as the higher charge states grow in. This would have been the case if one started with a fixed, small amount of gas, under good vacuum conditions. However, the CEBIS I nitrogen injection clearly showed a leveling off of each successive charge state, which is consistent with what one would expect for a continuous and undepleted supply of neutral nitrogen, clearly the case at an internal source pressure of ~10^{-9} torr. Figure 6 shows the result of a computer calculation of the charge state evolution for the case of constant neutral density.[9] Qualitatively, the agreement is good.

When argon was injected into the source some surprisingly good results were obtained. CEBIS 1 produced clean argon charge state spectra reaching up to Ar^{16+} with no background gas contamination, fig. 7a. The electron beam energy of 2.85 keV is about three times the ionization potential for Ar^{15-16+}, but well below the threshold of 3.947 keV necessary for further ionization of Ar^{16+}.[10] Figure 7b shows the time evolution of the argon charge states. Since the ion pump efficiency for noble gases is poor, argon gas can be injected infrequently and yet retain a sufficient background level for several days of operation. As was expected, at short confinement times (~50 mSec) the background gases dominated the spectrum; however, at long confinement times (~500 mSec) not only do the argon ions dominate the spectrum, but the background ions begin to disappear completely from the spectrum.

If one examines the evolution of a light background gas such as oxygen in the presence of the heavier argon gas, one sees that the behavior of the oxygen is radically different from that without the argon gas. Consider the oxygen evolution data in figure 8 with the nitrogen data of figure 5. At first glance the oxygen evolution data appears to follow the charge state evolution for a fixed amount of gas injected into the beam.

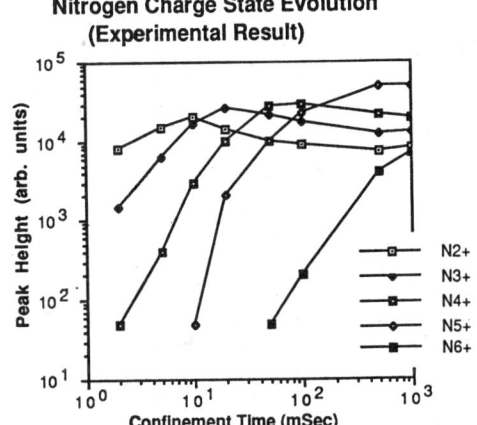

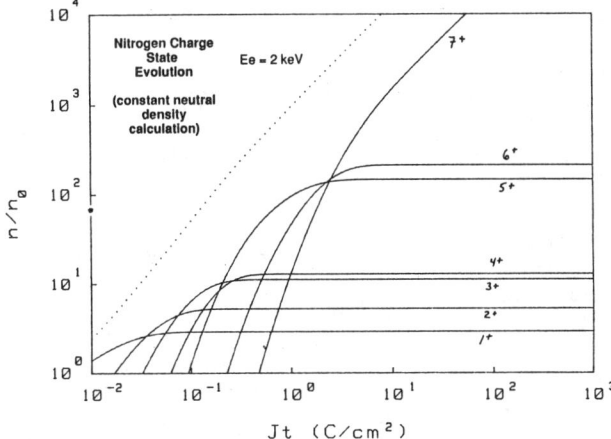

fig. 5 (above)

Observed nitrogen ion charge state evolution in CEBIS I.

fig. 6 (left)

Calculated nitrogen ion charge state evolution in an EBIS under similar beam conditions, assuming a constant neutral density of nitrogen.

However, the higher oxygen states are not building up at the expense of the lower charge states; rather, all the charge states are approaching a time limit at which they appear to be removed from the trap. There must therefore exist a mechanism which causes the background gas ions to be expelled from the trap. A plot of the highest argon, carbon, and oxygen charge states extracted from the trap as a function of confinement time, figure 9, shows that the high charge states of argon evolve at the expense of the background gas ions, an observation which is consistent with the process of ion cooling described by Levine et. al. to explain the Electron Beam Ion Trap (EBIT).[11] At the longest confinement time recorded in this series, the argon high charge states are still rising. However, doubling the confinement time at this point leads to an overall degradation of the spectrum, indicating that once the background coolant is gone the heavier ions also escape from the trap.

Further evidence for ion cooling can be obtained from the evolution of background gases, argon, and xenon charge state spectra when all are simultaneously trapped in the well. This data, figure 10, was taken with the aid of an analyzing magnet instead of the TOF. At short confinement times (50 mSec) one sees high charge states of the background gas ions, mid-charge states of argon, and low charge states of xenon (fig 10a). As the confinement time is lengthened, the background gas ions disappear and there is a considerable shift to higher argon and xenon charge states (fig 10b). Finally, at a long confinement time, one sees a pure xenon spectrum with only a small amount of H_2^+ remaining. The broad xenon peaks are due to the many unresolved isotopes in natural xenon. Placing slits before and after the bending magnet allows one to resolve and identify the most abundant isotopes, figure 10d. Again, the lighter, lower charge state ions are lost prematurely from the trap; i.e., they don't achieve the high charge states they would have in the absence of high Z ions, and once the coolant ions are gone the remaining ion species spectrum tends to deteriorate.

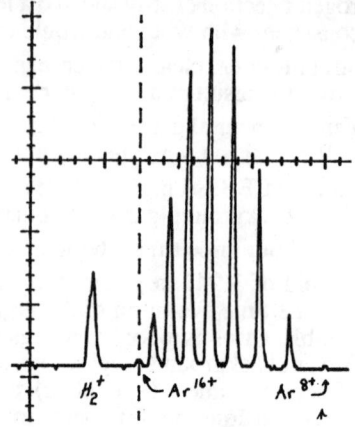

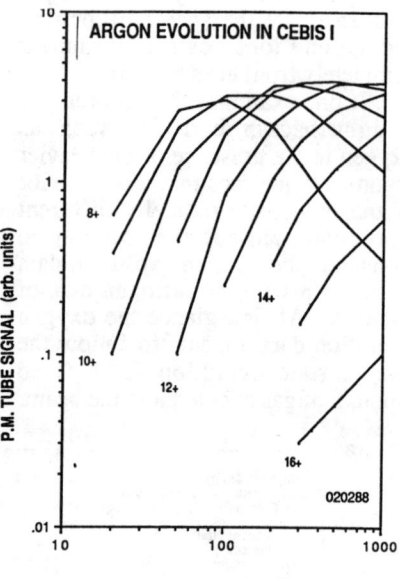

fig. 7 a. (top)   Oscilloscope trace of Argon TOF spectrum for a 900 mSec confinement. Electron beam conditions: V=2.85 kV, I=10.5 mA

b.(bot.)   Summary of a series of such snapshots for confinement times of 20 - 1000 mSecs.

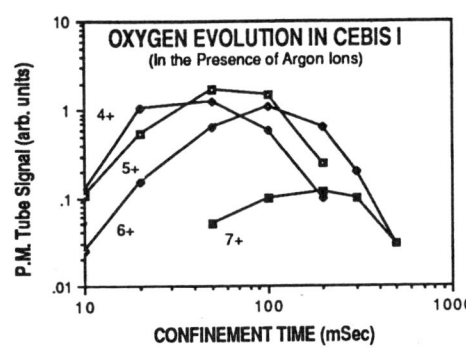

fig. 8

Oxygen charge state evolution in CEBIS I in the presence of Argon ions.

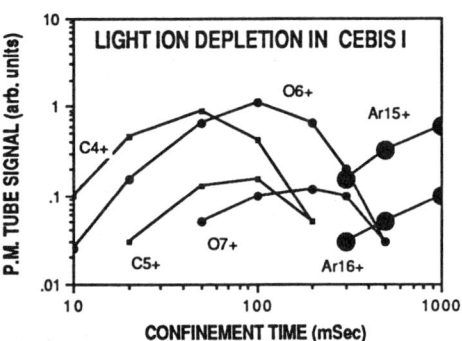

fig. 9

Light ion depletion in CEBIS I in the presence of Argon ions.

OBSERVATION OF ION HEATING

Ion heating may be observed by carrying out an energy analysis of ions coming over the axial electrostatic barrier. However, the problem is somewhat complicated by the partial compensation of the radial space charge potential well. The trap region of the beam becomes partially compensated, thereby reducing the potential depression, but the axial barriers remain uncompensated. As a result, if the trap and barrier drift tubes are held initially at a difference of 50 volts, the difference can be reduced to say 25 volts due to a partial compensation of the electron beam inside the well. The amount of energy a particle needs to gain through heating in order to escape over the axial barrier is:

$$E_{heat} = q \, \Delta V$$

where ΔV is the effective potential barrier height that the particle must surmount, and q is the ion charge (figure 11). Lower charge states need to accumulate less energy through heating in order to escape the trap. All particles escape with the same E/q, and if energy analyzed by an electrostatic analyzer, will pass through the analyzer at the same voltage setting. Particles escaping over the barrier will have a greater E/q than the particles extracted during the extraction period, i.e., particles coming from the trap.

EXPERIMENTAL ARRANGEMENT

The experimental arrangement consists of CEBIS I, a 90° analyzing magnet, and

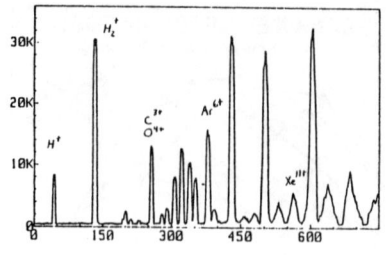
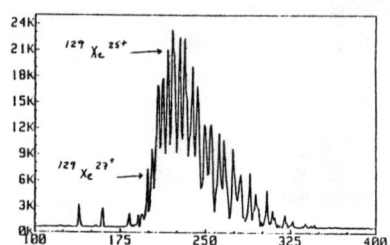
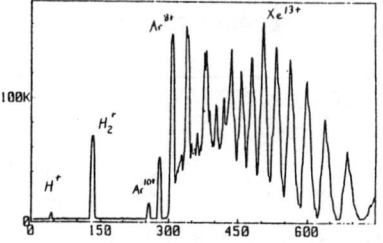
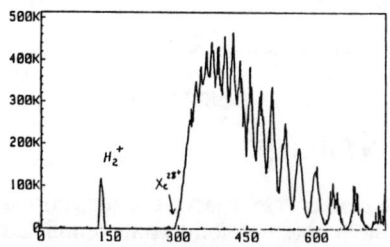

fig. 10 Magnetically analyzed Xenon spectra at electron beam conditions: V=3.5 kV, I = 7 mA.

a. (top left) Conf. time = 50 mSec. Note mixture of background ions, argon, and xenon.

b. (mid. left) Conf. time = 300 mSec. Now one sees only xenon and argon.

c. (bot. left) Conf. time = 1200 mSec. Pure xenon spectrum.

d. (top right) Xenon spectrum taken with slits in bending magnet to show some resolving of xenon isotopes.

a 127° cylindrical electrostatic analyzer with a channeltron ion detector (figure 12). The electrostatic analyzer is locked at the energy of a selected ion peak. (Particles of all q/m should pass through the analyzer at the same setting since they are extracted at the same E/q). A magnetic analyzer scan is then performed to record the relative abundance of each q/m species at the preset electrostatic analyzer setting. A schematic of the data acquisition system is given in figure 13. Two channels of data are recorded at each step of the scan. The first channel records pulses due to ions occurring during the ion extraction period only. The second channel records pulses continuously. A typical CEBIS I extraction period lasts ~50 microseconds with a typical confinement time of ~500 milliseconds. This represents a factor of 10,000 to one of confinement to extraction time. During this long confinement time ions coming over the barrier at a moderate rate could integrate significantly and obscure the ion signal during the extraction time; however, the use of the two data channels solves this problem. A relatively shallow radial trap was produced by a 15 mA, ~3.5 kV electron beam. The applied axial potential distribution on the drift tubes was such that the extraction end barrier tube was only 50 volts above that of the well floor.

ION HEATING RESULTS

To establish whether or not there were two groups of ions present, i.e. a burst during the extraction pulse with an energy associated with the well floor and a trickle of ions with an energy corresponding to the axial barrier, a sequence of electrostatic analyzer scans at successive magnetic analyzer settings was carried out in the vicinity of the C^{1+} peak. (This was necessary since particles of different E/q will not follow the same trajectory through the bending magnet and the electrostatic analyzer was not located at the focus of the analyzing magnet.) The result, figure 14, shows that indeed there were two groups of ions, separated in energy by about 27 eV. Furthermore, the detected number of ions coming over the barrier was about a factor of ten greater greater than that coming from the trap. Also, the measured relative energy spread, $\Delta E/E$, for

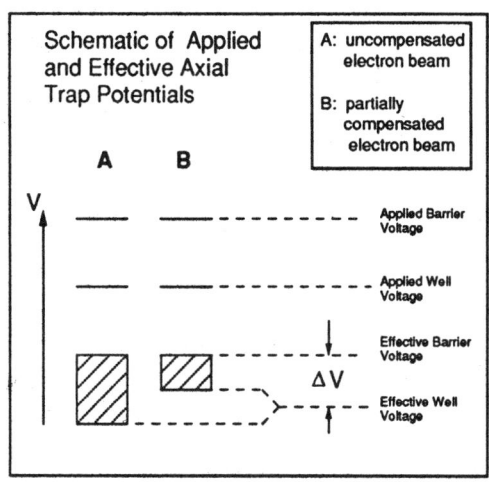

fig. 11 Schematic defining ΔV, the effective potential an ion must overcome to escape the axial barrier.

the trapped ions was ~1%; whereas, the ions coming over the barrier had a measured energy spread of ~0.5%. The difference in axial energy spread can be explained as due to the different radial energies the ions can have in the electron beam. In general one can expect this spread in energies to increase with electron beam current for the same beam voltage.

TIME DEPENDENCE OF IONS COMING OVER THE AXIAL BARRIER

Due to the relatively high background and operating gas pressure in CEBIS I, it is first necessary to establish that the ions coming over the barrier were indeed heated out of the trap and not created in a single pass on top of the barrier electrode. Accordingly, a multichannel scaler, triggered by an ion extraction pulse, recorded the total ion output from the source from one extraction cycle well into the next confinement period (refer to figure 13). Figure 15a shows the spectrum of extracted ions confined under moderate ionizing conditions for the gas mixture present. The setting of the energy analyzer corresponds to the energy of peak "A" in figure 14. (Under these same conditions but at longer confinement times one would obtain a

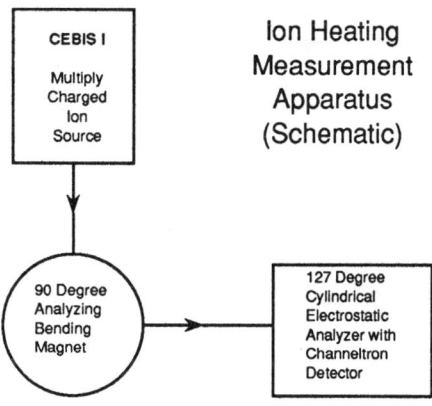

fig. 12 Schematic of Ion Heating Experimental Setup

pure argon spectrum.) Figure 15b shows the spectrum of ions detected during the confinement cycle, i.e. coming over the barrier. Note that for the lower charge states the number of ions in each peak is one to two orders of magnitude greater than for the corresponding trapped ions. Figure 16a shows the time dependence of C^{2+} ions coming out of the source during the confinement cycle. If the ions were being produced solely on top of the barrier one would expect a fairly constant ion output.

For the ion output observed, fig. 16a, there is a time period during which few ions have reached the energy necessary to

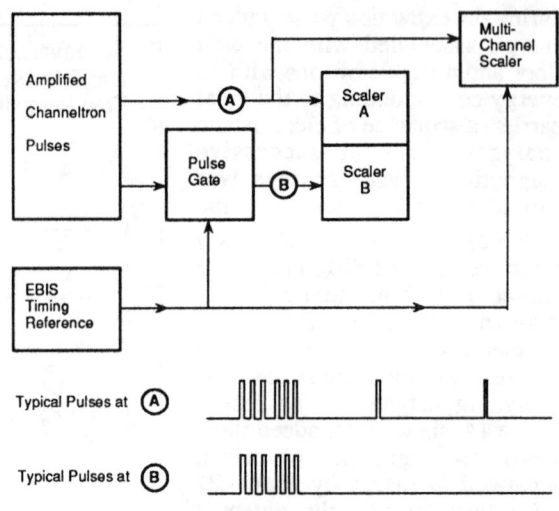

fig. 13 Data aquisition block diagram. Channel "A" records all pulses. Channel "B" records pulses which occur during the ~50 microsecond long ion extraction only.

overcome the axial barrier. The maximum in the C^{2+} detection rate and subsequent decline could reflect the evolution of ion charge state concentrations before equilibrium is reached. Such an equilibrium depends on the C^{2+} concentration, the ratio of high charged ions to low charge ions, and the rate of production of new low charge ions. The source parameters were changed by lowering the electron beam energy, current, and ion confinement time; thereby reducing high charge state ion production. The resulting time integrated spectrum of ion charge states over the barrier was similar to 15b. However, the time dependence of a given ion species ions was quite different. Figure 16b shows that at reduced ionizing conditions, a much

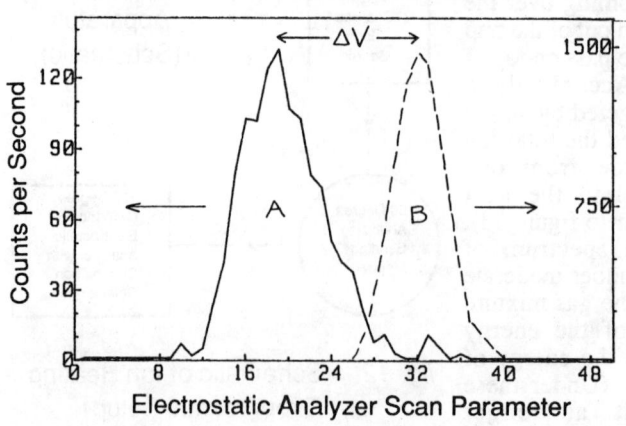

fig. 14 Two peaks of distinct energy ($\Delta V \sim 27$ volts) corresponding to:

A Trapped C+ ions $\Delta E/E \sim 1\%$

B Barrier C+ Ions $\Delta E/E \sim 0.5\%$

(note separate scales)

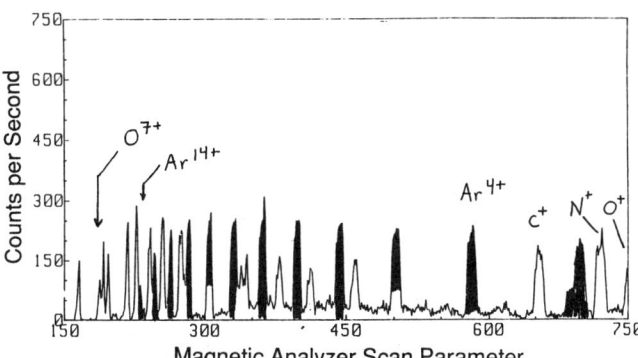

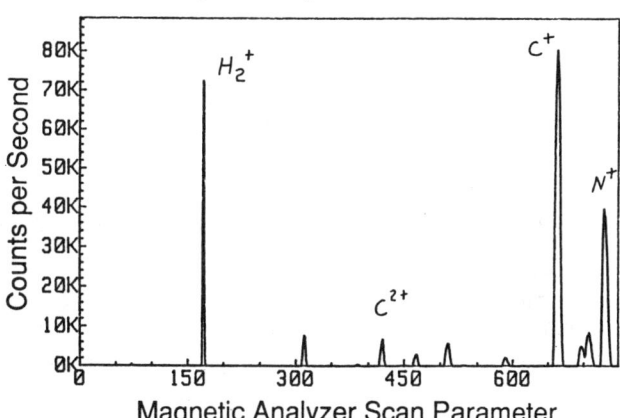

fig. 15

a. (top) Spectrum of extracted ions confined for 200 mSec in a 3.6 kV, 16 mA electron beam.

(Darkened peaks are argon)

b. (bot.) Ions coming over the axial barrier for the same beam conditions and confinement time. Note that the magnitude of the low charge state back-ground ion peaks is one to two orders greater than that of the confined peaks.

longer heating period is required, and equilibrium is reached without the above noted decrease in the C^{2+} ion escape rate over the axial barrier. For these source conditions the spectrum was dominated by background gas ions; the high charge state argon ions are not present for energy exchange. These results are preliminary and a more complete, systematic investigation will be made.

CONCLUSIONS AND CONSEQUENCES

In CEBIS I, the production of high charge state ions coincides with the depletion of the lower charge state background gas ions. In the absence of high atomic number (i.e., high Z) ions, high charge state nitrogen ions were observed to reach and maintain a constant abundance; however, in the presence of high charge state argon ions, the abundance all nitrogen ions was rapidly depleted. Furthermore, the charge state spectrum of a high Z ion such as xenon is seen to be depleted from the low charge state end when "cooling" lower charge state background gas ions are not available. The time dependent nature of low charge state ions escaping over the axial electrostatic barrier has been recorded and is consistent with the hypothesis that such ions are generated inside the trap rather than on top of the barrier. Furthermore, it has been observed that these ions can exhibit a lower spread in axial energy than the trapped ions.

fig. 16

a. (top) Time dependence of C2+ ions escaping over the axial barrier for electron beam conditions:
V = 3.6 kV, I = 16 mA
and an ion confinement time of 200 mSec.

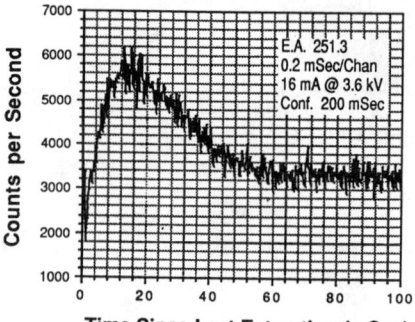

b. (bot.) Time dependence of C2+ ions escaping over the axial barrier for reduced ionizing conditions. The electron beam is:
V = 3.2 kV, I = 12.8 mA
and the ion confinement time is 150 mSec

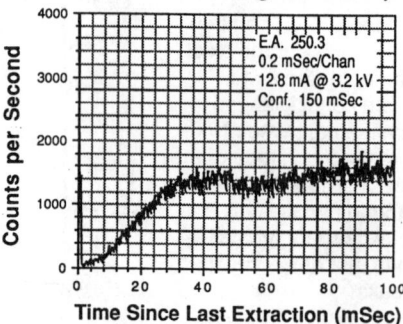

This work indicates that it may be possible to take advantage of two distinct modes of EBIS operation: In the "containment mode", the intended operating mode of CEBIS I, the highest charge states of ions can be produced and extracted in intense ($>10^5$ ions/pulse) but short (~50 microseconds) pulses. It may be possible to extend the confinement time, thereby increasing the ratio of high to moderate charge states in the extracted ion pulse, by injecting a suitable amount of low Z ions into the source as the naturally available "cooling" ions are depleted. In the "continuous flow mode", low Z ions of low and moderate charge state stream over the barrier during the confinement period of the high charge state ions. Such ions have been observed to have a lower energy spread, are well spread out in time, and yield an average available particle current one to two orders of magnitude greater than in the containment mode. This mode of source operation is somewhat similar to the TOF mode of the Frankfurt EBIS[12] and the NICE[13] source in Japan. Such source operation is convenient for many atomic physics experiments since it avoids the pile up of signals associated with the intense bursts of the containment mode. In addition, it provides a capability for setting up experiments without an auxiliary source; one could switch to the high charge state ions of the confinement mode after initial testing has been done with the distributed, more abundant ions of the continuous flow mode. Furthermore, the studies with argon and xenon indicate that one can determine and enhance the flow of low charge state ions over the barrier by choosing and introducing a suitable high Z ion species into the trap.

ACKNOWLEDGEMENT

I would like to thank V.O. Kostroun for his support and discussion. I would also like to thank James Perotti and Steve Sackett for their many critical contributions to the mechanical and electronic components of the source, respectively.

This work was supported in part by the US Department of Energy, Office of Basic Energy Sciences, Division of Chemical Sciences.

REFERENCES

1. V. O. Kostroun, E. Ghanbari, E. N. Beebe and S. W. Janson, Physica Scripta T3, 47 (1983).
2. E. Ghanbari, Ph. D Thesis, Cornell University, 1984
3. E. N. Beebe, Proceedings of the Third International EBIS Workshop, p. 259, Cornell University, May 20-23,1985.
4. V. O. Kostroun, Private Communication
5. V. O. Kostroun, Proceedings of the Third International EBIS Workshop, p. 55, Cornell University, May 20-23, 1985.
6. S. Janson, Proceedings of the Third International EBIS workshop, p. 205, Cornell University, May 20-23, 1985.
7. E. N. Beebe and S. Janson, Private communication
8. E. N. Beebe, M. Eng. Project, School of Applied Physics, Cornell University, 1982.
9. V. O. Kostroun and R. Yao, Private communication
10. T. A. Carlson, C. W. Nestor, Jr., N. Wasserman and J. D. McDowell, Atomic Data, 2, 63 (1970).
11. M. A. Levine, R. E. Marrs, J. R. Henderson, D. A. Knapp and M. B. Schneider, Physica Scripta T22, 157 (1988).
12. M. Kleinod, R. Becker and H. Klein, Proc. of the Third International EBIS Workshop, p. 17, Cornell University, May 20-23, 1985.
13. H. Imamura, Y. Kaneko, T. Iwai, S. Ohtani, K. Okuno, N. Kobayashi, S. Tsurubuchi, M. Kimura, and H. Tawara, Nucl. Inst. and Meth. 188, 233 (1981).

DISCUSSION

Levine: Do you resolve the nitrogen line width to see the temperature of the nitrogen ions coming out? I was wondering if you could change the barrier height, say half way through, to see how that temperature varies with the trap height.

Beebe: I can alternate it from pulse to pulse. I am looking for any input here. If anyone has any ideas, I have a lot of flexibility in this source for about another month.

Becker: I have a proposal for your flexibility. If you want to cool ions by themselves, there are two things that I could imagine. One is you raise the current in the beginning and then lower it at the end of the ionization time. Therefore, you force the ions out of the trap.

Beebe: I might try the idea of pushing the bottom of the well up again because there is quite a difference working with lower currents than with higher currents. Now that the source is well tuned, one can turn it on and off, and do almost anything with it and it still works.

Levine: I want to comment on that. The loss rate is proportional to the exponential of the charge state times the well potential over kT. If you are losing your high charge states ions after you have formed them, it might be that you can stretch them out by just raising the well potential so that the low charge ions carry away more energy.

Beebe: So just gradually increase it?

Levine: Yes.

Kleinod: I would like to point out that the neon spectrum we had shown, in the continuous extraction mode, was free of any residual gas. It proves here that by having the right conditions, you can get rid of impurities. You can even have a higher percentage of the 22 isotope than you will find in nature. So even in this little difference in mass we could see, and I think it is the same point, the A=22 ions were cooler than the A=20.

Beebe: I would like to thank everyone at this conference. I have really learned a lot. The EBIS field has certainly advanced since the last conference when many of us were still worrying about leaky magnets, cooling lines, etc. Really good results have been presented on electron beam propagation and the achievement of high charge states. It appears that the ion heating/cooling observations might provide some clues for more successful source operation.

PANEL SESSION: e-BEAM PLASMA INTERACTIONS IN AN EBIS TRAP;
EBIS PHYSICS; DIAGNOSTICS

N. Rostoker (Moderator), V. Kostroun, R. Marrs,
R. Schmieder, H. Tawara

Rostoker: The first set of business is to define the scope. The topics are: (1) instabilities; (2) ion heating and cooling; (3) some questions about the electron beam profile; and (4) diagnostics. Let me start by making a couple of remarks about instabilities. I know quite a bit about instabilities but not so much about this device. But, frankly, I think most people have the same problem. First of all, what is the criterion for when the thing starts to behave like a plasma and has collective instabilities? Offhand I think it is when the Debye length gets smaller than the characteristics dimensions. What dimensions? That is the one we always talk about in plasma physics — what dimension? The length? The radius? Probably the radius. Oh yes, the beam quality is important: the main thing is that you have to make density and you have to compress it. The Debye length stays constant while you do this. It is sensitive to the square of the emittance divided by the line density. It would stay the same. It is a matter of compressing it to get the density you want. Does the radius stay larger than the Debye length? There is another possible point of view: the transit time should be less than the oscillation frequency of the beam plasma, so there is no time for any instabilities to grow. That is a pretty severe criterion, and so is the one on the Debye length. Probably these criteria are usually violated unless you want to really strap yourself down. I am not sure; if those are the criteria, I think you really have to look at it carefully because this is not an ordinary plasma; it is non-neutral. There are all sorts of possibilities of the ion oscillations coupling with the diocotron frequency or with some other electron oscillation frequency. I would like to point out that there is literature on this subject. This is not the first time anyone has tried to do this sort of thing, to make highly stripped ions with electrons. This is just the first time it succeeded, which is unusual! But there is a paper from the early sixties, I think by Levy, Bethe, and others on a device called HIPAC. Then there were many experiments after that for many years at the University of Maryland with mirrors and also in my laboratory. There the idea was to accomplish this by having a confined electron plasma in a torus or in a mirror, and put in ions. The ions would be trapped by the space charge of the electrons and there is sort of nτ ion confinement, and that is a lot tougher. By giving up the confinement of the electrons, you make great progress. This method did not work because it is hard to confine electrons; there is a sort of nτ for every ionization state you need. No one ever got big enough nτ's for the high ionization states. But on the other hand, it was a weakly supported program, both at AVCO and everywhere else it took place. I remember that at the Department of Energy I talked with someone about getting support for this. This is when the first budget cuts took place and all they said was, "God help me from a new idea! I can not support the ones I have."

But I bring this up because there is literature on this subject, at least lots of theories and calculations, about all kinds of instabilities and problems that are common with this method, although this method does have a distinct advantage in that it has been successful! I did want to say one other thing about avoiding instabilities. I am not sure that there is a criterion for instabilities. But if there were, I know of electron sources that could avoid them. There is one analyzed for small deflections by transfer of magnetic field. This is the electron source used in an electron microscope, which is a small current but I think it is the brightest source there is. I remember when I was at Cornell a long time ago, Ben Segal worked on this and he could focus this beam to 10^6 A/cm^2. He only had about tens of nA, but, nevertheless, when you think about this you can get high density with this beam and it is a very small radius. And the Debye length is always going to be larger than the radius, I think. It may be possible to avoid plasma effects with this. I mean the signal will be smaller, so will the noise, and it is dc, so it might be useful to think about this source, which is well-documented. I have some references if anyone wants it. Well, I think that is all I have to say to start with and I think it would be a good idea if each panel member said what he has to say. Please interrupt from the audience anytime you want to say something.

<u>Kostroun</u>: Well, I think when we started eight years ago and even before that, there was no thought of plasma instability, so I guess the people that were involved did not have a plasma physics background. I remember walking around and talking to my colleagues about this problem. No one said off-hand, "You know this is not going to work." We were all basically motivated by the work at Dubna. At that time Donets had a device that was working. He was, even in 1980, producing very high charge states and trapping them in a beam that was about 1 m long. I think that when we went into it, we found out that all of the instabilities, noise and misbehavior was ultimately tied to bad launching or misalignment, things that we felt we had control over. I think the questions of instabilities came up when Mort Levine, Ross Marrs, and Bob Schmieder did some work on the Berkeley source, the LLL source, and they came out with a paper on instabilities. My own personal feeling was that that source was misaligned badly. However, in the long run, whether there are instabilities or not, the idea of building a short device has had tremendous success. I think the next question is whether or not our 30 cm or 1 or 2 m long devices behave in a similar fashion.

<u>Marrs</u>: You can imagine several types of instabilities. There is one I do not understand at all, so I will bring it up in case

some other people do. There must be some halo or secondary electrons, or Penning electrons, around the beam that presumably couple to it in some way that I do not understand. I do not understand where they go if they are there, or if they are a problem. Perhaps they contribute to the ionization in some way in removing the initial electrons, or they contribute to some sort of oscillation that then couples the free energy of the beam into the ions. I am not sure what the experience is in the travelling wave tube industry, but I would feel better if we could understand this better.

Rostoker: There is an experiment that has been going on in Novosibirsk for 12-15 years in which they have been trying to heat a plasma with an electron beam. There were experiments ten years or more ago in this country along this line. These are pulsed beams and are intense. The experience was that they did not heat worth a damn. The coupling was very weak. The people in Novosibirsk have a quasi-thermonuclear, if you pardon the expression, project based on this. But they persisted and they finally have obtained strong coupling and ion heating. It may be related to this. The important thing was the magnetic field, getting it large enough. It really is not understood very well. If you asked them to explain it, they will give you an awesome array of nonlinear equations and mode coupling and you will go crazy. But, there are some perceptions that seem to be correct: (1) the beam has to be cold; (2) a strong magnetic field helps it stay cold, the way it is prepared; and, (3) the magnetic field where it starts to couple is about the kind you fellows work with. So I think that one should be concerned about the possibility of coupling and ion heating. They have a big theoretical group that writes papers mainly on this.

Becker: You said that the electron beam is cold. Essentially, if you compressed it to a high degree you get high transverse beam temperature. As we have found in actual temperature, we find this high energy. So you need several kilovolts to stop all the electrons, which is surprising because if you accelerate electrons it should flatten the distribution; you should be in the millikelvin range, as they do on electron cooling devices. But if you compress a beam highly in the radial direction, they you have some oscillation with transference of energy to the axial direction.

Rostoker: If the beam has a temperature, it changes the interaction from hydrodynamics to kinetics. And the kinetic one is weaker; the hydrodynamic one is much stronger.

Marrs: We found the opposite effect actually. We do not see any evidence for any super energetic electrons or any electrons with energy with more than 50 or 100 volts different from the null energy. I do not know if this is just because we have a different machine, whether it is shorter, or different diagnostics.

Kostroun: We have a similar result on our EBIS. The end of the bremsstrahlung spectrum, as the electron beam hits the collector, corresponds to the cathode energy, more or less, and the potential

potential that is required on the extractor to stop the beam is also the same.

Marrs: In our case we measure cross sections which are highly resonant and have essentially no yield off resonance. In fact, there is no yield when the normal energy is off resonance.

Knapp: It is worth pointing out that in EBIT we are just now beginning to make detailed measurements of the beam energy widths. In fact, the dielectronic recombination is sort of model dependent, but the width we observed for the beam energy is wider than you would anticipate. There is no real explanation for that; it really is only about 50 volts wide but it is still wider than you would expect. And we have not tried varying very many things to see what it depends on.

Levine: Could I make a comment about the modified two stream instability? I followed the work of Litwin, Vella, and Sessler, and actually calculated for the EBIS geometry. Donets comments in his paper that if he goes over running about 100 mA; i.e., if his charge density in the ions corresponds to more than about 2% of the electron charge density, then he does see an anomalous heating of the ions. He measures using the collision cross section and essentially what he is measuring is the number of ions outside the beam. We tried to make a measurement of this as carefully as we could in the Berkeley EBIS where we could effectively vary the length of the trap. And we were careful to make sure that we did not have any of the instabilities present that come from the streaming and entrapped electrons, even for our shortest trap. I mean it showed some of the characteristics of this instability but whether it was or not is an open question. I do not think we are seeing that sort of thing in EBIT, which is really short enough to suppress it.

Amboss: How short?

Levine: Nominally 2 cm.

Marrs: It is, say, 30 cm or so from the gun to the collector.

Levine: But the interaction region is nominally 2 cm, and that was set by the condition that plasma frequency could not grow in that length.

Hershcovitch: I have read many of the papers on electron beam instabilities relevant to EBIS. In none of these papers have I seen a reference to the parameters or whether all the equations do apply to an EBIS device.

Levine: Jean Faure may want to comment on that but, Tagger has redone the Litwin, Vella, Sessler calculation and he has found that that can be stabilized if the ions lie outside of the beam. So he has done it in the EBIS configuration. I have a preprint of that paper, but I have not seen where it has been published.

Amboss: The ions have to get heated to get outside the beam first?

Levine: Exactly. But then you are open to a lot of interaction with Penning electrons. I am not crazy about that solution.

Hershcovitch: My comment was not about that. I do not disagree with that, but I do disagree sometimes with the use of basic equations without looking at whether they are applicable or not to the parameters.

Levine: That is what I meant. Tagger exactly considered the EBIS geometry and it was for the right parameters and was geometry sensitive.

Amboss: I understood from the conversation that you actually get noise on the electron beam, never mind the ions, so you have a problem right off.

Levine: It depends on conditions. I think you can always make the beam noisy by doing the wrong thing.

Amboss: I was under the impression that the drift tube is very often segmented into a number of segments. Is that right?

Levine: Yes.

Amboss: What are you likely to get from a kind of resonant interaction? We would love to get that in travelling wave tubes. Is that a possibility?

Marrs: I should comment on people I have talked to and the transparencies I have seen here. Every one of them show some sort of noise, either on the beam itself or on other electrodes around the beam. This is related to the tuning conditions and how much current you scrape off, which I assume is some sort of strange oscillation or instability involving electrons scrapped off from the beam. This is something apart from having ions or not; it is just property of the beam itself. I do not know how harmful it is, but it certainly is there.

Kostroun: I do not think that the electrodes are really that sensitive in that sense of the word. We started out many years ago with meshed electrodes. The idea was that we would get better pumping. We were told by Donets, for example in Stockholm, that when he tried it, half of the power in his electron beam went into microwaves. At which point, I told him, "Well, If I was that successful I would go to the Air Force, forget the EBIS and go into the microwave generation." In any case, with our mesh electrodes, there is no problem. If there was a problem, then it was an alignment problem. We have not found any particular coupling, noise, or anything. Supposedly, meshes are the worst things that you would use.

Marrs: But you have noise on your beam.

Kostroun: I have noise if the beam is misaligned.

Rostoker: There should be a diocotron frequency one could observe anyway, and you can use this to measure the density. You can get some measurements from this, like the line density.

Becker: This has caused a lot of confusion in the EBIS community. Donets reported that he has tried mesh drift tubes and non-mesh drift tubes, and the mesh ones are very bad. But after a long discussion, and after Arianer made the same experiments, it came out: the mesh ones were simply sticking together, allowing coupling into the beam. And non-meshed ones have been machined to be interleaving, so in that case, you have a capacitance which shuts out the rf coupling to some extent. So I suppose the basic question is really about rf coupling to the beam, and I would not call this an instability. It is primary beam behavior and is probably enhanced if you have a displaced beam, because then all the modes are coupled better to the beam. Another thing is, what is happening with the temperature inside the beam and the secondary electrons, and probably also the ions, especially if you have two populations of ions. The secondary electrons are especially bad, and we know from the people doing the electron cooling business, that they see secondary electrons if you do not take care to get rid of them, because they want to do a stable dc compensation of the electron beam including space charge compensation by protons. This means you must get rid of the secondary electrons, otherwise you have a plasma that is causing a lot of problems. I want to emphasize the difference between instabilities of the electron beam, which to my understanding are rf coupling and instabilities which are really of plasma physics origin which must always have to do with a temperature distributed species.

<u>Rostoker</u>: The secondary electrons are the worst thing. You must get rid of the secondary electrons. You do that by controlling the background so that it is a lower density and the beam electrons blow away the secondary electrons. If you do not do that, if you have too high a density, it is a mess. Ion focussing will not work anymore.

<u>Hershcovitch</u>: How can you get secondary electrons to be trapped? You have a collector attracting them and you have magnetic field lines for radial confinement. There is a very high potential due to the primary beam and many devices have extremely high vacuum. Why would they stay in the trap and not go to the collector?

<u>Becker</u>: If you make collector extraction, then the potential from the gun goes down, it also goes down to the collector. Certainly, the perfect trap! In the radial direction you can only do Bohm diffusion, and that must take ions with them, but in the axial direction you trap them perfectly.

<u>Marrs</u>: So does anyone understand how they get out eventually? Clearly they do not build up forever.

<u>Becker</u>: The best thing is to put the collector on a high potential and dump the secondaries. Then you have a wonderful source.

<u>Schmieder</u>: I would like to offer an empirical summary of the state of the business as I see it, and invite your criticism. It seems to me that several laboratories, in particular Frankfurt, Cornell and Dubna, have gone to a lot of work and succeeded in upping the beam current density to 1000 or 2000 A/cm^2 and that this was quite an accomplishment. But it still falls far short of what people originally talked about in terms of 100,000 or 1,000,000 A/cm^2. Only Reinard Becker knows how to make 1,000,000 A/cm^2! However, when you look at the performance of the devices in terms of the

charge states that are obtained, there seems to not be such a great difference. That is, the devices with very high current density that have been a lot of trouble to make work are not so much better than devices that have relatively low current density and have been somewhat easier to make. In addition, the instability measurements that we made on the Berkeley EBIS were done at relatively high currents, i.e., several hundred mA total. And most of the best performance that, at least in terms of the high charge states, comes from relatively low currents: 15, 20, 30, 50 mA. In other words, it seems like low currents and low current density is just as good as high currents and high current density. And why is this? What seems consistent in my mind is that the ions are getting heated and are getting out. So if you think of this criterion of $J\tau$ and you want to maximize to J, it is trouble. What you want is to maximize τ; you want to confine them for a long time. In other words, we need to learn how to cool the ions. We may have pushed the state-of-the-art of compressing beams to the point where nonlinearities are taking over and will just stop us, and we now have to pay attention to trapping them for longer times, which means keeping them cool.

Marrs: I just might comment that at Livermore we find that as we turn the knob that increases the beam current density, the performance will stay the same only if we increase the ion collisional cooling power. So you have to balance the heating with the cooling power, then it seems alright.

Becker: May I suggest that this is not a question of instabilities?

Schmieder: Well, it might be, if the instabilities are produced at high current, or high current density, then those couple strongly with the ions. For instance, the experiments on the Berkeley EBIS at several hundred mA may have been producing convective instabilities and heating the ions. We have some measurements that are consistent with that. Therefore you want to back down from that current so that those growth rates are sufficiently low.

Becker: Maybe, but it also could be this was just a current that you had reached where the structure as a whole generated positive feedback. Again, an EBIS is a structure waiting for modulation to occur. And you always have a certain current when such a structure starts to oscillate. And Murphey determines the right frequency for 180° feedback. Each structure has its own characteristic current, and we have gone up to 500 mA and have not seen anything. Donets has a structure which was oscillating at about 180 mA. So it depends on how you build things.

Schmieder: I certainly do not disagree with you, but I do not think it impacts the point. Sooner or later you come into a current limit, or current density limit, no matter what the structure, and even if you make a series of irregular drift tubes to avoid the periodicity, you are going to run into that limit. It is my feeling, and I am offering it for criticism, that through the efforts of you and the others who have upped the current density, we are now sensing the limits at which you can run these beams through these tubes. But that does not end the development of EBIS.

Levine: I think that we ought to try to distinguish between current and current density. My understanding is that current density is not necessarily a problem; that things scale with current density and one can arbitrarily up the density. On the other hand, as you increase the current, you increase the number of Debye lengths radially across your beam. I think there is a difference between current and current density. For given current, the better density you can get, the better it is for EBIS because your τ scales with current density.

Schmieder: I guess I would like to politely disagree that current density can be upped without limit.

Levine: Well, certainly I do not think you can do it without limits, but the limits are not because of instabilities. The limits have to do with physical parameters: electron temperatures, etc.

Schmieder: You mean we are not yet sensing the limiting current density, although we may be sensing limits in total current?

Levine: I would think so.

Becker: It seems to me that puts the finger on the right mark, because current means beam loading or coupling, and these are the rf properties. And current density means density of the particles inside, and this means comparison with Debye lengths, transit time, etc.

Amboss: Your microwave tube instinct has convinced me that you could use lousy materials to damp out unwanted oscillation. Maybe you could coat your drift tubes to get rid of the rf instabilities.

Schmieder: Yes. In fact when we were doing experiments on the Berkeley EBIS, one of the things Mort suggested, and we did, was putting a resistor between the drift tubes to try and damp out these oscillations. I do not recall some real clear data whether that helped or not. Mr. Chairman, may I bring up another related topic? In the general field of letting it hang out and asking for criticism, I would like to make a minor proposal. The question is: How well does an EBIS compare to other kinds of sources? If you have an accelerator you want to inject into, or if you have an atomic physics experiment you want to do, is an EBIS a good idea? Is EBIS A better than EBIS B? Let me reduce an EBIS to the simplest configuration and propose this. Here is the beam with ions trapped in it. The beam gets collected out here and the ions get extracted. What we really have is a source of ions emanating from a small volume somewhere near the collector. They come out with some solid angle. They do this because of their transverse kinetic energy in the beam, and because as the electron beam expands from its own coulomb repulsion, the ions tend to go with it. Those of you who have done calculations of the ions being extracted from the collector region see how the ions get pulled out. So they certainly come out with a solid angle Ω. Furthermore, there is some relative volume, or area A, that they will come out from. So how about the following as a criterion: $B = \dot{N}U/A\Omega$. This is a kind of "brightness". $\dot{N}$ is the number of ions coming out per second. "U" is a quantity that represents those ions. I would like to propose that the important quantity for atomic physics experiments is U = the sum of ionization potentials in producing that ion; it is the total potential energy available. Now, for injection into an accelerator you may prefer to use the charge state Q in place of U, but the important thing is that we should not only count ions; we should count how good those ions are for what we want to do with them. For two general modes of operation, continuous and pulsed, maybe we have the following numbers; I invite you to criticize the numbers or substitute your own. Suppose we get a million ions per second of a certain kind of dc mode, and suppose they have 10 keV of potential energy, that is the sum of all the ionization potentials. Then they come out, at least in my code calculations, within maybe 10^{-4} sr. They emit from an area probably something like 10^{-4} cm^2. So, I get 10^{18} eV/s-cm^2-sr. In pulsed operation maybe they all come out in a microsecond, but the intensity over that time is much greater. Since we are confining them in pulsed operation, maybe we have higher charge states and 100,000 eV. Then with the same geometry, we get a correspondingly higher number, 10^{25} eV/s-cm^2-sr. Now there is a duty cycle in the pulsed operation; maybe you can get one pulse per ten seconds and the pulse lasts for microseconds, so that the duty cycle is 10^{-7}. So the brightness times it duty cycle, which would represent your ability to collect experimental data in your experiment, sort of comes down to the same: 10^{18} eV per second per cm^2 per sterradian, or perhaps a tenth of a Watt per cm^2 per sterradian. That is the end of my proposal.

<u>Kleinod</u>: The alphabet is rich enough. Please do not choose B and do no call it brightness. I have enough confusion about sources and about the definition of brightness and emittance!

<u>Schmieder</u>: Alright. If you would like to pick a letter, please feel free.

<u>Kleinod</u>: Capital "Q" is good; Q for "quality".

<u>Schmieder</u>: Q was used for the charge state. It is a kind of brightness after all, per unit area and solid angle. Perhaps we can choose "C", the letter following B for brightness.

<u>Kleinod</u>: But the word "brightness" is not good. People would feel induced to use it and compare it to other brightness values without reflecting that you have a different definition.

<u>Schmieder</u>: Okay. Should we invent a new English word then?

<u>Kleinod</u>: Go ahead!

<u>Rostoker</u>: A combination of quality and brightness - "Quightness"!

<u>Schmieder</u>: "Quightness"! Then C = Quightness is defined as the number of ions emitted per second per cm^2 per sterradian times the total potential energy of the ions summed over all charge states. Thank you!

ELECTRON BEAMS

REFLECTIONS

THERMIONIC SOURCES FOR HI-BRIGHTNESS ELECTRON BEAMS

R. E. Thomas

U. S. Naval Research Laboratory, Washington DC 20375

ABSTRACT

This paper surveys the capabilities and limitations of modern thermionic electron sources for producing high emission density (>10 A/cm^2)-high brightness beams. The emphasis is on dispenser cathodes. The capabilities of existing commercial cathodes as well as the potential for future cathode improvements as demonstrated in various prototype structures are described.

INTRODUCTION

The field of thermionic electron emission is often considered a "mature" technology in which only incremental progress, if any, can be made. However in recent years with the advent of modern surface analytical capabilities that enable one to better understand the Physics and Chemistry of emitter surfaces. Considerable improvements have been made in thermionic electron-sources. Figure 1 serves to illustrate the changes that have taken place in the emission capabilities of these devices over the past seventy years.

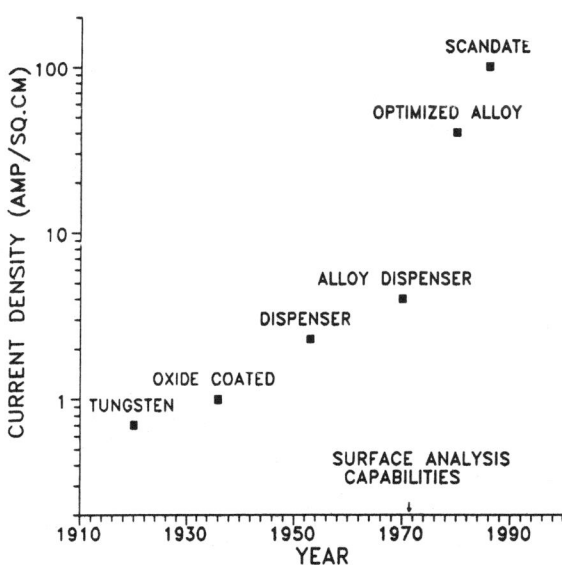

Fig. 1 Historical perspective of thermionic cathodes.

The current density shown for each cathode type has been obtained for at least a few thousand hours of life. In the case of the scandate cathode (whose development is the most recent), life in a device has not been demonstrated, but stable operation in close spaced diodes at 100 A/cm^2 has been demonstrated. As can be seen, more than just incremental improvements have taken place since the early seventies.

A primary purpose of this paper is to describe how the structure, operating mechanisms, and to some extent the fabrication methods of modern thermionic e-sources are related to the capabilities and limitations of the cathode as an electron source in applications, e.g., Free Electron Lasers (FEL), Accelerators, EBIS, and other high power e-beam devices which need high brightness e-beams. Emphasis will be placed on describing those characteristics of cathodes that determine their brightness as a source, as well as those properties that determine how well it can operate in a hostile environment, and factors that limit its operating life.

In addition to describing the commercially available (off-the-shelf) cathodes, we will discuss prototype cathodes that have been evaluated and have demonstrated the potential for improving the capabilities of high brightness electron sources. The very limited amount of data that exists on the operation of cathodes above 10 A/cm^2 will also be discussed.

II - EMITTER TYPES

A considerable variety of thermionic emitters are available. The most commonly known types are

A - Pure Materials
   1. Refractory metals (e.g. Tungsten)
   2. Compounds (e.g. LaB$_6$)

B - Oxide Cathodes
   1. Spray coated (Ba, Sr, Ca, oxides)
   2. CPC (oxide coated Ni particles)

C - Dispenser Cathodes
   1. Tungsten matrix (type B, S)
   2. Coated W matrix (type M)
   3. Mixed metal matrix (MMM)
   4. Controlled porosity (NRL prototype)

D - Cermet Cathodes
   1. Scandate (W+Sc$_2$O$_3$ emitting surface)

E - Thorium Base Cathodes
   1. Thoria Coated tungsten or iridium
   2. Thoriated tungsten

From the above listing only the dispenser and cermet structures will be described in any detail. Pure materials, e.g., tungsten, as well as thorium based cathodes require too high an operating temperature to be feasible in most applications requiring high brightness. LaB$_6$ has been the object of recent work [1], and may be useful in some devices, especially for small area (< 2 mm Dia.) applications. However larger cathodes have been difficult to fabricate, and with the high operating temperatures (>1500 °C) required, considerable problems with thermal and mechanical stability have been encountered.

Oxide cathodes are perhaps the best known of all cathodes. Today they are used mainly in display tubes. Although their work function is low ($\approx$1.5eV), they are not suitable for high current density (>10 A/cm^2), high brightness applications. This is because of the relatively high resistivity of the coating (giving rise to a voltage drop and Joule heating at high current density), along with their slow recovery from poisoning. Detailed descriptions of the properties of these cathode can be found in previous review articles, e.g., [2].

III - CATHODE LIMITATIONS

The usefulness of a cathode as an electron source in a high brightness application is determined primarily by four factors:
   1) Current density ($J \approx 100$ A/cm^2 is desired in many applications).
   2) Beam quality (emittance, or angular divergence of emitted electrons).
   3) Poisoning resistance (ability to operate in non-ideal vacuum).
   4) Life (greater than a few hundred hours is required in most applications).

None of the above factors is independent of the others. For example, the current density is related to the operating temperature (T) and electronic work function (ϕ) through the Richardson equation.

$$J = 120T^2\exp(-\phi/kT) \qquad (1)$$

The operating temperature determines the evaporation rate of activating material (e.g. Ba) from the cathode. Thus, life is also dependent on J. Also ∅ is affected by the presence of poisoning gases and the rate of supply of Ba to the surface (a temperature dependent factor). Hence the relationship between factors 1,3, and 4 is established. Beam quality is also affected by poisoning, but in a less direct way, which will be described below.

A) Beam Quality

As has been mentioned above, for high brightness the cathode must, in addition to providing high emission density, also not introduce effects that give rise to angular divergence of the electrons as they leave the cathode. Three factors giving rise to this latter effect are: a) Thermal velocity distribution of emitted electrons, b) Surface roughness, and c) Non-uniform (or patchy) emission. Evaluation of thermal velocities suggests that this factor is negligible compared to the other effects [3].

Figure 2 illustrates the effect of both surface roughness and patchy emission. Fig. 2a shows the effect of a geometric irregularity on the surface which gives rise to electric fields perpendicular to the beam direction, thereby giving rise to velocity components in that direction. This effect has been evaluated by Lau [4] for both temperature limited and space charge limited emission.

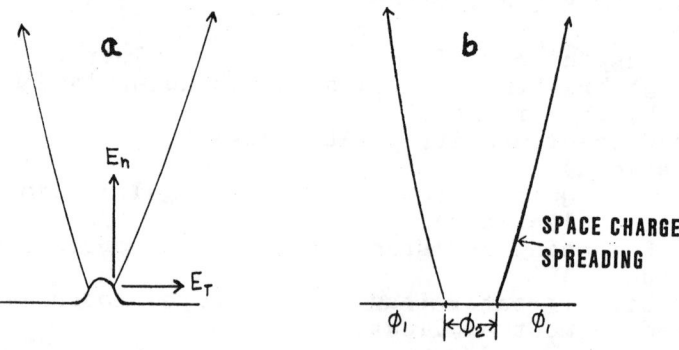

Fig. 2 Cathode sources of beam emittance.

Figure 2b illustrates the effect of patchy emission. The beam from a patchy cathode is comprised of any array of beamlets. Work function differences as small as 0.15 eV can give rise to an order of magnitude difference in emission. Thus, on a patchy cathode most of the emission comes from the low work function patches. In contrast to the surface roughness effect, where space charge tends to smooth out the roughness and reduce the emittance effects, emission from the low work function patches will tend to be space charge limited, and the surrounding higher work function regions temperature limited. Thus, the beamlet will have an electric field perpendicular to the beam due to the space charge gradient. This field will persist until the space charge gradient is smoothed out. Thus, the fields which give rise to beam emittance act on the electrons for longer distances than the surface roughness fields. The emittance contribution due to space charge effects in multiple beamlets has been treated by Wangler [5].

Unfortunately little or no experimental correlation has been obtained between the above factors and actual brightness measurements. However, based on the theoretical estimates, it is expected that control of these factors in the fabrication of cathodes will be critical to obtaining high brightness. In what follows we will describe those aspects of cathode fabrication technology that contribute to surface roughness and patchy emission.

IV - DISPENSER CATHODES

A - Impregnated Matrix Types

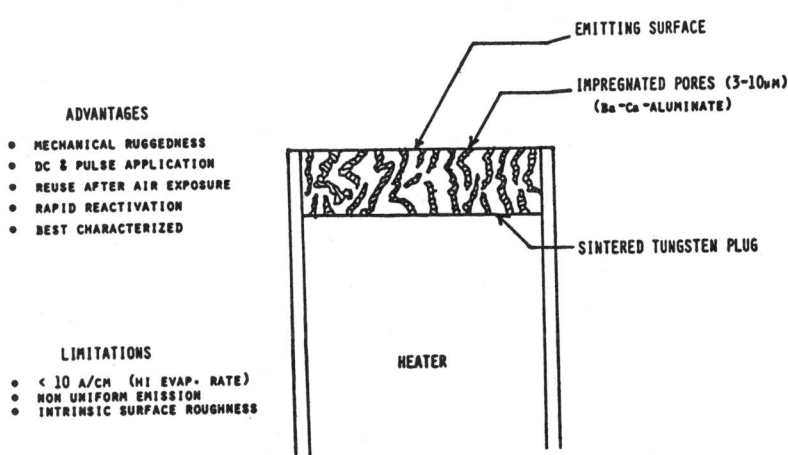

Fig. 3. Cross section of type B cathode

Figure 3 shows a schematic cross section of an impregnated dispenser cathode. The emitting surface and major part of the cathode body consists of a pressed and sintered porous tungsten plug (pore sizes 3-10 microns). The pores of the plug are impregnated with a compound of $BaO:CaO:Al_2O_3$. The ratio of the three oxides is varied to control the Ba dispensing rate (the ratio 5:3:2 is commonly known as the type B cathode). At operating temperatures (1000 - 1200 °C) the impregnant reacts with the tungsten to produce free Ba which migrates to the emitting surface. Approximately a monolayer of Ba + O is maintained on the emitting surface by this dispensing action. The reaction sustaining oxygen on the surface is not known. However it is tightly bound to the tungsten and desorbs very slowly compared to Ba [6]. This Ba,O monolayer forms an electric dipole on the surface lowering the work function from 4.5 eV of clean W to approximately 2.0 eV.

The advantages of this cathode structure are tabulated in Fig. 3. A major advantage is the ability to reactivate rapidly after poisoning, and maintain the dipole layer in a poisoning environment. This is by virtue of the short distance Ba must diffuse over the surface ($\approx$15 microns). However, as the cathode ages the pores are depleted of impregnant and the transport distance increases. A noticeable decrease in the ability of the surface to remain active begins to be seen when the depth of depletion approaches 100 microns.

The practical emission limitations of the type B cathode are due primarily to the work function which determines the operating temperature, which in turn determines the loss rate of Ba, and end of life due to Ba depletion. Life expectancy as limited by Ba depletion will be discussed below. Those aspects of the impregnated matrix cathode that affect beam quality are illustrated in figures 4 and 5.

Figure 4 shows a scanning electron microscope (SEM) image of different areas on the emitting surface of a type B cathode early in life (<100 Hrs.). The white regions are excess impregnant. The regions of excess impregnant have significantly lower ø ($\approx$1.75 eV) than the darker monolayer covered tungsten areas. Thus, even though the average emission is enhanced by the impregnant, the patchiness of emission is quite pronounced, and beam quality is expected to be degraded. Excess impregnant can remain on the surface for several hundred hours until it decomposes into volatile compounds and desorbs. Therefore the time during which it influences beam quality will depend on operating temperature. Commonly used methods for removing excess impregnant are: a) Picking it off with a sharp tungsten point, b) Using 400 to 600 grit alumina abrasive. c) Dipping the emitting surface into a chelating solution which disolves calcium oxides. a) and b) are often used together. Probably the cleanest approach is to use the pick technique on the larger impregnant regions, and then ion bombard the surface to remove the additional thinner regions of impregnant. This latter process is often used in preparation for coating the cathode with an emission enhancing metal, which will be discussed below.

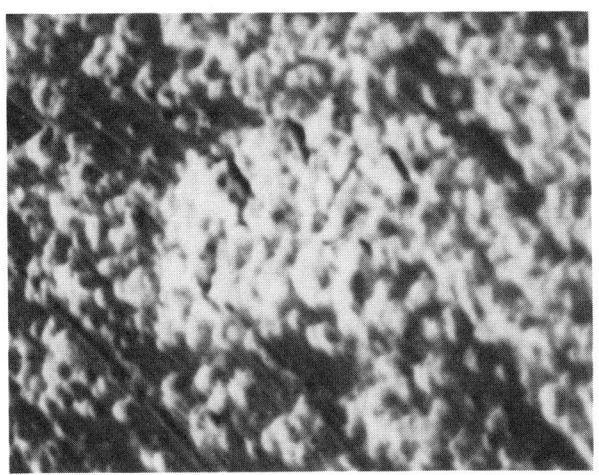

15μ

Fig. 4 SEM image type B cathode with excess impregnant.

Excess impregnant is not the only source of emission non-uniformity. Figure 5 shows both a SEM image (5a), and an emission image (the emission from the surface is projected via a electrostatic immersion lens onto a phosphor screen) of a type B cathode that had operated for about 1000 Hrs at 1050°C . In the SEM image it can be seen that the impregnant is depleted from the pores near the surface, and from the emission image the pores are seen to be non-emitting. In addition non-uniformities in emission can be correlated with differences in surface crystallography. For example, it is seen that the thermally facetted crystal A is emitting much more than the smooth crystal B. This correlation between the two images can be found on a large number of the exposed grains. The effect simply illustrates the dependence of work function on the underlying crystal orientation. Different orientations of the same material when activated with Ba and O give rise to several tenths of a volt work function differences [7,8].

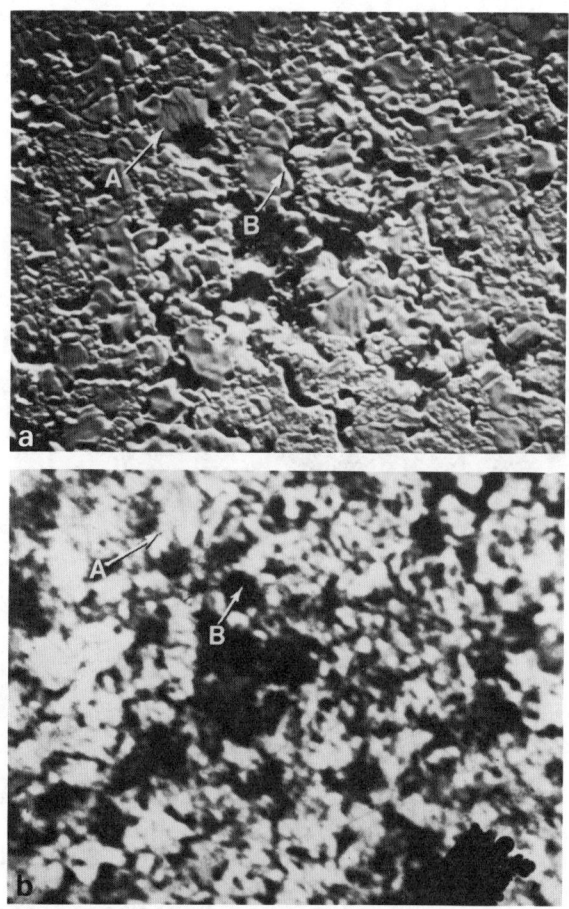

Fig. 5. Type B cathode a) SEM image. b) Emission microscope image of same area as in a).

From figures 4 and 5 it is seen that non-uniformity of emission can be caused by excess impregnant, non-emitting pores, and variations in crystal orientation on the emitting surface. In addition pores are seen to be sources of surface roughness. The cathode shown in figure 5 was polished prior to use. Consequently the normal roughness due to machining marks on the surface was not seen. These marks can give additional roughness effects with height variations up to 12 microns.

B - Coated Matrix (alloy) Cathodes.

A type of cathode which is finding increasing popularity for higher emission density applications is the type M. In it's simplest form it is a conventional tungsten matrix (type B, or S) that has been coated with approximately 5000 Angstroms of Os, Os-Ru, Ir, or some noble metal (Pt is not useful). Figure 6 illustrates the structure of this cathode as well as some of the problem areas associated with it.

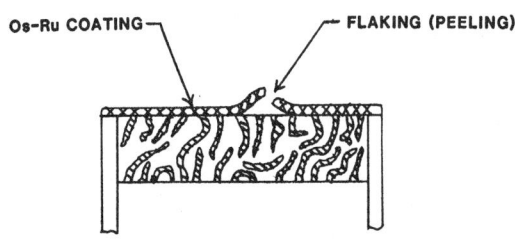

ADVANTAGES	LIMITATIONS
* LOWER WORK FUNCTION	* REPRODUCIBILITY
- (higher current density)	- (Surface preparation)
- (lower temperature)	- (Deposition control)
* BETTER EMISSION UNIFORMITY	* SURFACE COMPOSITION
* EXISTING MATRIX TECHNOLOGY	- (Time dependent)

Fig. 6. Cross section of M cathode.

Coating of the surface gives rise to lower work function (≈ 0.2 eV less than the uncoated matrix) allowing greater current density at the same temperature, or a lower operating temperature (thus longer life) for the same emission density. In addition to higher emission density, in practice it is found that the coated cathodes tend to have more uniform emission. The reason for better emission uniformity, compared to the uncoated structure, has not been unambigiously determined as yet. However, it has been found that under certain deposition conditions the coating tends to deposit in a preferred crystallographic orientation, and that the oriented deposits give greater emission density. If the surface maintains this orientation during operation, it would explain the improved emission uniformity.

These advantages are obtained without the necessity for modifying the conventional tungsten matrix technology. However, the coating process does give rise to some problem areas. In order to insure good adherence of the coating uniformly over the surface, the surface must be cleaned of excess impregnant or other contaminants prior to coating. Also the deposition rate needs to be well controlled to prevent excess stress build up in the films, as well as, obtaining the preferred orientation deposit mentioned above. If these factors are not well controlled non-adherence (flaking) can result in localized areas on the emitting surface. Figure 7 shows a three dimensional map of the emission from a type M cathode taken by measuring the emission through a 5 micron hole in an anode being scanned across the surface [9]. As can be seen a large part of the surface gives uniform emission. However many areas have variations in emission of a factor of two or three. Optical microscopic examination of this cathode showed a correlation between regions of non-uniform emission and areas where coating was flaking.

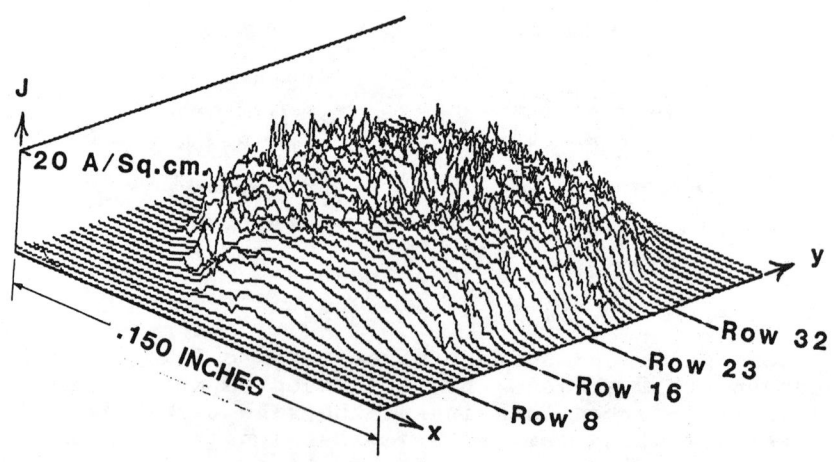

Fig. 7. Emission map of M cathode with flaking coating.

Another problem with the simple coating technique

for enhancing the emitting properties is that at normal cathode operating temperatures interdiffusion of the W substrate and the coating occurs, thereby forming an alloy surface whose composition is changing with time. As might be expected, it has been found that the work function of the alloys so formed is a function of their composition. Figure 8 shows the measured dependence of work function on composition for Ir/W and Re/W system. Similar results have been found for Os/W [10]. It is found that near 50/50 composition gives the lowest work function for all the alloys measured. Based on these measurements one might expect that in early life the alloy concentration would be too rich in the noble metal, and as the cathode aged it would go through a minimum in work function (maximum in emission) and then degrade in emission as the surface became W rich. This is precisely what is usually observed [11]. . The time for degradation is of course temperature dependent, but in attempting to operate at higher current densities degradation in a few hundred hours can be seen.

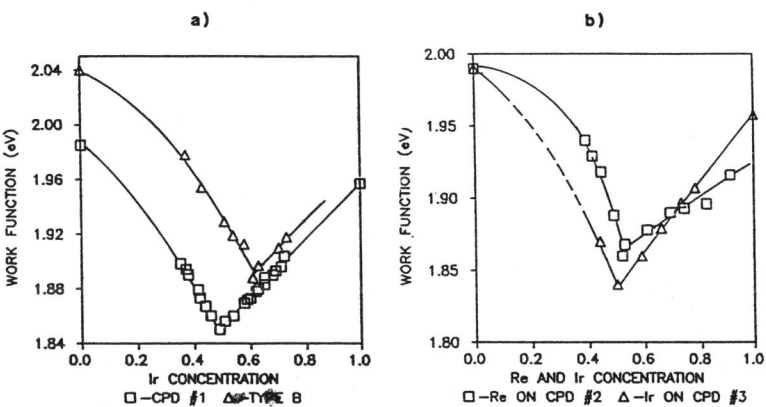

Fig. 8. Work function vs alloy composition for CPD and type B cathodes. a) Ir/W. b) Re/W and Ir/W.

Emission degradation by a factor of three or four (depending on temperature) can occur as the cathode reverts to a W rich composition, and can require a temperature increase of up to 100 °C to maintain emission. The other factors, non-adherent coatings and non-uniformity of orientation are expected to influence brightness by increasing emittance. In addition coated cathodes still have all the surface roughness effects common to matrix cathodes.

C. Alternative Fabrication for Alloy Cathodes

A number of fabrication techniques have been explored in an attempt to better control the properties of alloy cathodes. One such approach is called the mixed metal matrix (MMM) cathode [12]. Instead of coating a pure W matrix, the entire matrix is made from a sintered plug of W + noble metal (usually Ir) particles. The cross-section of such a plug is illustrated in figure 9. This fabrication, while providing the stable surface composition desired, requires very careful control of the distribution in size of the W and Ir particles for proper cathode performance. Thus, making the fabrication process difficult to control. These cathodes are available as special procurements, and are not considered off-the-shelf items.

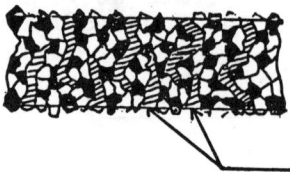

ADVANTAGES

- STABLE SURFACE COMPOSITION

LIMITATIONS

- CRITICAL PARTICLE SIZE & DISTRIBUTION

- DIFFICULT TO REPRODUCE

- SURFACE ROUGHNESS

Fig. 9. Cross section of mixed metal dispenser cathode

Other approaches to controlling the surface composition have been developed by the Varian Microwave Tube Div. One approach called the controlled doping (CD) cathode [13] is to deposit multiple layers of alloy material each of different composition. The compositions are adjusted to minimize diffusion into and out of the surface layer.

Another approach to stabilizing the surface composition also developed at Varian [14], is to add a small amount (a few percent) of the noble metal to the tungsten matrix. This tends to reduce the chemical potential driving the diffusion from the noble metal rich region to the tungsten matrix.

It is clear from the results reported by Varian,

that by proper fabrication techniques, concentration stability on alloy cathodes can be significantly improved over that found on conventional type M cathodes. In addition improved emission uniformity probably results if care is taken to control the deposition rates so as to obtain preferred orientation coatings. In fact test results at high emission densities have been reported [15] on a Varian fabricated cathode with a preferred orientation coating. These results will be discussed below.

D - Controlled Porosity Dispenser

One factor which can contribute to poor beam quality and which cannot be avoided in the matrix cathode is the random nature of the matrix pore structure. Non uniformity of emission often results from depleted pores which do not interconnect with other pores in the matrix. Also the rate of replenishment of Ba to the surface in general must decrease as the pores are depleted of impregnant during operation.

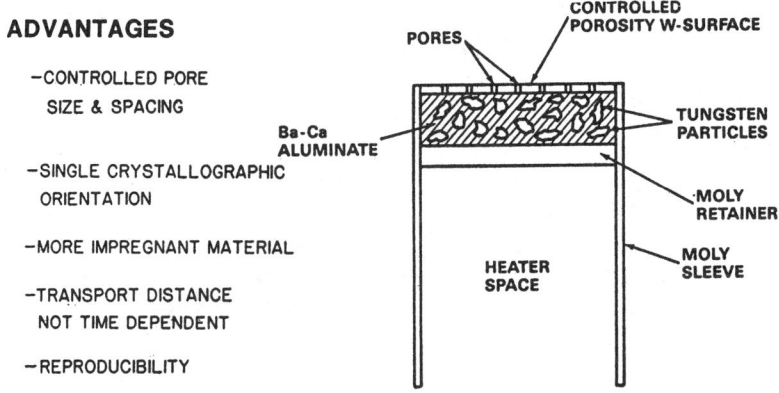

Fig. 10. Cross section of CPD cathode.

A controlled porosity dispenser (CPD) cathode structure designed to avoid these problems has been developed by NRL (via contracts with Varian and Hughes) over the past few years. The structure of this cathode is illustrated in figure 10. Instead of an impregnated matrix, a reservoir containing $BaO:CaO:Al_2O_3$ + particles of tungsten as a reducing agent provides the source of Ba. The CPD emitting surface consists of a thin sheet of W (or alloy, e.g., used for alloy cathodes) in which a uniform array of pores has been drilled or etched (Commercial dispenser cathodes using the reservoir structure but with a standard matrix as the emitting surface are made by Siemens, of Munich, W. Germany.).

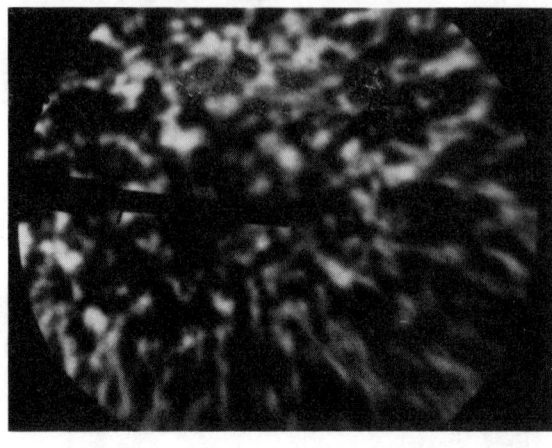

TYPE-B CATHODE
0.6 A/cm² PULSED

|←165 μ→|

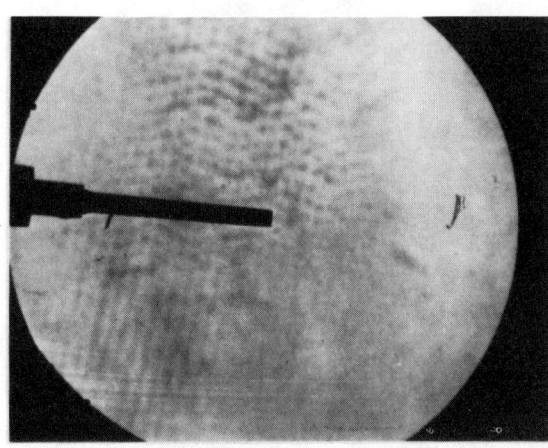

W-CPD
0.7 A/cm² PULSED

Fig. 11. Emission images. Top-type B. Bottom-CPD.

The advantages of the CPD structure are listed in figure 10. It should be mentioned that by chemical vapor deposition or sputter deposition preferred orientation surfaces can be fabricated [8]. Since the diffusion source for Ba is from the back of the emitting surface (due to the vapor pressure of Ba in the reservoir a thick Ba layer accumulates on the reservoir side of the emitting surface), the transport distance remains constant during life. Thus, the dispensing properties of this structure are much more uniform and reproducible than impregnated matrix structures.

The improved emission uniformity of the CPD relative to type B cathodes is illustrated in figure 11. As can be seen at similar current densities the CPD cathode is considerably more uniform. An advantage of being able to control the shape of the pores is illustrated in figure 12. Here is shown a plot of surface Ba concentration as a function of distance away from a pore on a simulated CPD cathode. The data was obtained using Auger electron spectroscopy [6]. As can be seen the gradient of Ba concentration from circular pores is much greater than that near a slot. Consequently control of pore shape provides another method of enhancing emission uniformity, or providing more rapid reactivation from poisoning environments.

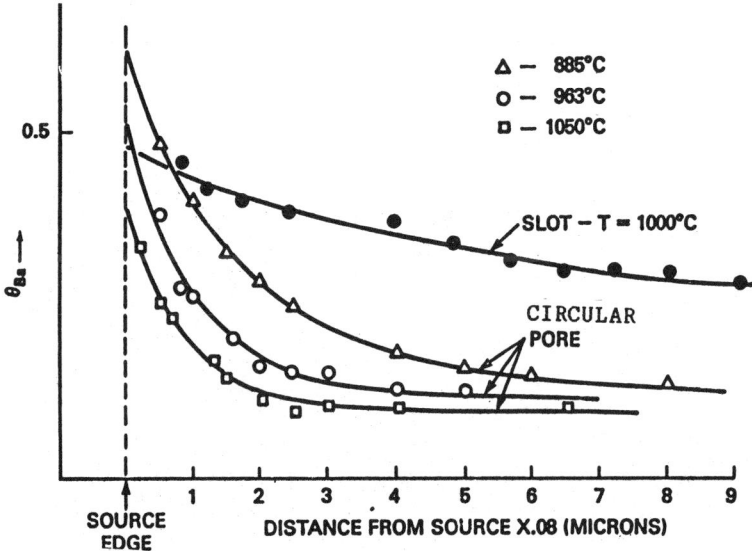

Fig. 12. Ba diffusion profiles. Slot vs circular pore.

Figure 13 is an example of a slotted CPD cathode surface made at the Hughes Electron Dynamics Div. The slots were fabricated with a laser drilling technique. Slots widths are about 5-7 microns. The large slots were introduced to provide non-emitting regions which would be in registry with a control grid electrode.

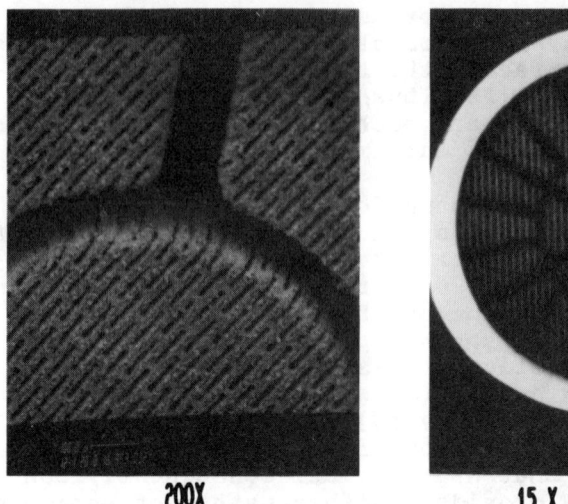 

CPD SPHERICALLY RADIUSED 8727H CATHODE ASSEMBLY, 0.222 IN. DIAMETER
GROOVED INTEGRAL GRID PATTERN
SLOTTED PORES (LASER DRILLED), 5 X 100 MICRONS, 30 MICRONS CENTER-TO-CENTER
SURFACE IS "AS-DEPOSITED" CVD TUNGSTEN

Fig. 13. Slotted CPD cathode structure.

Although the CPD approach has significant advantages for improving cathode reproducibility and performance, at present the difficulty of fabricating it in a batch processing mode (especially for concave surfaces) limits its availability and consequent usefulness. However, anticipating that methods will become available to enable its efficient fabrication, this type of cathode structure probably has the greatest potential for future improvement of dispenser cathodes.

V - SCANDATE CATHODES

Emitters referred to as scandate cathodes have been under study for a number of years. However, until recently their performance has not been very stable or reproducible. A recent version developed by Philips (Eindhoven) [16-18] has demonstrated consistently high emission densities at low temperatures for several thousand hours in diode tests. The basic structure of this cathode is shown in figure 14. The cathode is built on the standard W matrix structure, but the emitting surface is composed of a thinner plug of W mixed with Sc_2O_3 (~5 percent by weight). After sintering the top plug on the main W matrix, the entire matrix is impregnated with $BaO:CaO:Al_2O_3$, usually the 4:1:1 ratio. Stable emission densities of 100 A/cm^2 at 1225 °K for thousands of hours are reported.

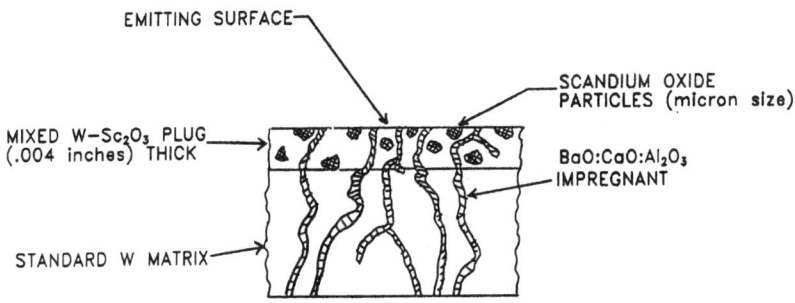

Fig. 14. Cross section of Philips "top layer" scandate cathode.

Present limitations of this cathode are non-uniform emission, difficulty of reproducibility, and emission degradation under back ion bombardment. This latter effect would make it unsuitable for DC or high duty operation. An emission map illustrating the non-uniformity is shown in figure 15. The emission mapping apparatus is the same as used for the M cathode data. The data of the lower map is temperature limited and shows RMS variations of about 70 percent. The upper map shows considerable smoothing due to space charge as the temperature is raised. However, the smoothing does not mean that beam quality is good, since the space charge spreading effect from the patches, discussed above, will degrade brightness.

A serious obstacle to further development of this cathode is the lack of understanding of the mechanism giving rise to the emission enhancement over that of the standard matrix cathode. It is obvious that the emitting surface contains Sc in addition to W, O, and Ba. However the exact chemistry of the emitting surface, as well as the distribution of the various species on the surface is not clearly established. Some models have been proposed [16-19], but more work is needed to establish an unambiguous picture. Because of the low operating temperature and stable high emission levels that have already been demonstrated, it is hoped that with a better understanding of the emission mechanisms, improved fabrication techniques can be developed, and even greater performance may be possible.

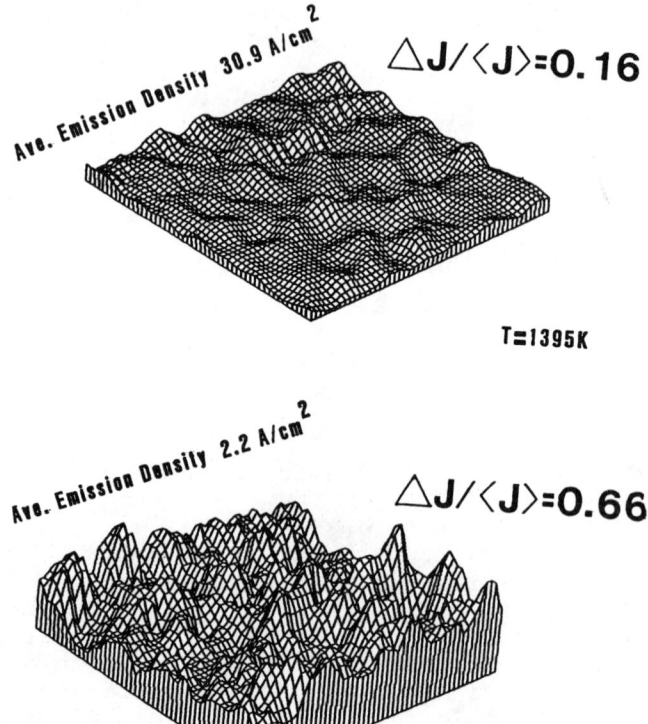

Fig. 15. Emission Maps of Philips "top Layer" scandate cathode (emission from a 0.5 X 0.5 mm area).

VI- HIGH EMISSION DENSITY TEST DATA

A - Emission Data

The data available on thermionic cathodes at emission densities above 10 A/cm^2 is quite meager. The main reason for this is the limited availability of high power modulators in combination with electron gun structures designed for these current densities. Most available data has been taken in close spaced diode (CSD) structures. Data obtained in CSDs can be suspect if care is not taken to determine whether anode interaction effects (e.g. reflux of Ba from the anode giving greater cathode activation) are present. Figure 16 shows a selection of data points for the different types of cathodes that have been discussed. The data were selected from tests where stable emission was obtained for a useful period of time. The solid lines are drawn to approximate the expected temperature limited (TL) data for the different cathode types. Some of the data within the M cathode set was from Ir based rather than standard Os based coatings.

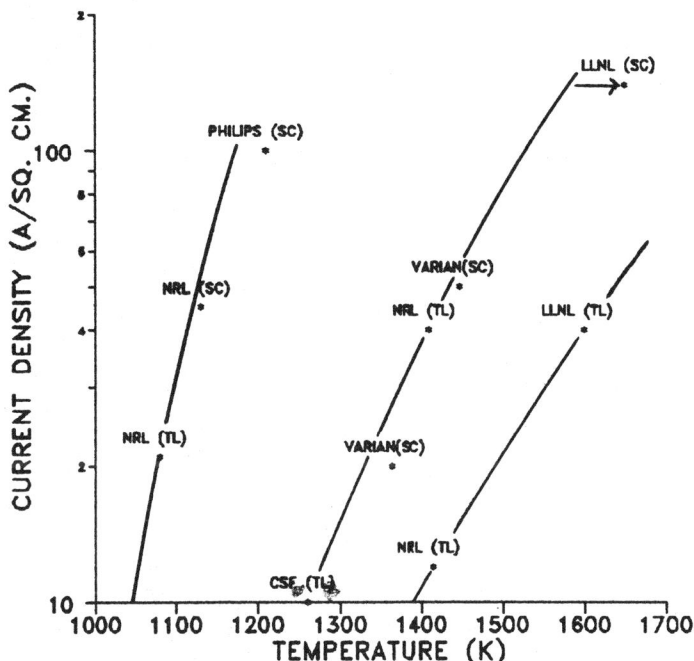

Fig. 16. Hi-emission data from different cathode types.

Data points from non-CSDs were obtained at the Varian Microwave Tube Div. and Lawrence Livermore (LLNL). The type B cathode at 40 A/cm^2 was operated temperature limited in an accelerator test station to end of life at approximately 40 hours. The LLNL data point for a type M at 140 A/cm^2 space charge limited (SC) was on the preferred orientation coating mentioned above [15] (coating parameters on the other coated cathodes tested were not specified). Operating time on this cathode was indicated to be only a few tens of hours. However this was not end of life. Both poisoning and brightness data, which will be shown below, were obtained on this cathode.

It should be noted that the scandate data is at significantly lower operating temperatures than any of the other cathodes. However, no data has yet been obtained in non-CSD test structures.

B - Brightness Data

Figure 17 shows beam brightness (from Ref. [15]) as a function of total beam current for several types of e-sources. The two thermionic sources (NRL LaB$_6$, and LLNL type M) both provide better beam quality than the velvet and graphite type plasma field emitters.

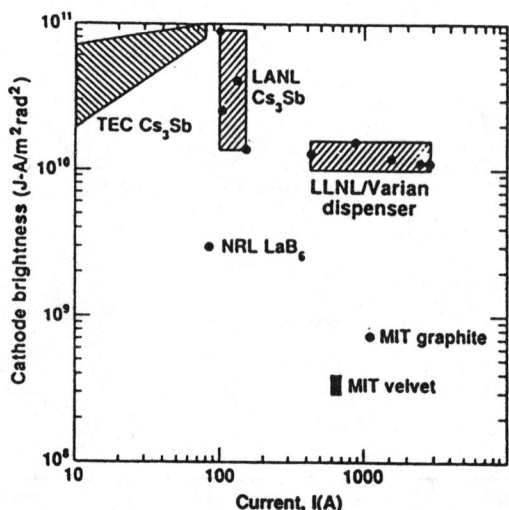

Fig. 17. Beam brightness from various emitters from Turner, et.al. [15].

The beam quality from the type M is somewhat less than that from the Cs$_3$Sb photocathode of Los alamos (LANL). This is somewhat unexpected since the distribution of initial electron velocities from a photoemitter is considerably larger than the spread from thermionic cathodes. This might suggest that, if surface roughness and patchy emission are not playing a role on the thermionic cathode, the beam brightness from the thermionic cathode may be limited by non-cathode related gun factors.

C - Operating Environment Effects

There are two primary mechanisms by which a partial pressure of a gas in an electron beam device can degrade the cathode performance. 1) If the gas is chemically reactive with the emitting surface, it can change the nature of the electric dipole which ordinarily lowers the work function. 2) Because of interaction with the electron beam positive ions can be formed which are accelerated back to the cathode, thereby sputtering away the emission enhancing layer (usually Ba-O). Both of these effects can be offset by dispensing more Ba-O to the surface. However this requires operation at higher temperatures, with a consequent reduction in cathode life. In a cathode having non-uniform dispensing over the surface, or variations in surface chemistry (e.g., crystallographic variations) the rate of poisoning can also be non-uniform. Thus emission non-uniformities are enhanced by poisoning.

In most work that has been done to determine the poisoning effects of gases on cathodes it is found (as might be expected) that oxidizing gases (e.g., O_2, H_2O, CO_2,... etc.) give rise to the greatest rate of poisoning for a given partial pressure in the device [20-22]. Fortunately the pumping speed of vacuum pumps and the gettering action of internal tube surfaces is also greatest for the more common oxidizing gases. In addition most high power vacuum tubes are thoroughly baked during the exhaust procedure, and then sealed off. Thus the partial pressures of oxidizing gases in conventional high power tubes is usually quite low after a short burn in period.

However, many experimental devices that need high brightness sources are in a demountable form, and background pressures from 10^{-7} to 10^{-5} Torr are not unusual. In these systems the partial pressure of H_2O can be significant. In addition, unusual gases such as fluoro carbons (used as cooling liquids) can also find their way into these systems. For these latter types of gases very little evaluation has been done to determine their effect on cathode operation. Also most evaluations of the effect of common gases mentioned above have been done at low current densities. At higher current densities residual gas ionization is greater and operating temperatures are higher, thus the net effect may be different.

Some poisoning data that have been obtained at moderately high emission density are shown in figure 18 as a plot of emission density vs the partial pressure of H_2O. The data was taken by LLNL [15] on the same Varian type M cathode discussed above. Total background pressure in the device was 10^{-7} Torr. The unpoisoned emission density is about 24 A/cm^2 (space charge limited). As can be seen, at the lower operating temperature (1120 $^\circ C_B$) the emission begins to degrade at a partial pressure of H_2O of 10^{-7} Torr. As expected, the pressure required to initiate emission decay increases with operating temperature. Tests on a fluoro-carbon (FC_{75}) used as a coolant were also done, showing at least an order of magnitude greater sensitivity to poisoning by this gas. This demonstrates the need for reducing the capability for such gases to be present in the device, as well as the need for more evaluation of poisoning effects from such gases.

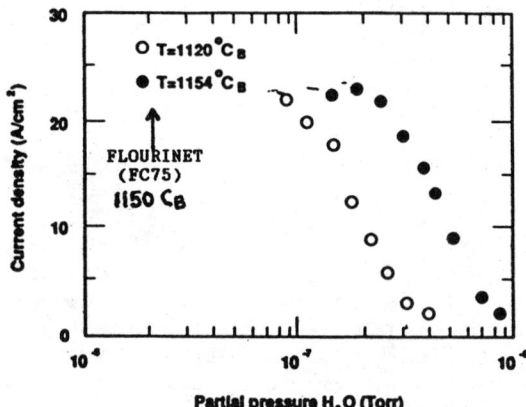

Fig. 18. Current density vs partial pressure of poisoning gas. From Turner, et al, [15]

VII - OPERATING LIFE LIMITATIONS

For dispenser cathodes operating life is a strong function of temperature. During operation the pores become depleted of impregnant, and the distance that the barium must travel to replenish the emitting surface increases. At some point the transport of Ba is insufficient to maintain an adequate Ba coverage on the surface, and emission begins to drop. One can increase the temperature to maintain the emission but this accelerates the degradation. The emission level chosen for end of life is somewhat arbitrary. A reasonable point is when the temperature limited emission is about half the initial emission level. For standard type B cathodes this will occur at a depletion depth of about 100 microns. If one assumes this to be the same for all dispenser cathodes, using evaporation rate data for type B cathodes, life as a function of temperature can be approximated for each cathode type. The approximation is a crude one because of differences in impregnant composition, pore structure, and surface composition, all of which affect the net rate of impregnant depletion.

Based on the above model, expected cathode life as a function of emission current density is plotted in figure 19 for the different cathode types. Each cathode type has a different curve because of the differences in work function. Temperature is shown on the right hand ordinate. The horizontal line at about 1590 °K is the temperature at which the standard heater assemblies are likely to be unreliable.

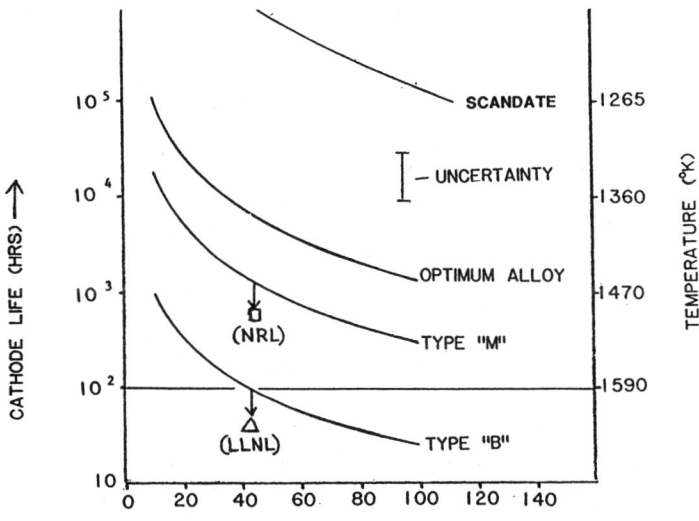

Fig. 19. Predicted cathode life vs current density. Based on Ba depletion model.

There is very little existing life data at high current densities to compare with these predictions. One type B cathode tested at LLNL, and a type M at NRL, both at about 40 A/cm² are in reasonable agreement with predicted life. Thus the curves may be used, with perhaps a factor of 2 or 3 uncertainty. The scandate cathode end of life was assumed to be also due to impregnant depletion. However, at this time it is not clear what other mechanisms may limit life in this cathode. Nonetheless, it was included for completeness, and to illustrate the exceptional potential it may have for long life at high current densities.

VIII- SUMMARY

Based on our current understanding of life limiting mechanisms and the available emission data, it seems clear that 100 A/cm^2 thermionic cathodes are presently available with life of perhaps a few hundred to a few thousand hours. Available data on a specially fabricated M cathode [15] indicates very acceptable beam quality at these current densities. However, in order to assure low emittance care must be taken in the fabrication process to minimize surface roughness, and in the case of coated cathodes to provide coatings that will be adherent and of uniform crystallographic orientation to minimize work function variations. It is probably not reasonable at this time to expect off the shelf cathodes to reproducibly yield optimum beam quality. Prototype cathode structures (e.g., the CPD) have demonstrated the potential for very uniform emission with minimal surface roughness. Thus, for applications requiring very high brightness sources, the expense of fabricating such structures may be justified.

The life of presently available dispenser cathodes is probably limited mainly by impregnant depletion, with alloy stability becoming an important factor for coated cathodes. However, in practice, it may be found that in many devices the operating temperature required for the needed emission will be higher than expected (with consequently shorter life) because of the presence of poisoning gases in the device. In such cases reservoir type cathodes, e.g., the Siemens MK, or CPD structure may provide additional life because of their greater Ba supply capability. Care must be taken to minimize the presence of flourine containing gases ($P<10^{-8}$ Torr is suggested by existing data).

It is often pointed out that in practical beam forming gun structures current densities are limited to approximately 100 A/cm^2 or less because of voltage breakdown effects. However, higher current densities may be useful in devices operating with pulse lengths sufficiently short (< 1 microsec.) so that an arc does not have sufficient time to propagate between the cathode and anode during the on-time of the pulse. No existing thermionic cathodes have yet demonstrated the capability of supplying current densities much greater than 100 A/cm^2. However, because of its low work function and consequent lower operating temperature, the scandate type of cathode (or some other cermet structure) has demonstrated considerable potential and may be a candidate for such applications.

References

1. P. Loschialpo and C.A. Kapetanakos, "High-Current Density, High Brightness Electron Beams from Large-Area Lanthanum Hexaboride Cathodes, NRL Memorandum 6119, Naval Research Laboratory, December 13, 1987.

2. G.A. Haas, "Thermionic Electron Sources", Methods of Experimental Physics, Vol 4, Part A, Academic Press, New York, 1967.

3. J. D. Lawson, "Physics of Charged Particle Beams" Clarendon Press, Oxford, (1978), p 201.

4. Y.Y. Lau, J. Appl. Phys. 61 (1). 1987, 36.

5. T. P. Wangler, IEEE Trans. Nucl. Sci., 32, 5, 2196 (1985).

6. R. E. Thomas, Appl. Surf. Sci., 24 (1985), 538.

7. T. Pankey Jr., and R. E. Thomas, Appl. Surf. Sci., 8 (1981) 50.

8. C.R.K. Marrian, and R.E. Thomas, Appl. Surf. Sci., 17(1984) 285.

9. J.W. Gibson, and R.E. Thomas, Appl. Surf. Sci., 24 (1985) 518

10. R.E. Thomas, and J.W. Gibson, Appl. Surf. Sci. 29 (1987) 49.

11. L.R. Falce, "A Study of M Cathode Aging", Internal Report, Hughes Electron Dynamics Div., Torrance, CA (1984).

13. M.C. Green, "An Optimized Dispenser Cathode", Final Report, RADC contract F30602-82-C-0069. Varian Microwave Tube Div. (Sept. 1985).

14. G.A. Goeser, "A Phase Stabilized Coated Cathode for 4 A/cm^2 100,000 Hour Service", 1988 Tri-Service Cathode Workshop, Asbury Park, NJ, Mar. 22, 1988.

15. W.C. Turner, Y.J. Chen, WE. Nixon, G. Miram, M.C. Green, and A.V. Nordquist "High-Brightness, High-current Density Cathode for Induction Linac FELs", Lawrence Livermore Laboratory, Report UCRL-99042, Sept 1988.

16. A. Van Oostrom and L. Augustus, Appl. Surf. Sci. 2, (1979) 1752.

17. J. Hasker and H.J.H. Stofflen, Appl. Surf. Sci. 24, (1985) 330.

18. J. Hasker, Van Esdonk and J.E. Combeen, Appl. Surf. Sci. 26, (1986) 173.

19. J.W. Gibson, G.A. Haas, R.E. Thomas, IEEE Trans. Elect. Devices (To be published Jan. 1989).

20. C.R.K. Marrian, G.A. Haas, and A. Shih, Appl. Surf. Sci., 16(1983) 73.

21. C.R.K. Marrian, G.A. Haas, and A. Shih, Appl. Surf. Sci., 24(1985) 391.

22. C.R.K. Marrian, and Arnold Shih, IEEE Trans. Elect. Dev. 33, 11 (1986) 1874.

DISCUSSION

<u>Amboss</u>: When you measure space charge limited emission, Longo has shown there is a formula: $1/J$ meas = $1/J$ space charge limit + $1/J$ temp. limit. In these cathodes, which have very uniform emission, do you still get this formula obeyed, or is there a difference?

<u>Thomas</u>: Well, I have not looked at Longo's formula too carefully. I think that he is talking about a cathode that has a combination of temperature and space-charge limited regions on the surface and, therefore, you combine them in that fashion. If you do not have those variations on the surface, then you expect the whole thing to be either temperature or space-charge limited. In other words, I would not expect it to apply.

<u>True</u>: Rodney Vaughan published a paper a few years ago in the Transactions on Electron Devices in which he showed that as the cathode got better and more uniform, the knee did get sharper, so that would support your comment.

<u>Thomas</u>: Yes, that is observed quite often. If you plot a roll-off curve of the temperature of the current density as a function of temperature, there is a knee. As it becomes temperature limited, that knee is sharper depending upon how patchy the cathode emission is.

<u>Becker</u>: I bought several cathodes with Scandium impregnation from Spectra-Mat, so they are available commercially.

<u>Thomas</u>: Yes, we have not had very much success with those cathodes. They tend to deplete fairly quickly.

ADVANCES IN E-BEAM FORMATION, FOCUSSING AND COLLECTION

Richard True

Litton Systems, Electron Devices Division
San Carlos, California 94070

ABSTRACT

This paper presents computer codes and methods useful in the design of high quality guns of the type used in electron beam ion source (EBIS) machines. It is known that magnetic compression is a good way to achieve ultra-high density beams, consequently, the paper considers techniques useful in the design of immersed flow focussing systems. The final topic considered is the modelling of multi-stage depressed collectors (MDC's) used for efficient spent E-beam collection.

GUN DESIGN

Presented in Fig. 1 is a computer simulation of a diode Pierce gun focussed by a magnetic field (1), (2). Deformable mesh code DEMEOS (3) was used to generate the plot.

In design of such guns, code TMLBMC can be used to obtain a preliminary design. A printout from this code is shown in Fig. 2. The code obtains a solution to the conical flow gun problem by the method of successive bisection, and prints out all salient Pierce gun parameters including the cathode-to-anode spacing, plus nonthermal and thermal beam envelope data through the beam waist. Alternately, it is possible to achieve a preliminary gun design by synthesis (4). The meaning of optical compensation is discussed in the paper by Vaughan.

Next, electrode contours external to the beam (focus electrode and anode with hole) can be obtained by the technique of Fig's 3 and 4 (5). Electric fields are matched at the cathode center and along the beam edge from solutions of Laplace's equation in combination with data from TMLBMC. Use of the deformable mesh code is quite convenient in this regard as Neumann boundaries (N) can conform to the beam edge (they do not have to be parallel or perpendicular to the axis). In the plots, B-boundaries are Dirichlet, and G's are fixed in space for mesh density control (G-type mesh points are treated as normal interior nodes otherwise).

After this step, DEMEOS can be used to simulate the whole gun (Fig. 1 without magnetic field). Thermal beam dilation, and sigma (the standard deviation of the transverse velocity distribution), can be determined from cold beam DEMEOS results plus results from TMLBMC, or a thermal beam model can be used in DEMEOS (2).

© 1989 American Institute of Physics

BEAM FOCUSSING

Presented in Fig. 5 is a block diagram showing the interrelationship between the beam formation, focussing, and collection problems. The theory of the author (2) intercouples the various problems via a parameter called tunnel emittance (where T-emittance is the product of sigma and beam filling factor, r/a). Among other things, it enables one to optimally match the gun to the focussing system for maximization of beam transmission to the collector (minimization of T-emittance growth).

Figure 6 shows a simulation of a shadow gridded gun focussed by a double period magnetic field (6). The rms magnetic field level for the beam envelope was chosen per (2) taking into account the variation of magnetic field with radius (7). It should be pointed out that the halo of wild electrons outside of the main beam core originate from near the shadow grid wires in the gun and carry but a small fraction of the total beam current.

Presented in Fig. 7 is a plot of the beam filling factor versus distance (over the full tube length) and Fig. 8 presents a plot of T-emittance over the same distance. It can be seen that the beam filling factor remains essentially constant as the beam flows downstream. On the other hand, T-emittance first increases then levels off. Such a growth in T-emittance appears typical insofar as it has been observed before (2).

It is known that higher levels of beam compression can be achieved with with immersed flow focussing (8) in comparison to Brillouin flow (zero flux at the cathode). Thus this method of focussing appears ideal for use in EBIS machines.

Figure 9 presents a computer simulation for an immersed flow focussed beam from a shadow gridded gun. In design of such a focussing system, it is essential that the beam and magnetic field be properly matched (in both the gun and beam region). Code POISSON (solenoids) or PANDIRA (permanent magnets) are useful in design of the magnetics (9). Values of the magnetic field along the axis (computed or measured) can be passed to DEMEOS (which then uses a second order power series expansion for fields off-axis), or numerical values of the magnetic field over the whole region from POISSON or PANDIRA can be passed to DEMEOS (which eliminates the power series expansion) (10).

The next two plots (Fig's 10 and 11) show the beam filling factor and the T-emittance for the gun of Fig. 9. It can be seen in Fig. 10 that the beam filling factor is decreasing with distance. This is due to the fact that the main field is tapered by 14 percent over the distance (roughly) 0 to 4 times the plasma wavelength.

It can be seen in Fig. 11 that the average value of T-emittance in the focussed beam is less than that in the gun with zero magnetic field (see B=0 case data in Fig. 9) and that the T-emittance is not increasing as in the periodically focussed case of Fig. 8. Linking flux through the cathode tends to quiet the beam (reduce the value of sigma) and increase beam stiffness. It has been found that improper matching between the gun and the magnetic field tends to increase T-emittance over the case shown (like periodically focussed beams (2)).

BEAM COLLECTION

The final topic considered in this paper is beam collection. It is possible to model the whole beam flow region (that is the gun, beam transport region, and single or multi-stage depressed collector) in one shot using DEMEOS. A plot of such a case downstream is presented as Fig. 12. The beam in this case originates from an immersed shadow gridded gun. It can be seen that as the collector is depressed from ground potential that the beam expands more rapidly in the collector.

It is possible to set up a statistical model of the spent beam based on the grown value of T-emittance and the energy distribution in the spent beam as explained in (2). This method is particularly useful in design of multi-stage depressed collectors because it is easier to set up and is more computationally efficient than the full simulation.

REFERENCES

1. R. True, IEEE Trans. Electron Devices 31, 353 (1984).
2. R. True, IEEE Trans. Electron Devices Part II 34, 473 (1987).
3. R. True, IEDM Tech. Dig., 257 (1975).
4. J.R.M. Vaughan, IEEE Trans. Electron Devices 28, 37 (1981).
5. R. True, Fifteenth IEEE Intnl. Conf. on Plasma Science, (1988).
6. R. True and W.B. Reyes, IEEE Microwave Power Tube Conf., (1988)
7. R. True, IEEE Microwave Power Tube Conf., (1988).
8. K. Amboss, IEEE Trans. Electron Devices 16, 897 (1969).
9. A.M. Winslow, J. Comp. Phys. 2, 149 (1967).
10. G.P. Scheitrum, private communication.

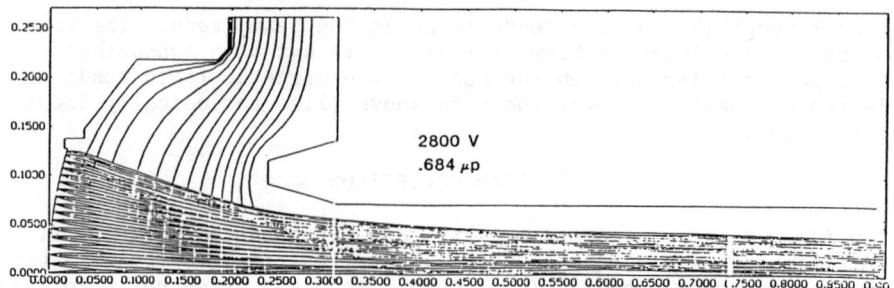

Figure 1.

GUN WITH ENTRANCE FIELD PER TRUE 1985, AND MAIN FIELD B$^+$ FROM B$_b$ $[F(1+197.6\sigma^2/P\mu)]^{1/2}$ WHERE F = 1.65 AND σ = .0151.

```
TMLBMC: Conical Flow, Vers. 2.2, 4-6-88
Basic pgm for Microvax II by R. True
DRS cap gamma =1.25

Case date =08-Nov-88
Units: enter 1 for cm, 2.54 for inches [default 2.54]?

* Case title? L-2087 WITH HYPO CKT

enter disk rc, th(deg), V0 (kv), pmu? .125, 17.4, 9.6, .54

enter optical comp factor [default .9]?
uncompensated th = 15.6129 deg used below

i0 = .507926 a     jc = 1.57426 a/cm^2
cath sph radius = .464448
anode sph radius = .213307     cath-anode dist = .251141
```

Fig. 2. Sample printout from Pierce gun design code.

```
enter mean grid spacing to cathode [0 to bypass]? .015
grid sph radius = .449448    vg = 102.117 volts

enter thickness of grid [0 to bypass cutoff calc]? .002
enter width of grid wires? .002
enter short pitch (grid aperture plus one width)
[note: aperture distance = minor dia in hex grid]? .032
enter long pitch [if enter 0 long set equal to short]?
effective linear grid pitch = .226274E-01
grid cutoff voltage =-86.0906


enter 0 for thermal beam analyis; 1 for new grid
spacing; 2 for new problem; and 3 to stop? 1
enter mean grid spacing to cathode [0 to bypass]? .251141
grid sph radius = .213307    vg = 9600.18 volts


enter 0 for thermal beam analyis; 1 for new grid
spacing; 2 for new problem; and 3 to stop?

enter Tc(deg C), a [circuit i.r.]? 1130, .082
check pmu = .539996
```

z	re	re/a	sig	r95	r95/a	r95-re
0.25114	0.05741	0.70011	0.00174	0.05915	0.72133	0.00174
0.27626	0.05525	0.67376	0.00181	0.05394	0.65775	-0.00131
0.30137	0.05318	0.64851	0.00189	0.05203	0.63455	-0.00114
0.32648	0.05120	0.62442	0.00198	0.05023	0.61253	-0.00097
0.35160	0.04932	0.60151	0.00208	0.04852	0.59173	-0.00080
0.37671	0.04755	0.57982	0.00219	0.04692	0.57218	-0.00063
0.40183	0.04587	0.55941	0.00231	0.04542	0.55394	-0.00045
0.42694	0.04431	0.54031	0.00244	0.04404	0.53705	-0.00027
0.45205	0.04285	0.52256	0.00257	0.04277	0.52155	-0.00008
0.47717	0.04151	0.50620	0.00272	0.04161	0.50747	0.00010
0.50228	0.04028	0.49128	0.00289	0.04058	0.49487	0.00029
0.52740	0.03918	0.47782	0.00306	0.03967	0.48377	0.00049
0.55251	0.03820	0.46585	0.00324	0.03889	0.47421	0.00069
0.57762	0.03734	0.45542	0.00344	0.03823	0.46623	0.00089
0.60274	0.03662	0.44653	0.00365	0.03771	0.45984	0.00109
0.62785	0.03602	0.43922	0.00388	0.03751	0.45750	0.00150
0.65297	0.03555	0.43349	0.00411	0.03729	0.45470	0.00174
0.67808	0.03521	0.42936	0.00436	0.03721	0.45378	0.00200
0.70319	0.03500	0.42683	0.00463	0.03729	0.45472	0.00229

```
enter 0 for thermal beam analyis; 1 for new grid
spacing; 2 for new problem; and 3 to stop? 3
```

Fig. 2. Sample printout from Pierce gun design code.
(continued)

224

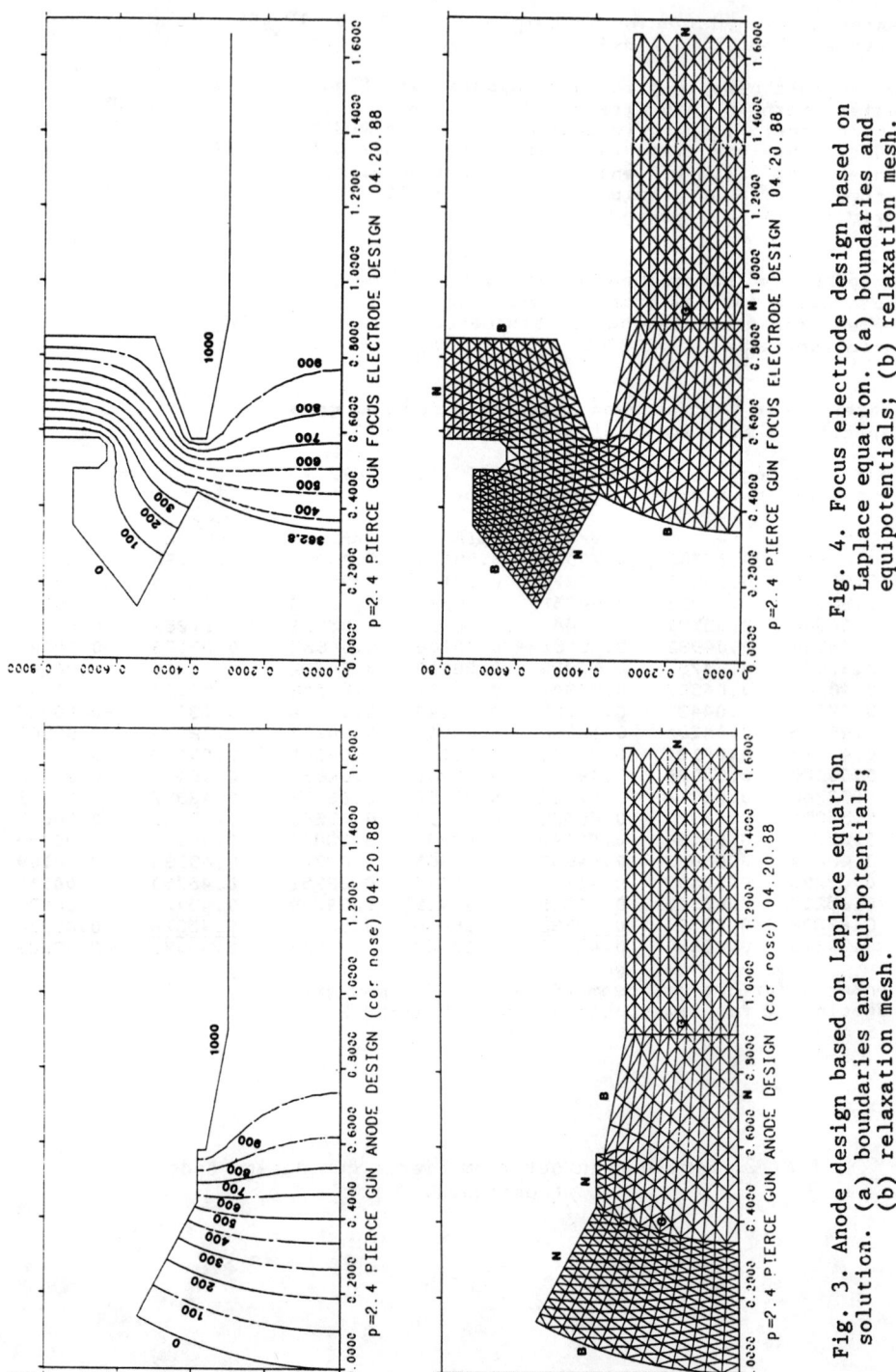

Fig. 3. Anode design based on Laplace equation solution. (a) boundaries and equipotentials; (b) relaxation mesh.

Fig. 4. Focus electrode design based on Laplace equation. (a) boundaries and equipotentials; (b) relaxation mesh.

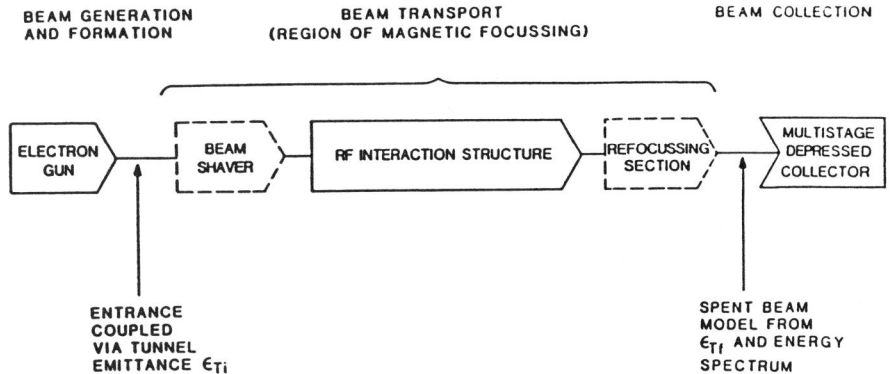

Fig. 5. Framework of generalized focussing theory in linear electron beam device.

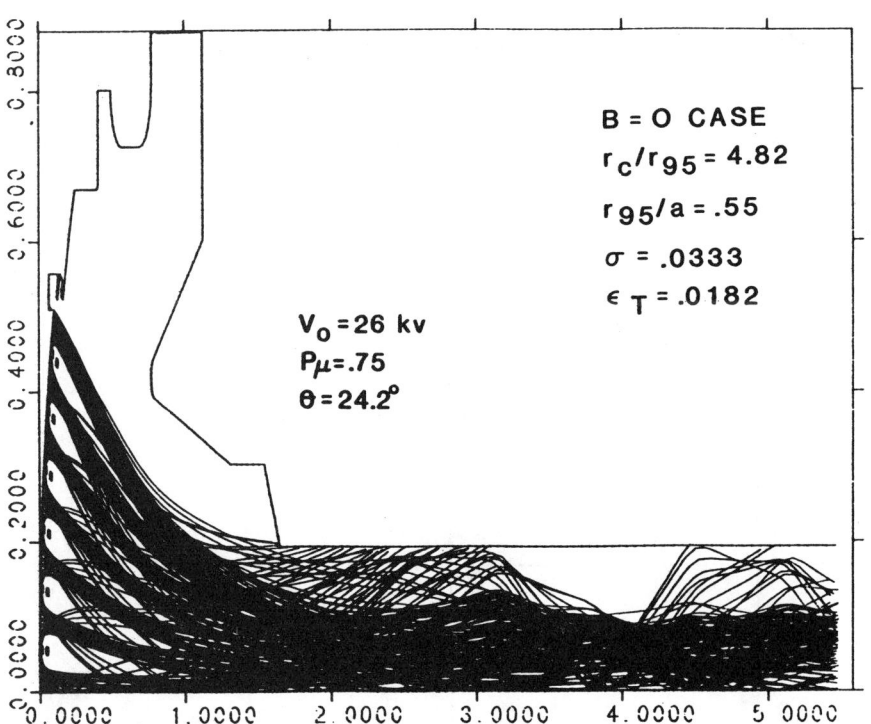

Fig. 6. Simulation of L-5637 gun used in emittance growth study (thermal beam model). Beam focussed by double period magnetic field (through 5 peaks).

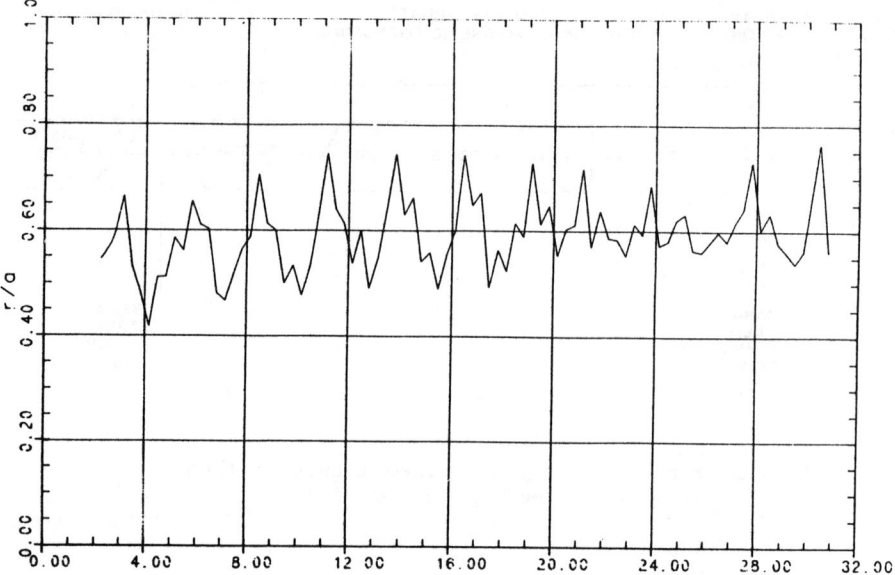

Fig. 7. Beam filling factor versus axial distance over the full tube length (thermal beam simulation).

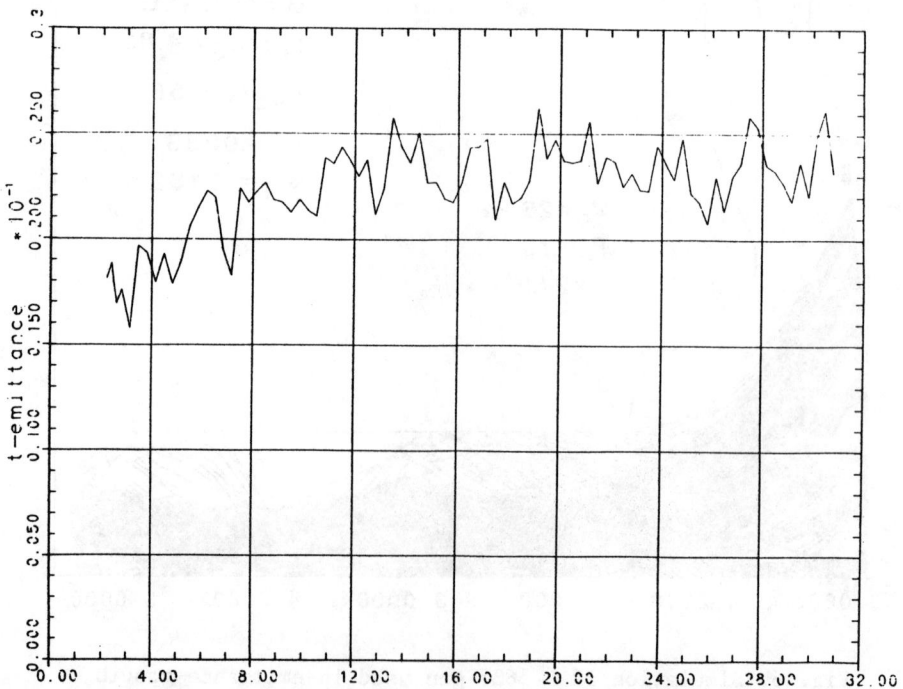

Fig. 8. Tunnel emittance versus axial distance over the full tube length (thermal beam simulation).

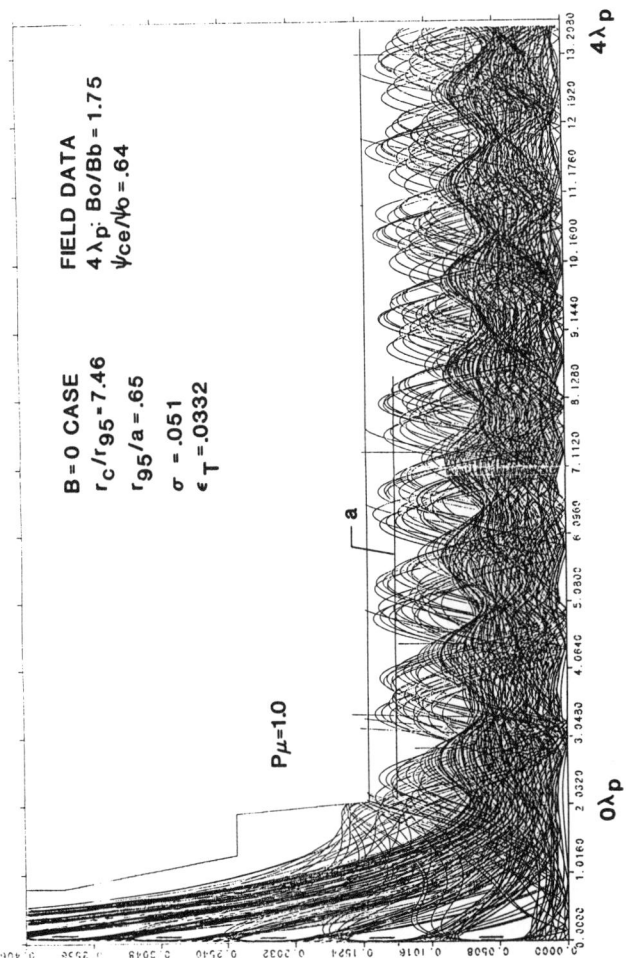

Fig. 9. Simulation of space-charge balanced flow focussed beam from shadow gridded gun (thermal beam model).

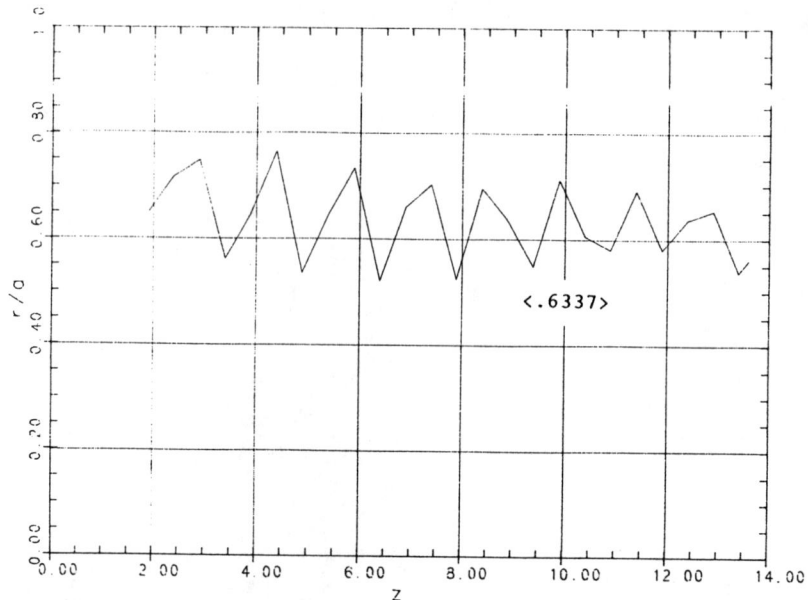

Fig. 10. Beam filling factor versus axial distance for space-charge balanced flow case.

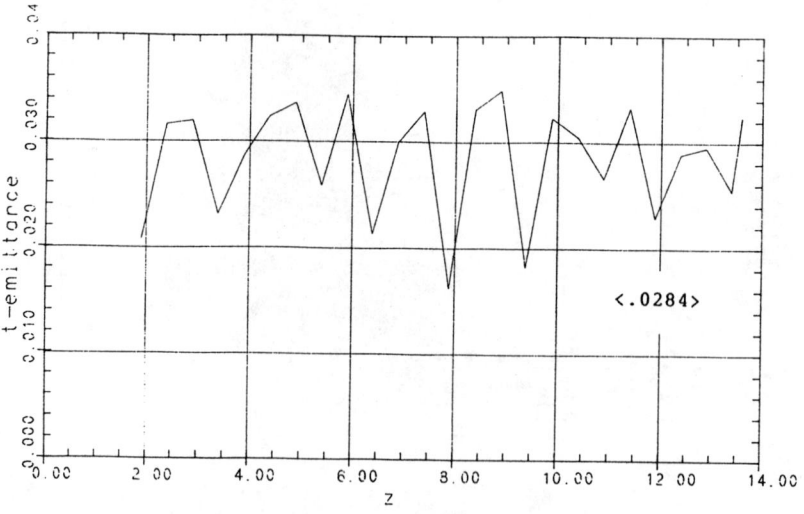

Fig. 11. Tunnel emittance versus axial distance for space-charge balanced flow case.

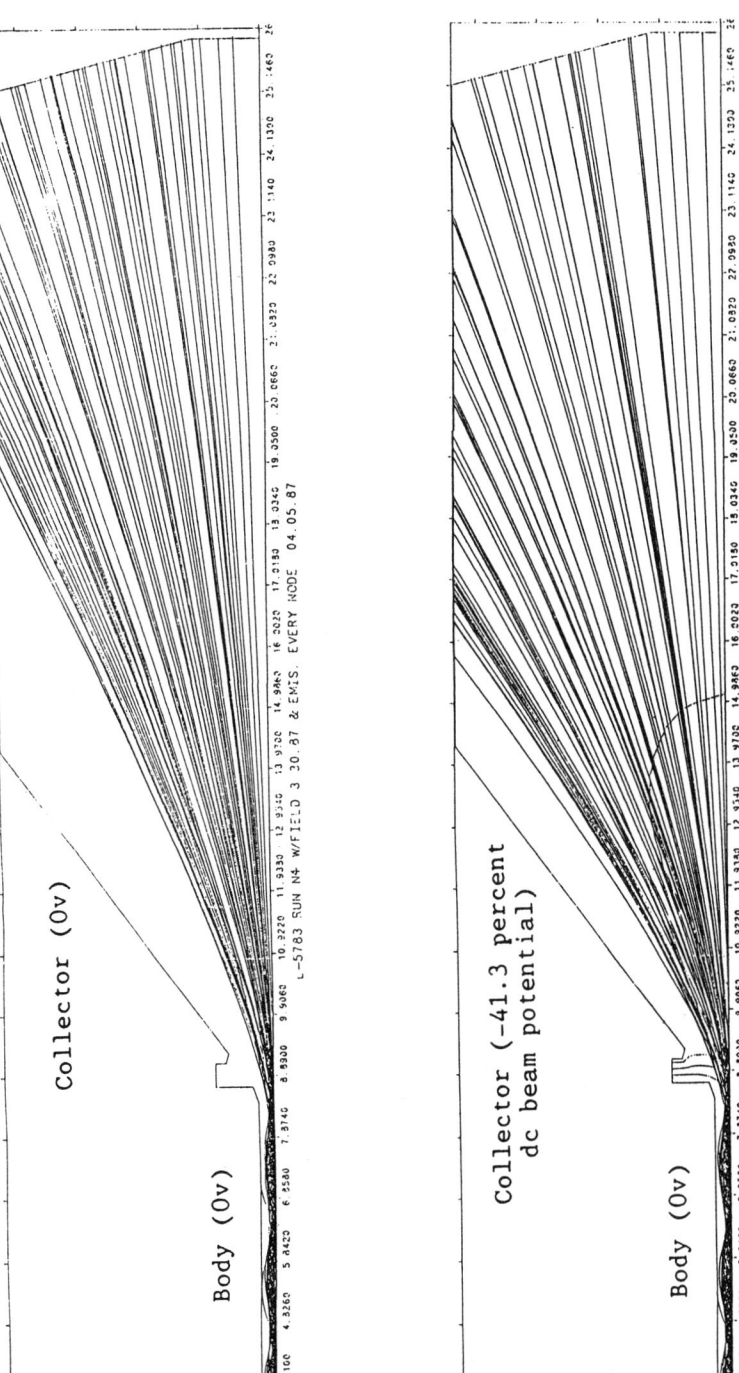

Fig. 12. Downstream view of end-to-end simulation (shadow gridded gun-to-collector) of space-charge balanced flow focussed beam.

DISCUSSION

<u>Kostroun</u>: You mentioned that Amboss, in some classic paper, says that you can get greater compression of the beam in an immersed flow. Did I understand that correctly?

<u>True</u>: Yes.

<u>Kostroun</u>: This puzzles me because I always thought that you can get greater compression from a shielded cathode than from an immersed cathode.

<u>True</u>: I knew that there would be some controversy because you have been using Brillouin flow.

<u>Kostroun</u>: It is sort of related to Busch's theorem too.

<u>Amboss</u>: No, it is the paper on the verification of the Herrmann theory. This paper relates to low perveance thermal beams in solenoid fields. There it is shown that a small amount of cathode flux prevents the Gaussian tail in the current density from developing so that you end up with a smaller diameter beam. But the amount of cathode flux is limited, otherwise the beam gets bigger.

<u>True</u>: Yes, in other words you can wind up with basically a higher overall beam compression since flux through the cathode tends to quiet the beam down and keep the electrons in line. So it has certain advantages. But the difficulty is that you can lose big if you do not do it right. And that is the other point. You can gain by doing it, but if you do not do it right, then it is worse, much worse.

<u>Amboss</u>: How far downstream do you trust your computer modelling? I mean we work in the design of travelling wave tubes with beam lengths which are about, say, 100 beam tunnel radii. Have you ever done any computation carrying a beam that far down to the collector?

True: I have. You have to watch out for instability problems. In PPM systems in particular, the beam tends to become unstable, and you have to break it into pieces. It is important that you overlap your solutions. I gave a paper on this at the Monterey Power Tube Conference this year. You run the problem downstream for a certain distance and then you restart each continuation frame from a position upstream of the downstream Neumann boundary. This is done to eliminate errors from non-axially-uniform space-charge distributions across the downstream Neumann boundary. In order to obtain very accurate solutions far downstream, you must use the frame overlapping technique above, and you have to crank the convergence tolerances way up high so as to obtain a highly converged solution. For immersed flow focussed beams, the extra beam stiffness shows up in the computer calculations. That is, solution stability downstream is higher and you can carry the simulation forward over a much greater distance. In terms of a length-to-circuit inner diameter of 100 to 1, sometimes the calculations are stable and sometimes they are not. In this case you are on the ragged edge of stability. When the solutions settle down and do converge, you get a good reliable solution. You can see it because the beamlet trajectories look reasonable and do not vary from iteration to iteration, and changes in potential are small. But if it is non-convergent over that distance, there is not much you can do. You just have to break the problem into pieces.

Amboss: Do you approach the Pierce-Walker theory at all in your computations and has anybody looked at this? Are you familiar with it?

True: I am not familiar with it. But based on a conversation I had yesterday with Reinard, I believe the material I have presented probably provides a similar result.

Becker: I have been asked to comment on this but I wanted to put up a question. Perhaps I will reverse it. Now, Pierce-Walker's theory, I suppose, is the real truth of beam focussing, and it has been nicely reproduced by the people looking at ion beam transport, for instance, in Berkeley for this heavy ion fusion stuff. So Dick has done a wonderful job showing how it can really move from a current dominated beam with a rectangular shape to an emittance-dominated beam with a more Gaussian-like profile. And in all cases this was a matched beam in the sense that it was just fixed to the aperture and the focussing capability of the channel. So I suppose Pierce-Walker is true, no question about it. Bernd Fogen did the work on comparing Pierce-Walker's theory with Herrmann's theory. As a feature, you get a relation on how to increase the magnetic field in order to focus the beam. If you compare this with Pierce-Walker's theory, it agrees, but you have to do an arbitrary definition of a radius in a Pierce-Walker beam. If this contains 67 or 68% of the current, then it matches.

True: Right, and if you take a Gaussian, the 95% radius and the 1.65 factor, it is like you are bringing yourself in towards that point.

Becker: OK, my question to the experts now is, "What is the real difference between a Brillouin beam with some cathode flux and immersed flow, and is there a transition between them or not?" The Brillouin flow, in general, is considered only to be a perfect Brillouin flow, but when you do a real Brillouin flow then you have a little cathode flux that corresponds to some cathode immersion. It means that for the differential equation, term with $1/R^3$ is showing up. So the beam has a hard, stiff core inside.

True: I think that there are two idealized limits. One is Brillouin flow with zero flux at the cathode which is a mathematical statement - it can never be exactly zero, but close; you can make it pretty close. And then there is confined flow where the field is very high and you have virtually 100% cathode immersion. In this case the beam is very stiff and is kept ramrod straight at 3 or 4 times Brillouin. In between you have, what I call, immersed flow or space-charge balanced flow where you have less than 100% immersion of the cathode. In the case I showed, the relative cathode flux was 0.64, whereas at higher main field levels over Brillouin, this number might be 0.7 to 0.8. In the immersed flow case, beam stiffness is higher in comparison to Brillouin flow and the electrons take different paths (they make little cycloids rather than coming down near the axis). So the basic flow pattern is different. In terms of quieting the beam, limiting emittance growth, and providing for potentially greater compression, it appears that immersed flow focussing might prove useful. On the other hand, you can not knock success. People here have achieved fantastic compressions with Brillouin flow. For instance, Val Kostroun did it with a very carefully controlled field in the cathode region. Whether immersed flow focussing will lead to higher levels of performance in EBIS devices, remains an open question.

THE EFFECT OF SMALL TRANSVERSE MAGNETIC FIELDS ON ELECTRON BEAM TRANSMISSION

K. Amboss

Hughes Aircraft Company, Electron Dynamics Division, Torrance, Ca. 90505

ABSTRACT

The paraxial ray equation for the beam axis of a space charge electron beam in the presence of a small transverse magnetic field is presented and solved for the simple case of a uniform axial focusing and uniform transverse perturbing magnetic field. The solution is used to explain the interception generally observed when the amplitude of the field is changed and the sense of the magnetic field is reversed. Values for permissible transverse fields in electron beam ionization sources (EBIS) devices are discussed.

INTRODUCTION

Very small departures from rotational symmetry of the focusing fields used in a variety of electron optical devices can have a major impact on their performance. In electron microscopes, resolution is reduced; in cathode ray tubes, picture quality is degraded; in traveling-wave tubes, power output is adversely affected; and in EBIS machines, beam alignment and trapping are compromised. The effects of departures from rotational symmetry were first studied in the context of transmission electron microscopy. In particular, the work of Sturrock[1] on microscope lenses can be applied directly to the problem of the transport of space charge electron beams as was shown by Amboss.[2]

Sturrock expresses the departure of the magnetic field from rotational symmetry in a Fourier series in the azimuthal coordinate. He finds, in the paraxial approximation, the lowest order field term to result in a transverse displacement of the beam axis from the nominal axis of rotational symmetry. The beam remains rotationally symmetric and centered on the displaced axis in the paraxial limit, even in the presence of space charge.

The equation of the beam axis is the familiar paraxial ray equation of space charge optics but without the space-charge term and with a term containing the transverse magnetic field on the right hand side. The equation is derived in Reference 2 and presented in Section 2.0. The equation is specialized to the relatively valid simplified case of a uniform axial and uniform transverse magnetic field and is solved to allow discussion of the approximate motion of the axis. This solution could of course have been obtained directly from the Lorentz force equation for this simple case.

The solution is used in Section 3.0 to discuss the often-used diagnostic techniques of solenoid current reversal and magnitude change.

An estimate of the transverse fields in EBIS machines is made in Section 4.0.

© 1989 American Institute of Physics

THE EQUATION OF THE BEAM AXIS

In discussing the motions of electrons in fields that depart to some degree from rotational symmetry, it becomes necessary to set up an appropriate coordinate system. For this discussion, the z-axis is conveniently taken to be the center of the drift tube through which the beam passes, and complex coordinates

$$u = x + iy = re^{i\theta} \tag{1a}$$

$$\bar{u} = x - iy = re^{-i\theta} \tag{1b}$$

centered on this axis are used to describe the motion and the fields. An off-axis expansion of the components B_z and $B_\perp$ about this axis takes the form:

$$\begin{aligned} B_z(u, \bar{u}, z) &= \left(B_o(z) - \frac{1}{4} u\bar{u} \frac{d^2 B_o(z)}{dz^2} + \cdots \right) \\ &+ \left(u \frac{d\bar{B}_1(z)}{dz} + \bar{u} \frac{dB_1(z)}{dz} + \cdots \right) \\ &+ \frac{1}{4} (u^2 \bar{B}_2(z) + \bar{u}^2 B_2(z) + \cdots) \end{aligned} \tag{2a}$$

$$\begin{aligned} B_\perp(u, \bar{u}, z) &= \left(-\frac{1}{2} u \frac{dB_o(z)}{dz} + \frac{1}{16} u^2 \bar{u} \frac{d^3 B_o(z)}{dz^3} - \cdots \right) \\ &+ \left[B_1(z) - \frac{1}{8}\left(u^2 \frac{d^2 \bar{B}_1(z)}{dz^2} + 2u\bar{u} \frac{d^2 B_1(z)}{dz^2} \right) + \cdots \right] \\ &+ \frac{1}{2}(\bar{u} B_2(z) + \cdots) \end{aligned} \tag{2b}$$

Since $u\bar{u} = r^2$, the $B_0(z)$ term and derivatives can be readily identified with the rotationally symmetric part of the field; by setting $u = \bar{u} = 0$, $B_1(z)$ is seen to be a transverse magnetic field on the axis, and $B_0(z)$ is the axial field on the axis. $B_1(z)$ is therefore a part of the paraxial field and must be included in the paraxial ray equation.

The paraxial motion in a magnetic field is most conveniently solved by a transformation to the Larmor frame, in which the transverse coordinates $w, \bar{w}$ are related to the laboratory frame coordinates $u, \bar{u}$ by the transformation (non-relativistically for simplicity):

$$w = ue^{i\chi} \tag{3}$$

where:

$$\chi = \int_{z_A}^{z} \sqrt{\frac{\eta B_0^2}{8V_0}}\, dz \tag{4}$$

and where $V_0(z)$ is the potential on the axis and $\eta = e/m$, the charge-to-mass ratio of the electron. z_A is the axial coordinate at which boundary conditions are applied. The paraxial ray equation of the axis $k = k_x + ik_y$ has been shown to be

$$k'' + \frac{1}{2}\frac{V_0'}{V_0} k' + \left(\frac{1}{4}\frac{V_0''}{V_0} + \frac{1}{8}\eta\frac{B_0^2}{V_0}\right) k = -i\sqrt{\frac{\eta B_1^2}{2V_0}}\, e^{-i\chi} \tag{5}$$

where a dash denotes differentiation with respect to z.

Equation (5) is just the paraxial ray equation with the space charge term:

$$\frac{\rho_0}{e_0 V_0} k \tag{6}$$

absent and with the B_1 term on the R.H.S. added. Equation (5) can be solved directly by computer for given specific fields and boundary conditions. However, since the motion in the region of interest is in a uniform field at constant potential, a solution can be obtained directly, assuming B_1 also to be uniform. Under these conditions, Equations (4) and (5) reduce to

$$\chi = \Gamma_0(z - z_A) \tag{7}$$

$$k'' + \Gamma_0^2 k = -2i\Gamma_1 e^{-i\Gamma_0(z-z_0)} \tag{8}$$

where:

$$\Gamma_0^2 = \frac{1}{8}\eta\frac{B_0^2}{V_0} \tag{9}$$

$$\Gamma_1^2 = \frac{1}{8}\eta\frac{B_1^2}{V_0} \tag{10}$$

The homogeneous part of Equation (8) describes simple harmonic motion, i.e.:

$$k = Ae^{i\Gamma_0(z-z_A)} + Be^{-i\Gamma_0(z-z_A)} \tag{11}$$

It is, however, more convenient to use the linearly independent solutions g(z) and h(z), used in electron optics, which obey the boundary conditions in the Larmor frame in some initial plane $z = z_A$, where $k = k_A$ and $k' = k'_A$

$$g(z_A) = 1 \; , \; g'(z_A) = 0 \tag{12a}$$

$$h(z_A) = 0 \; , \; h'(z_A) = 1 \tag{12b}$$

and which are calculated to be:

$$g(z) = \frac{1}{2}\left(e^{i\Gamma_o(z-z_A)} + e^{-i\Gamma_o(z-z_A)}\right) \tag{13}$$

$$h(z) = \frac{1}{2i\Gamma_o}\left(e^{i\Gamma_o(z-z_A)} - e^{-i\Gamma_o(z-z_A)}\right) \tag{14}$$

In terms of the values for k and k' in the plane $z = z_A$

$$k = k_A g(z) + k'_A h(z) \tag{15}$$

The solution of Equation (8), including the particular integral, which can be obtained by variation of parameters, is

$$k(z) = \left[k_A + 2i \int_{z_A}^{z} h\Gamma_1 \, e^{-i\Gamma_o(z-z_A)} dz\right] g$$

$$+ \left[k'_A - 2i \int_{z_A}^{z} g\Gamma_1 \, e^{-i\Gamma_o(z-z_A)} dz\right] h \tag{16}$$

Evaluation of the integrals with g and h given by Equations (13) and (14) and transformation to the laboratory frame by the operation

$$u = k \, e^{i\Gamma_o z} \tag{17}$$

gives

$$u = u_A + (z - z_A)\frac{B_1}{B_o} + i\frac{\sqrt{2\eta V_o}}{\eta B_o}\left(e^{i\frac{\eta B_o}{\sqrt{2\eta V_o}}} - 1\right)\left(\frac{B_1}{B_o} - u'_A\right) \tag{18}$$

This equation describes an initial displacement u_A of the beam axis from the axis of the system that grows linearly with distance $(z-z_A)$ and a rotation with a cyclotron radius R_c:

$$R_c = \frac{\sqrt{2\eta V_o}\left(\frac{B_1}{B_o} - u'_A\right)}{\eta B_o} \tag{19}$$

Figure 1 shows the motion of the axis for B_1 and u'_A real. It is a helix, ngent to the field line B.

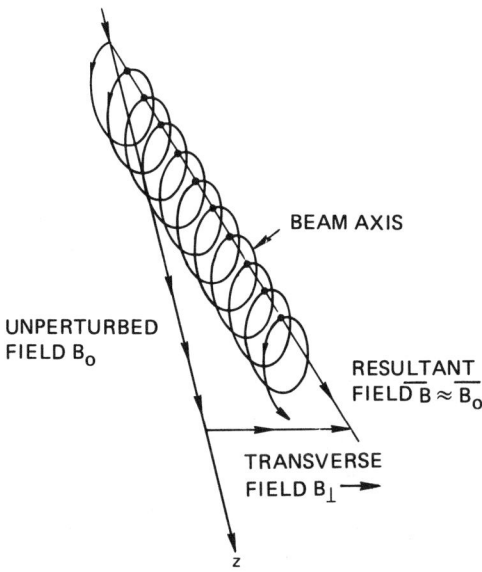

Fig. 1   Motion of the beam axis in a uniform magnetic field, B_o, perturbed by a small uniform transverse magnetic field, $B_\perp$, directed along the x-axis.

DIAGNOSTIC TECHNIQUES

Useful information about the alignment and the transverse field can often be obtained by observing the variation in beam interception as a function of magnetic field strength, that is, solenoid current, for both senses of the magnetic field. One direction of solenoid current generally produces lower interception than the reverse direction. Figure 2 shows data from a traveling-wave tube that exhibit this phenomenon. It also shows that the interception varies sinusoidally and decreases with increasing solenoid current.

A difference in interception on current reversal can be explained by Equations (18) and (19) with the aid of the diagrams in Figures 3 and 4. Assuming that any iron in the solenoid magnetic circuit has been brought into the cyclic state by repeated current reversal so that hysteresis effects are small, current reversal will reverse the sense of both B_0 and B_1 and, therefore, change the direction of any field line by 180°. The cyclotron radius, R_c, given by Equation (19) is unaltered in magnitude; however, the center about which the beam axis rotates has moved through a distance $2iR_c$, and the direction of rotation is also reversed. Figures 3(a) and (b) show the paths of the beam axis on current reversal near the entrance plane z_A. Figure 3(a) shows the gun axis coincident with the drift tube axis in the plane z_A; in this case, the excursion of the beam axis from the drift tube axis is the same for either sense of the magnetic field, and no difference in interception would be noted. In the case illustrated in Figure 3(b), the gun axis is displaced by u_A in the entrance plane; in this case, one field direction leads to higher interception than the other. Figure 4(a) and (b) illustrate the motion for some distance in the field with both an initial displacement u_A and slope u'_A in the entrance plane. The gradual separation of the beam axis from the drift tube axis is increased periodically by the cyclotron motion, which leads to a larger excursion in the case of Figure 4(b).

A periodic variation in beam interception when the solenoid current is varied is noted in Figure 2 for both senses of the magnetic field. Two qualitative explanations are offered for this periodicity with the aid of Figures 5 and 6. Figure 5 shows the edge of an undulating beam passing by some local constriction in the diameter of the beam tunnel, assumed to be the result of a manufacturing error. In a weak magnetic field, the constriction intercepts some of the beam; as the field is raised the depth of undulating increases even as the maximum radial excursion decreases. A condition is reached where the constriction coincides with the trough of scallop with resultant minimum interception. As the field is increased further the scallop wavelength decreases and a peak of the scalloping beam is intercepted by the constriction. However, since the maximum radial beam excursion is smaller for the larger field, the interception peak is less.

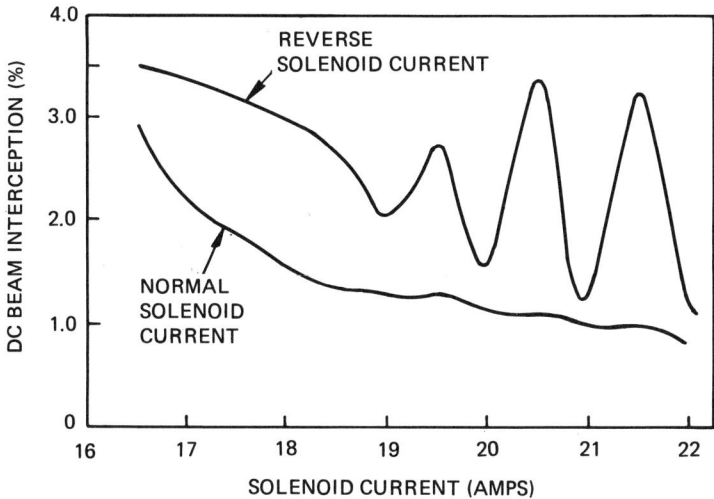

Fig. 2 Beam interception as a function of solenoid current measured on a traveling-wave tube for both senses of the solenoid current.

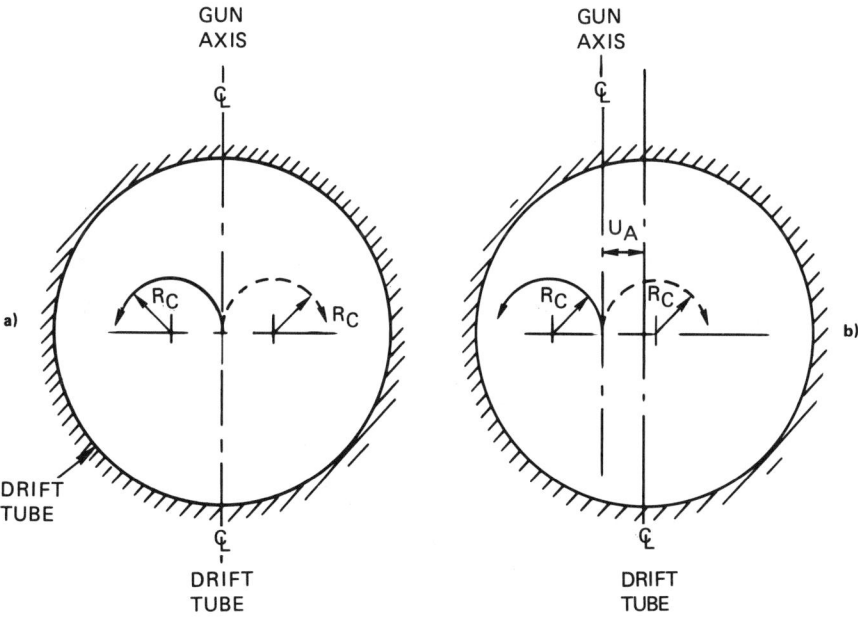

Fig. 3 Motion of the beam axis on solenoid current reversal near the entrance plane $z = z_A$. (a) Gun axis coincides with the drift tube axis in the plane z_A. (b) Gun axis displaced from the drift tube axis by a distance u_A along the negative x-axis in the plane z_A.

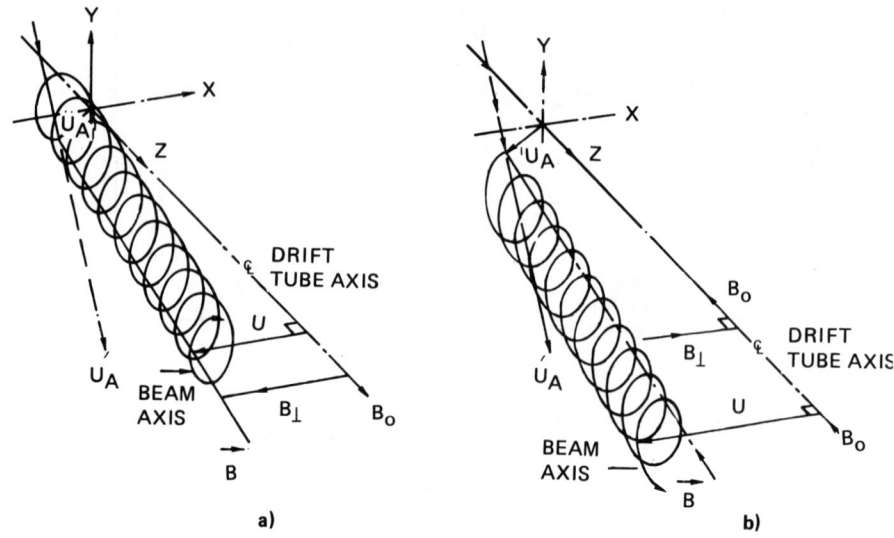

Fig. 4 Motion of the beam axis on solenoid current reversal with displacement u_A and slope u'_A of the gun axis to the drift tube axis in the entrance plane, z_A. (a) Current direction "normal"; (b) current direction "reversed."

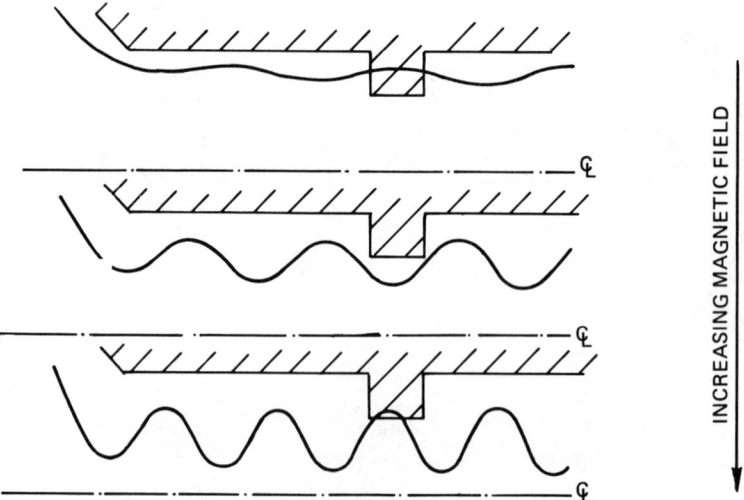

Fig. 5 Beam edge of an undulating beam near a constriction in the wall of the drift tube for three cases of gradually increasing magnetic field.

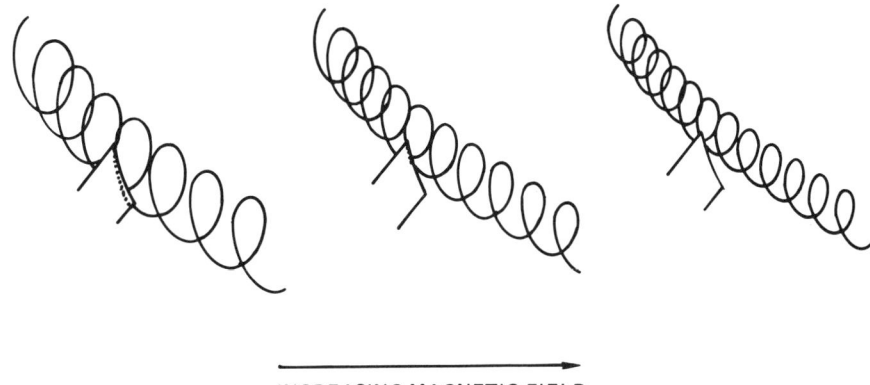

Fig. 6 Path of the spiraling beam axis in an increasing magnetic field, with the relationship to a constriction in the drift tube wall shown schematically. (The interception is on the beam edge.)

A somewhat similar periodic interception mechanism can be postulated for the transverse field. In this case, the radius R_c of the helix of the spiraling beam axis and the pitch of the helix decrease with increasing magnetic field and the beam axis passing the local constriction periodically has a gradually decreasing periodic displacement from it. The effect is illustrated in Figure 6.

ESTIMATE OF THE TRANSVERSE FIELDS IN EBIS MACHINES

The electron beams used in the various EBIS devices are highly compressed and, it has been shown, have an almost gaussian current density distribution. An approximate value of the transverse magnetic field can be calculated by assuming the beam interception to be the result of a single asymmetric impingement of a gaussian beam on a circular aperture, as sketched in Figure 7. Also shown in Figure 7 are graphs of computed percent interception as a function of transverse displacement of the beam axis, assuming the radius R_{99}, which encloses 99 percent of the beam current to be equal to the drift tube radius a and 0.66 a.

EBIS machines have confinement drift tubes that are typically about 0.8 or 0.9 cm in diameter and are about 1 m long. The magnetic fields used vary from instrument to instrument and range between 1 and 5 Tesla and are uniform to about 1 percent. Beam voltages range from 10 to 30 kV.

Using the extreme values of B_0 and V_0 in Equation (19), which maximizes the cyclotron radius, and realistically large values for B_1/B_0 and u'_A of, say,

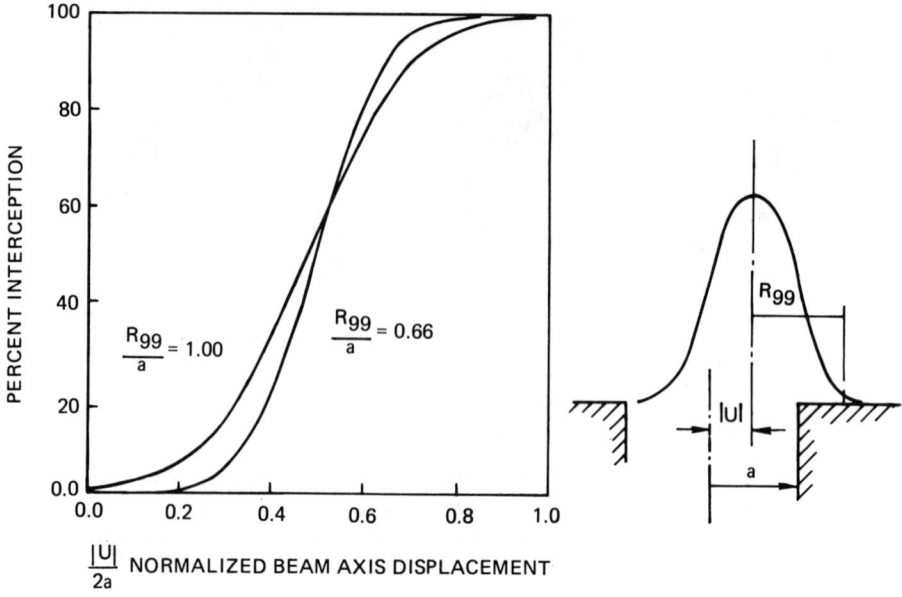

Fig. 7 Interception of a gaussian beam on a circular aperture of radius a as a function of transverse displacement |u| of the beam axis.

0.1 a each, one finds the cyclotron radius to be small enough to neglect for EBIS devices. The axial electron is therefore "tied" to a flux line and the tilt of the magnetic field is the primary defect.

Inspection of Figure 7 shows beam interception to increase by only 1.0 percent, from 1.0 to 2.0 percent for a displacement of the beam axis of 0.1 a. For the case of a 0.80-mm diameter confinement tube, this amounts to 0.04 cm. If the electron beam is to pass through a 1.0 meter long confinement tube with only a 1.0 percent increase in interception, then neglecting beam entrance misalignments and assuming a uniform transverse magnetic field, one finds from Equation (18) that B_1 must be less than 0.0004 B_0.

Further inspection of Figure 7 shows interception to rise rapidly for normalized displacements |u|/a exceeding about 0.50. For |u|/a = 0.42, about 50 percent of the beam current has been sheared off. In the case of a 1 m long confinement tube, the corresponding transverse field amounts to about 0.0037 B_0, less than half a percent of the axial field.

Solenoids are, of course, always provided with adjustments that allow a limited translation and tilt of the nominal magnetic field axis with respect to the drift tube axis. These adjustments allow compensation, generally only partial, of the transverse magnetic field. A parallel translation has, of course, no effect in the region of uniform field but produces a transverse field at the ends of the solenoid. This can be seen from Equation (2) by making the expansion about a coordinate system $v, \bar{v}, z$ which is displaced a parallel vector distance Δ from the $u, \bar{u}, z$ coordinate system. Thus, with

$$u = v + \Delta$$

Equation (2) becomes

$$B_z(v, \bar{v}, z) = B_o(z) - \frac{1}{4}(v + \Delta)(\bar{v} + \bar{\Delta})\frac{d^2 B_o(z)}{dz^2} + \cdots$$

$$+ (v + \Delta)\frac{d\bar{B}_1(z)}{dz} + (\bar{v} + \bar{\Delta})\frac{dB_1(z)}{dz} + \cdots \qquad (20a)$$

$$B_\perp(v, \bar{v}, z) = -\frac{1}{2}(v + \Delta)\frac{dB_o(z)}{dz} + B_1(z) \qquad (20b)$$

To the first order in v and Δ, Equation (20) becomes:

$$B_z(v, \bar{v}, z) = B_o(z) - \frac{1}{4} v\bar{v} \frac{d^2 B_o(z)}{dz^2} - \frac{1}{4}(u\bar{\Delta} + \bar{u}\Delta)\frac{d^2 B_o(z)}{dz^2}$$

$$+ u\frac{d\bar{B}_1(z)}{dz} + \bar{u}\frac{dB_1(z)}{dz} + \Delta\frac{d\bar{B}_1(z)}{dz} + \bar{\Delta}\frac{dB_{sub1}(z)}{dz} \qquad (21a)$$

$$B_\perp(v, \bar{v}, z) = -\frac{1}{2} v \frac{dB_o(z)}{dz} + B_1(z) - \frac{1}{2}\Delta\frac{dB_o(z)}{dz} \qquad (21b)$$

REFERENCES

1. P.A. Sturrock, Phil. Trans. 243, p. 387 (1951).
2. K. Amboss, IEEE ED-12, p. 322 (1965).

DISCUSSION

<u>Becker</u>:   I suppose one of the things which is most important with respect to transverse fields for the EBIS community is something which is considered in travelling wave tubes but which you have not addressed so far. This is the coupling of microwaves to the beam through electrodes at different potentials and decoupling from it. Essentially, EBIS devices are travelling wave tubes waiting for modulation. And you always have cross-talk through the leads. Therefore it is important that you are in the center of a beam tube. If you are off-axis, then coupling to the beam and from the beam is much easier. We always see that the beam gets noisy if it is displaced.

<u>Amboss</u>:   Well, we like to get coupling. That is our bread and butter. I have seen, in effect, that if you have the beam off-axis, then the bunching forms in an asymmetric fashion. I do not know if you can verify that, Dick. And you do not get the power output because you do not get the nice bunching that is expected in microwave tubes. So, in your case, you may get more noise.

CHARACTERISTICS OF TYPICAL PIERCE GUNS FOR PPM FOCUSED TWTs

R. Harper, M. P. Puri
Raytheon Company, Waltham, MA 02254

ABSTRACT

The performances of typical moderate perveance Pierce type electron guns which are used in periodic permanent magnet focused traveling wave tubes are described with regard to adaptation for use in electron beam ion sources. The results of detailed electron trajectory computations for one particular gun design are presented.

INTRODUCTION

The microwave tube industry has designed and produced many electron guns for use in linear beam devices such as TWTs. The availability of such electron guns, most of which are relatively inexpensive, can be of great benefit to an experimental project, since resources can be devoted to the major work of the project rather than to electron gun design. The performance of a typical TWT gun and the usual parameters which are used to specify a TWT gun are described in this paper to aid in finding an existing gun which can fulfill a given need. This discussion is restricted to conventional Pierce guns designs intended for DC operation. Gridded guns used for pulsed operation are not discussed.

TWT ELECTRON GUNS AND BEAMS

A typical TWT requires an electron beam with 20 to 100 amperes per square centimeter current density which is of high quality so that it can be magnetically focused to pass through a metal tunnel (the microwave interaction circuit) with a length to diameter ratio which usally exceeds 200/1. A periodic permanent magnet (PPM) focusing system, consisting of ring magnets of alternating polarity which produces a sinusoidally varying axial magnetic field, is used to maintain the beam at the injection diameter throughout the length of the interaction circuit. Figure 1 shows a simplified diagram of the electron beam flow from the electron gun through the PPM focusing system. Note that these TWT electron guns are convergent Pierce guns designed according to the principles originated by J.R. Pierce[1]. Since the desired beam current density usually is considerably larger than the emission capabilities of the cathode, the convergence has been chosen to obtain the required beam current density while still having acceptable cathode current density.

These electron guns are designed by a process of synthesis, choice of gun electrodes followed by detailed analysis and optimization using an electron ray-tracing computer code. The synthesis procedure can be accomplished using the approach described in Pierce's book and/or several papers[1,2,3] using hand calculations and graphical methods. Alternatively, one can use a method described by Vaughan[4] which is readily adaptable to small computers. The analysis, which requires the use of a main frame computer code (such as ones described by Kirstein and Hornsby[5], Herrmannsfeldt[6], True[7] or Dionne[8]), must be done to optimize the design because the simple Pierce gun theory does not take into account effects such as the anode hole. A well designed electron gun has required many days of work to achieve optimum performance.

Before discussing the characteristics of some typical TWT electron guns we should define the parameters involved. Table 1 lists the quantities usually specified by the user of the electron gun and then defines the parameters usually used in gun design work. Figure 2 serves to define the various radii at different locations on the electron beam and also includes an indication of the effects of thermal spreading of the beam caused by the finite velocity of emission of the electrons from the cathode. This thermal spreading is commonly stated in terms of the radius which will enclose 95 percent of the beam current as shown in figure 2. Whether or not one must be concerned about thermal spreading is determined by the magnitude of the parameter at the end of the list in table 1 which was defined by Herrmann[3]. If this parameter is greater than 3.2×10^{-6} thermal spreading of the beam will be less than 10 percent and can be ignored. For values of (PV/T) less than 0.01×10^{-6} thermal spreading will be severe (greater than 50 percent) and measurements of the electron beam diameter are necessary before deciding how useful the gun will be for a particular application.

Table 1
ELECTRON GUN SPECIFICATIONS

° Requirements as specified by user
 - Voltage
 - Current
 - Cathode current density
 - Beam diameter at waist
° Restatement of requirements
 - Perveance   P $= I/V^{3/2}$
 - Area convergence C $= (R_c/R_o)^2$
 - Voltage   V   volts
 - Cathode temperature T degree Kelvin
 - Thermal spreading parameter $\dfrac{PV}{T}$ $\dfrac{\text{amperes}}{\sqrt{\text{volts}} \cdot \text{degrees}}$

The effects of electron optical aberrations (which can be predicted by the ray tracing computer codes) and thermal spreading (which is not well predicted by the ray tracing codes) cause the performance of most electron guns to differ from that predicted by simple theory; and the optimization of an electron gun design to achieve a high quality laminar flow beam is to some extent an art, based on the science of electron optics, using detailed ray tracing computations.

CHARACTERISTICS OF A TYPICAL TWT ELECTRON GUN

The characteristics of a particular TWT electron gun will be described to illustrate our thesis. This gun, which was designed some years ago for use in low and medium power CW TWTs, was used in the recent EBIT experiments done at Lawrence Livermore Laboratory. It was designed to be relatively inexpensive, rugged, and easily manufactured; and it is manufactured in quantities of several thousand per year. Although the basic structure of the gun, which is shown in figure 1, differs somewhat from a typical Pierce gun in that the beam forming (focus) electrode does not have the usual slant angle with respect to the beam edge, it is a Pierce gun. The focus electrode is designed to operate at a negative potential with respect to the cathode, and this causes the zero equipotential to assume a shape closely approximating the usual Pierce focus electrode. This design approach gives a focus electrode which is easily machined, and which can be precisely located by suitable assembly fixtures. The result is ease of manufacture with control of performance characteristics; and minor sample variations can be compensated for by slight adjustment of the negative grid bias.

The electrical performance characteristics of this gun are tabulated in table 2. When operated with 4800 volts applied to the anode the beam current is 150 milliamperes at a cathode current density of slightly over 2 amperes per square centimeter, well within the emission which can be obtained from this type of cathode. This gun can be operated at beam currents up to 200 milliamperes (2.8 amp/cm^2) with an anode voltage of 5800 volts. Obviously lower beam current operation is possible by using a lower anode voltage. The thermal spreading in this gun is negligible so that the area convergence and beam waist diameter are essentially unchanged for anode voltages greater than 2000 volts.

TABLE 2. ELECTRON GUN PERFORMANCE CHARACTERISTICS

Beam Characteristics	
Perveance	0.45×10^{-6}
Cathode Diameter	0.119 inch
Beam Diameter	0.028 inch
Area Convergence	18.06
Typical Operation	
Anode Voltage	4800 volts
Focus Electrode Bias	-10 volts
Beam Current	0.150 A
Cathode	Impregnated Matrix
Cathode Temperature	1040°C Br
Cathode Current Density	2.09 A/cm^2
Thermal Spread Parameter (PV/CT)	$.091 \times 10^{-6}$

In the original design of this gun a computer code written by Boers[9] was used to trace the beam trajectories in the electrostatic case (without magnetic field) to optimize the design.

Recent simulations of this gun have been made using the latest version of Dionne's code, and the predictions agree quite well with the original simulations. Figure 3 shows the plot of trajectories for the electrostatic case and figure 4 shows a plot of the same trajectories with a two times expansion of the radial dimensions to give a clearer view of the trajectories. These plots show the radius of rings of charge as they leave the cathode and travel through the electron gun into the beam tunnel.

In a TWT the presence of the magnetic focusing field has a significant effect on the trajectories; and, in this particular case, a small amount of magnetic field is deliberately imposed at the cathode to give more stable focusing under conditions of rf modulation of the beam. Figure 5 shows the trajectories computed from the cathode through the fourth magnet of the PPM focusing systems. An expanded radial dimension plot of the same trajectories in figure 6 clearly shows the trajectories in a PPM focusing system. This beam is well behaved with only minor periodic rippling of the beam diameter. For comparison, the behavior of the beam when the PPM focusing system is shifted to be 0.125 inch further from the electron gun is shown in figure 7. It can be seen that the beam has much more rippling of the diameter such that the maximum diameter points are more than twice normal[10]. This is a clear illustration of the need to properly design the matching of the electron gun to the magnetic focusing system.

Another system for focusing a beam uses complete magnetic shielding of the cathode with an abrupt transition into the axial magnetic focusing field. The magnetic field strength must be the correct value to satisfy the Brillouin flow conditions in which the beam travels in a uniform axial magnetic field with a rotation such as to give a balance between inward and outward components of force. With care the Brillouin focusing conditions can be satisfied and a beam with uniform diameter and rotation can be obtained. This focusing may be desirable for experimental usage of these guns, but is seldom used in standard TWTs because of its sensitivity to perturbing influences. If the magnetic field is gradually increased, the beam diameter can be reduced while still maintaining the Brillouin flow conditions. This magnetic compression to obtain a smaller diameter beam has been used in some special TWTs.

Verification of the computed predictions of beam diameter for this electron gun has been obtained by use of a beam analyzer in which a small probe containing a collector cage bucket behind a pin hole aperture plate is moved across the beam to measure the distribution of current within the beam. Figure 8 shows a plot of the current density in the beam as it emerges from the anode. The slight peak of current density at the beam edge as seen here agrees with that predicted by the computer simulations. Such beam analyzer plots have been used for many years to check the performance of electron guns; but only recently, with the availability of desk top computers which are easily programmed for data acquisition has it been practical to generate plots such as this along the entire length of a PPM focused beam.

Recently we have built a computer controlled beam analyzer in which the probe is small enough to measure the beam inside the PPM focusing system. Figure 9 shows a series of plots of beam current distribution taken at regular intervals of one half a cell length along the beam in a PPM focused TWT. Figure 10 is a continuation of this series of plots. The beam shows some periodic scalloping of its diameter, which was predicted by the simulation; but a careful study of these plots reveals some interesting characteristics which were not predicted by the computer simulations. Specifically the beam, apparently in response to small transverse magnetic fields created by non-uniformities in the permanent magnet rings, moves on and off axis periodically as it traverses the PPM focusing system. The movement of the beam never is enough to cause interception of any electrons on the tunnel, so that, in terms of TWT performance, it is not noticeable in the testing of this TWT. In TWT manufacturing occasional tubes are encountered in which the beam transmission does vary with rotation of magnets; and it must be assumed that these are extreme cases of what has been observed in this beam analyzer. The usual electron ray tracing computer code assumes axial symmetry; and, therefore, it cannot predict these anomalies.

This rather detailed discussion of the performance of one particular TWT electron gun and PPM focusing system has been presented to show that good electron guns are readily available; and to show that computer simulations and perhaps beam analyzer measurements are available to describe completely the performance of these electron guns. If you can define the performance needed from an electron gun it may be worthwhile discussing the requirements with a TWT manufacturer. A considerable saving in time and money may result.

RANGE OF TYPICAL TWT PIERCE GUN PERFORMANCE

Available Pierce guns used in TWTs for dc operation cover the range of perveance from 0.100×10^{-6} to 2.0×10^{-6} with area convergence up to 50 to 1. The gun described in the preceding section and several others are listed in Table 3 to illustrate the available range of performance.

TABLE 3. CHARACTERISTICS OF REPRESENTATIVE GUNS

Gun	1	2	3	4
Perveance	0.45×10^{-6}	1.1×10^{-6}	0.12×10^{-6}	0.45×10^{-6}
Cathode Diameter	0.119 inch	0.134 inch	0.92 inch	0.080 inch
Beam Diameter	0.028 inch	0.040 inch	0.013 inch	0.015 inch
Area Convergence	18.1	11.2	50.0	28.4
Anode Voltage	2000V-5800V	1300-2770V	7000V-10000V	1650V-3420V
Beam Current	40-200 mA	50-160 mA	70-120 mA	30-90 mA

The maximum beam current obtainable from a gun is limited by the emission capability of the cathode (usually a maximum of 2.8 to 3.0 amperes/cm^2 for a normal impregnated cathode). Higher current for a given beam diameter requires either a larger cathode diameter or a cathode with higher emission capability. Both options have their advantages and disadvantages. Larger cathode diameter means higher area convergence which increases thermal spreading effects and also makes the electron optical design more difficult. Higher cathode emission density can be obtained by the simple method of raising the cathode temperature but at the cost of reduced life. New types of cathodes give promise of higher emission capability but less experience with these cathodes means some uncertainity. One possible type is the scandate cathode which has recently become generally available.

OTHER COMMENTS ON PIERCE GUNS

The TWT guns described here all have separately connected focus electrode, cathode, and anode supported by ceramic insulators which can standoff full beam voltage. This gives freedom to adjust the beam parameters by varying the focus electrode and/or anode voltages with respect to the cathode. There is a possible problem, however, which can occur unless care is taken in routing of the lead wires to the gun. These guns have V/I characteristics similar to those of an old fashioned triode vacuum tube, with an amplification factor similar to that of a triode. Hence, if the lead wires have sufficient inductance due to their length, resonant circuits, formed by the lead inductance and stray capacitances, can cause parasitic oscillations at frequencies in the range of tens to hundreds of megahertz. Elimination of these oscillations can be accomplished by any of several measures:
- Reduction of lead wire length.
- Bypass capacitors close to the gun.
- Non-inductive resistors (10 to 1000 ohms) in series with the focus electrode and/or anode leads close to the gun.

These all are standard techniques used in vacuum tube amplifier design, and experimentation is the best way to determine the best approach for a particular gun.

CONCLUSIONS

There are a number of electron guns originally developed for use in TWTs, that are readily available at low cost because of the continuing production of TWTs. The characteristics of some of these electron guns may be useful for EBIT and other experimental work. Many of these guns have been carefully optimized, and analytical and experimental characterizations of these guns usually are available to aid in determining whether they can be useful.

REFERENCES

1. J.R. Pierce, Theory and Design of Electron Beams, 2nd edition, 1954.
2. Danielson, Rosenfeld, Saloom, B S T J, vol. 35, p. 375-420, March 1956.
3. G. Herrmann, JAP, vol. 28, no. 4, April 1957, p. 474-478.
4. J.R.M. Vaughan, IEEE Trans Electron Devices, vol. ED-28, no. 1, January 1981, p. 37-41.
5. P. T. Kirstein, J. S. Hornsby, Report No. 63-6, CERN, Geneva, 1963.
6. William Herrmmansfeldt, Stanford Linear Accelerator Center Report No. SLAC 166.
7. R. True, 1975, IEDM Technical Digest paper no. 12.1 also, R. True "Space Charge Limited Beam Forming Systems Analyzed on a Deformable Relaxation Mesh", PhD Thesis, U. Conn, 1972.
8. N. Dionne, H.-J. Krahn, R. Harper, December 1980, IEDM Technical Digest, paper no. 18.4.
9. J. Boers, University of Michigan, Electron Physics Lab, Technical Report No. RADC-TR-68-175, April 1968.
10. M. Chodorow, C. Susskind, Fundamentals of Microwave Electronics, McGraw Hill, 1964, p. 42-52.

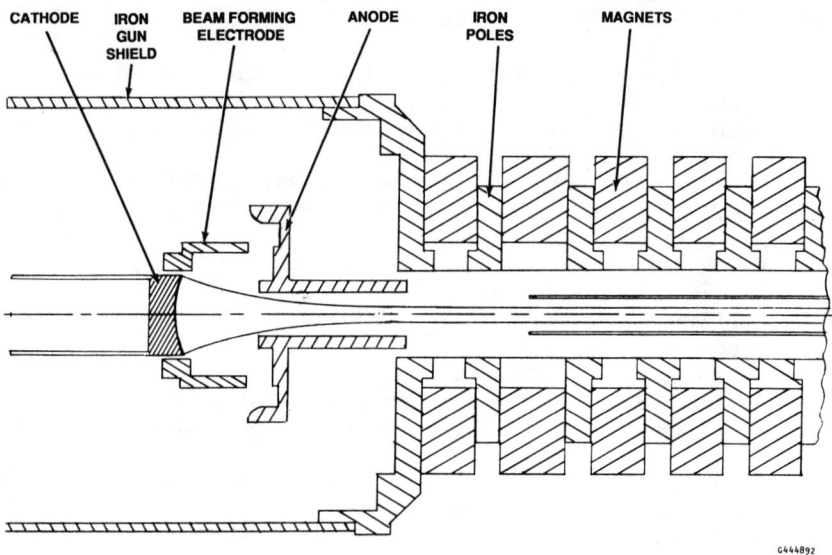

FIGURE 1. ELECTRON GUN AND PPM FOCUSED BEAM.

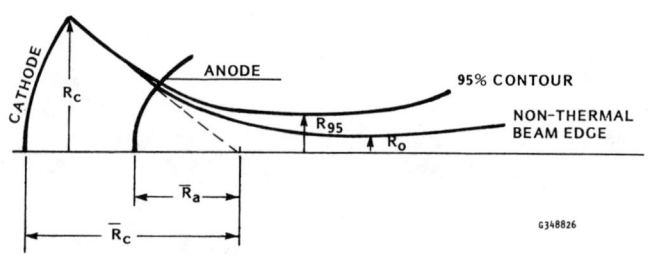

FIGURE 2. ELECTRON GUN PARAMETERS.

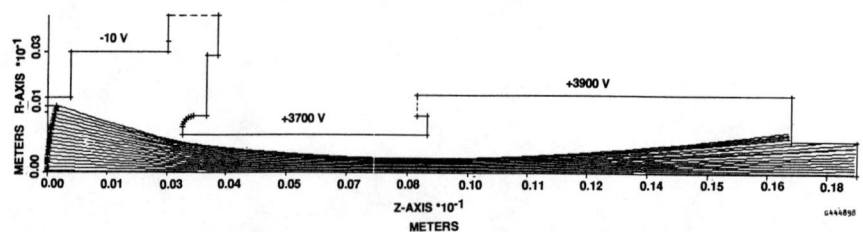

FIGURE 3. PLOT OF ELECTRON TRAJECTORIES. NO MAGNETIC FIELD.

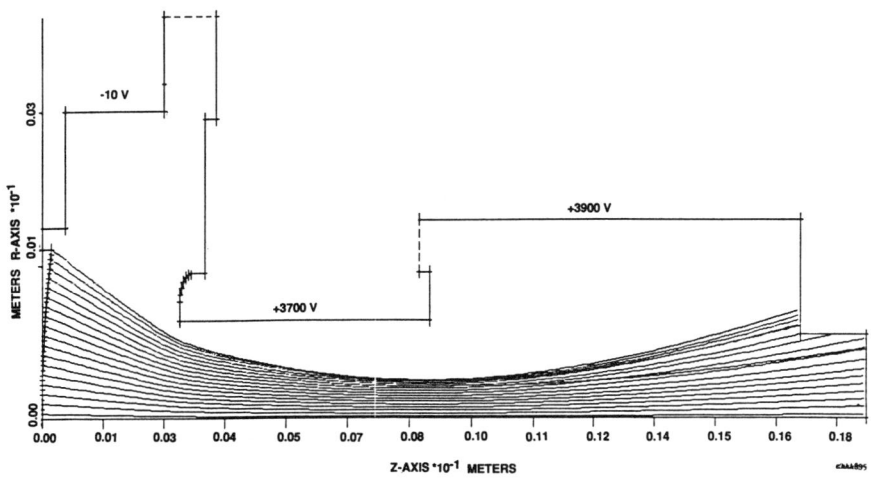

FIGURE 4. RADIALLY EXPANDED PLOT OF ELECTRON TRAJECTORIES. NO MAGNETIC FIELD.

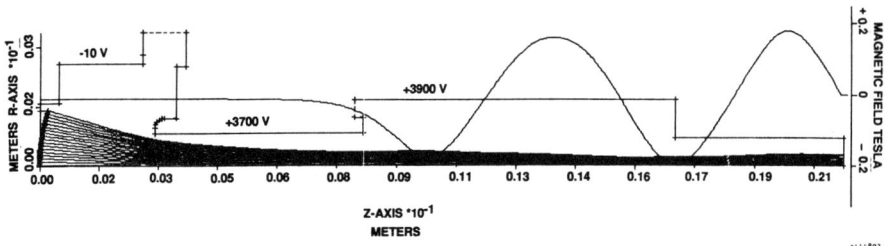

FIGURE 5. PPM FOCUSED BEAM TRAJECTORIES.

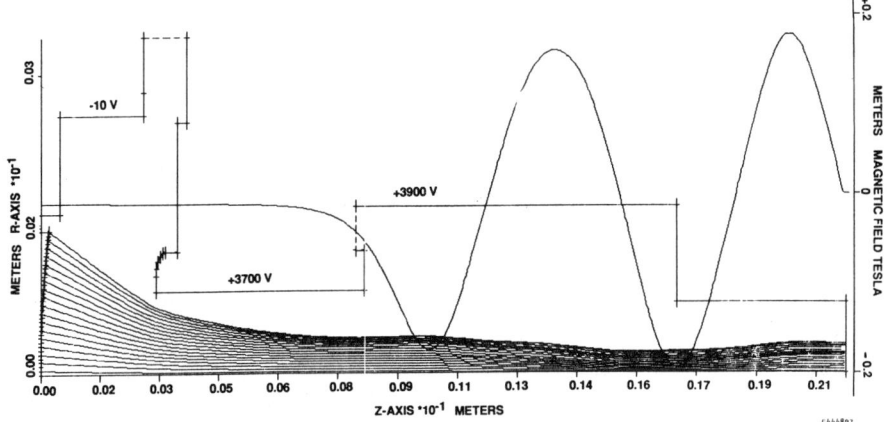

FIGURE 6. RADIALLY EXPANDED PLOT OF PPM FOCUSED BEAM TRAJECTORIES.

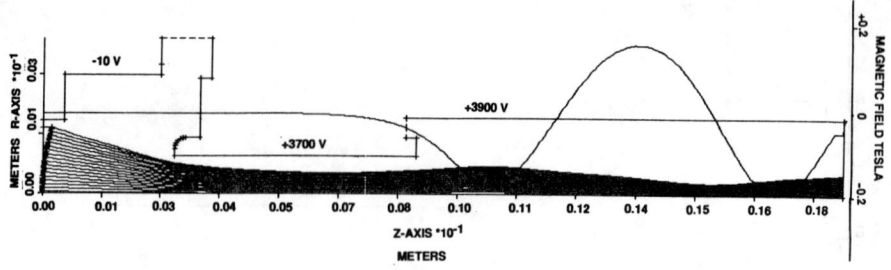

FIGURE 7. PLOT OF TRAJECTORIES IN PPM FOCUSING SYSTEM LOCATED 0.125 INCH BEYOND CORRECT POSITION.

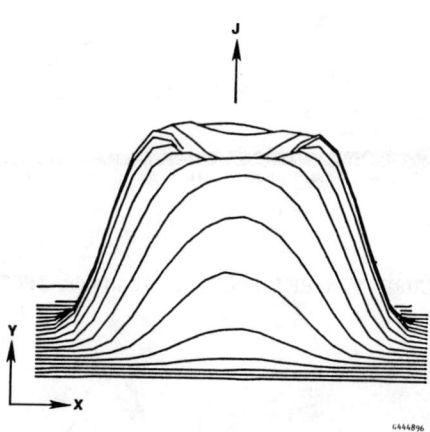

FIGURE 8. SURFACE MAP PLOT OF BEAM CURRENT DISTRIBUTION EMERGING FROM ANODE.

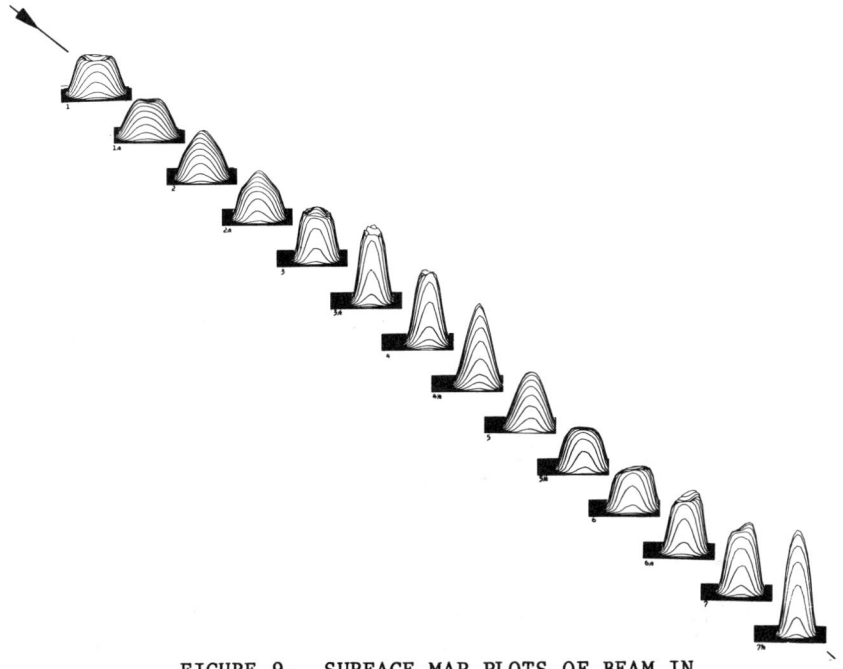

FIGURE 9. SURFACE MAP PLOTS OF BEAM IN FIRST SEVEN MAGNETS.

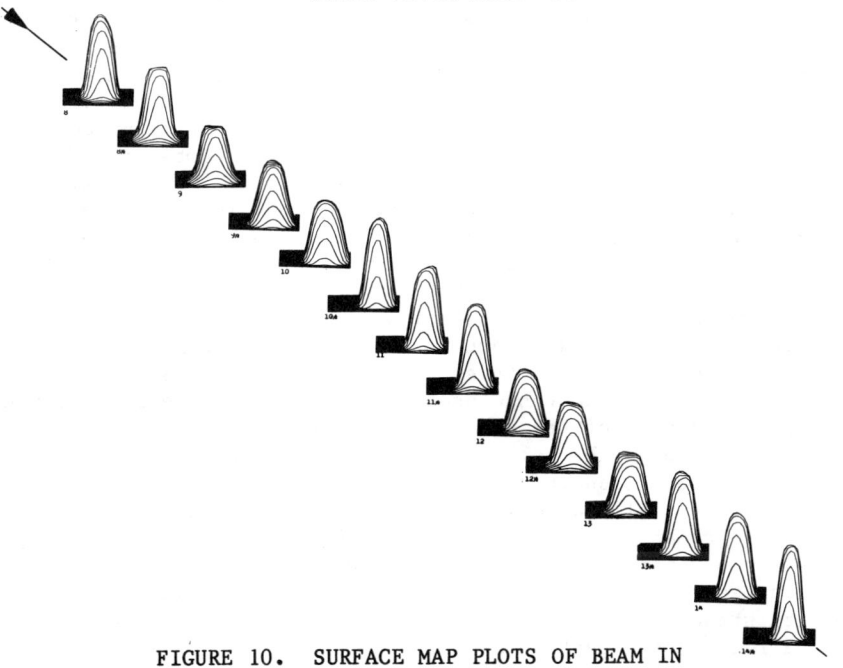

FIGURE 10. SURFACE MAP PLOTS OF BEAM IN NEXT SEVEN MAGNETS.

DISCUSSION

Stockli: Your beam profiles are very impressive, especially considering the very limited space inside a TWT. Could you sketch the mechanical arrangements for this probe?

Harper: I will try. Basically, the analyzer uses a long piece of metal tubing small enough in diameter to fit inside the PPM focussing structure with enough room to allow transverse movement. Welded across the end is a tungsten disc with a 0.002 inch diameter hole. Some of the slope at the edge of the plots is caused by the diameter of that hole because it obviously imposes a limit of the smallest size which we can resolve in terms of changes in the beam current density. We are trying to make one with a smaller diameter hole. Then behind the disc there is simply a long Faraday cage supported by insulators with a wire coming out for a connection. This is all inside the beam tunnel and can be moved back and forth.

Stockli: How thick is the tantalum aperture and how do you move it in all directions?

Harper: The plate is 0.006 to 0.007 inch thick, and it is tungsten. Be aware that, for these measurements, the beam has to be pulsed at a very short pulse length because of the rapid heating. It is not just a matter of conducting the heat away from there; it is the instantaneous heat. When you turn the beam on, the thermal time constant of the tungsten plate is such that in more than about 5 or 6 microseconds the plate will begin to melt. Therefore, these measurements were made with the beam current pulsed at a rate of about one pulse per second, with pulses one microsecond long. This is controlled by a small desktop computer which pulses the beam once every second and measures the collector current and stores the value in an array. The probe is moved in a regular pattern to scan across the beam and each number in the array represents the current density at one point. Then, after scanning the entire square pattern, we have 300 to 400 values stored in the array, and we can go back later and make these plots. We have an advantage in constructing this probe in that we have experience making things such as this. Essentially, this is like a very small TWT helix assembly. The only difference is that, instead of a helix, there is a small piece of metal tubing inside the probe shell and this is supported by four support insulators exactly like those used to support a TWT helix. If you look at this sketch of the end view you see the small piece of tubing (which forms the collector) supported in compression within the outer piece of square tubing by the four support insulators. Thus we have a small Faraday cage collector behind a hole in a plate, and it can be moved back and forth inside the beam tunnel. The design of this beam analyzer was the subject of a paper at the 1988 Microwave Power Tube Conference and there are slides which show its structure. Unfortunately, I did not bring them with me because I did not expect to discuss the beam analyzer design today.

Stockli: How did you scan X and Y?
Harper: There are linear motors made by the Daedal Company, which are controlled by the computer that move the probe.
Stockli: So it was controlled from the outside?
Harper: Yes. Without a detailed picture it is rather difficult to describe. A mechanical engineer worked for quite a few months on the detailed design of this structure. The basic choice which had to be made was whether to do the movement inside or outside of the vacuum. We originally started designing with the motors inside the vacuum and then abandoned that approach and went to a bellows arrangement with the motors outside the vacuum. The probe is mounted in a vertical position on a very firm baseplate and the gun and focussing structure moves. There is a very good reason for that. This probe is very long and very thin, and if you move it in a series of steps, the probe will vibrate, hence the choice of the concept of having the probe stationary and the structure move. Even then we had to program the computer to make a measurement (after a movement to a new position), make two more measurements and, if they are the same, then store that value in the array. If the collector current has not settled down, more measurements are taken until it has stopped varying at that position. Sometimes when you are measuring right at the edge of the beam a very small amount of vibration produces a large change in measured current. So, it may be necessary to make three or four measurements until the vibration stops. Then you know you have the correct value. When there is no vibration, the computer can take measurements at successive locations very quickly.
Levine: We tried something like this and had trouble with secondary electrons.
Harper: Yes. It is true that you must always worry about secondaries. Are there secondaries going from the target disc to the collector, or are they going the other way? We did not coat the disc. We made it out of pure tungsten and made it very clean. The secondary emission ratio of pure clean tungsten is fairly low. Then we operate them both at the same potential. The experimenter must always decide whether to make the collector positive or negative with respect to the target, which will have an effect on which way the slow secondaries go. We found very little difference, which indicates that there were few secondaries.
Becker: At what distance from the cathode did you really start your trajectories? I am a little bit anxious about this sharp corner of your Wehnelt electrode.
Harper: Let me go back to one of my computer plots.
Becker: I believe you either have the Pierce angle or aberration. You should take the computer simulation of the gun. If it is curved in the vicinity of the cathode, then there will be some influence downstream. But your outermost trajectory really looks linear.

Harper: What you are saying is, how do we handle the trajectories close to the cathode?
Becker: Yes.
Harper: In the computer code, they are launched perpendicular to the cathode surface and they are carried one mesh box out from the cathode before they are allowed to deflect. You have to do something of that sort. You can see the mesh box size by the tick marks along the cathode. Each tick is at one mesh box. Thus, each trajectory is carried a very short distance out from the cathode before it is allowed to be deflected by the electric field. I do not know if you are familiar with the Kierstein-Hornsby program or not.
Becker: Sure, we had that before.
Harper: This launching concept evolved from that program.
True: You might be able to cut down your secondary emission, if there is a problem, by carbonizing the tungsten to get tungsten carbide. It is very easy to do and tungsten carbide has a very low delta.
Harper: I do not believe we had a secondary emission problem in this probe, but it is a worthwhile comment. Thank you.

A NOVEL ELECTRON SOURCE FOR EBIS MACHINES

Richard True

Litton Systems, Electron Devices Division
San Carlos, California 94070

ABSTRACT

This paper describes a scheme for production of very dense rotating E-beams. Such beams appear well suited for use in electron beam ion source (EBIS) machines. The basic idea is to propagate a thick hollow beam (it can be solid) from a magnetron injection gun (MIG) through a magnetic field reversal followed by magnetic compression of the rotating beam. A small MIG systhesis program (TRMIG) was used in the preliminary design calculations and the paper will include descriptive remarks about this code. The paper includes the triple pole piece field reversal element of Scheitrum and True (1981) which has been shown to improve post-reversal beam quality.

GUN DESIGN

Table I includes a set of preliminary specifications for an EBIS machine capable of stripping the heaviest of atoms (1). In the table, multiplying cathode current density (20 a/cm^2) by magnetic compression (100) yields an objective beam current density of 2000 a/cm^2.

It is known that dc operation of a dispenser cathode at 20 a/cm^2 is beyond conventional cathode technology. It would be desirable to lower the level of cathode loading to be able to use industry standard dispenser cathodes which have proven to be rugged, reliable and capable of very long life.

One way to reduce cathode loading is to use a MIG as shown in Fig. 1 (2). In this gun, an axial magnetic field constrains the beam and prevents it from flowing over to the positive modulating anode (above the cathode). The beam compresses axially in this case and the current density of the ring beam (to the right) is substantially higher than the current density along the cathode (in Table II it is 5.71-8.51 times higher).

It should be mentioned that MIG guns can be made to generate solid beams in the limit by extending the cathode down to the axis. The standard hollow beam MIG, however, offers the potential advantage of ion injection and extraction from opposite ends of the EBIS (the non-emissive plug and cathode can have a hole). Further, reduction of heavy electronic charge along the axis might provide a quiet potential well into which the heavily stripped ions can collect.

Figure 2 presents a simulation of a gyrotron MIG gun (3) from code DEMEOS (4), (5) and Fig. 3 is a plot of the relaxation mesh on which the problem was solved. Figure 2 shows beam flow into the drift tube and illustrates magnetic beam compression. This MIG gun is configured to provide a beam having a high degree of rotation about a non-axis encircling guiding center (unlike the gun of Fig. 1).

Preliminary design of MIG guns can be accomplished using TRMIG (6). A sample printout from this code is presented in Fig. 2. It can be seen that microperveance, mod anode voltage, and cathode geometry (length, rear diameter, and angle) are entered. The code then calculates various quantities such as cathode current and loading (a/cm^2), magnetic field at the cathode (BHC), and inner and outer diameter of the hollow beam (see Fig. 1). Magnetic field in the drift tube (BDT) is calculated to provide a given size beam in the drift tube (adiabatic compression assumed). In the case of Fig. 4, multiplying area compression (498.3) by cathode loading (2.149 a/cm^2) yields a compressed beam current density of 1071.1 a/cm^2 (JB in Table II).

TRMIG also can be used to calculate coordinates of the electrodes (not shown). A correction for cylindrical geometry is included in the procedure. By entering different values of PHI, theoretical equipotential contours other than that corresponding to the mod anode can be obtained.

Table II summarizes four hypothetical MIG designs based on results from TRMIG. It can be seen that a net beam current density of roughly 1000-2000 a/cm^2 is possible from cathodes having a loading level of 1.43-6.08 a/cm^2, which makes it possible to use a dispenser cathode to do the job.

The magnetic field at the cathode is relatively high in these designs and radial beam convergence is moderate which serves to enhance beam stiffness and stability in the compressed beam region. It should be pointed out that a current density of roughly 2150 a/cm^2 in the drift tube does not represent an upper limit. If desirable, higher levels can be achieved using this basic approach.

MAGNETIC FIELD REVERSAL

It is known that when a beam passes through a magnetic field reversal (cusp) it is caused to rotate at the cyclotron frequency (see Fig. 5). Careful shaping of the field through the reversal is necessary to obtain a smooth post-reversal beam, whence this can be accomplished using a triple-pole-piece field reversal element (TPPFR) shown schematically in Fig. 5 (7).

Presented in Fig's 6 and 7 are a set of calculations which illustrate basic principles. At the exit of the field reversal (situated near the exit of the MIG), Fig. 6 shows that the beam possesses a small amount of rotational energy prior to beam compression. As the beam is compressed, the beam acquires a high degree of rotational energy at the expense of longitudinal. Fig. 6 also discloses that a higher applied voltage is necessary in the 2141.6 a/cm^2 design to prevent mirroring of the beam.

Figure 7 includes a relativistic correction in the calculation for the 2141.6 a/cm^2 case. In this case, the electrons cannot acquire as much rotational velocity as in the non-relativistic case and as a result, parameter alpha is reduced from 2 to approximately 1.5. Alpha values of 1.5 are known to provide stable beams from a focussing point of view.

It is realized that an electron cyclotron resonance (ECR) instability may occur in such system. Fortunately, there is a high degree of axial velocity shear in the thick hollow beam and it is anticipated that such an instability will not occur. If it does, it can be controlled by a suitable choice of design parameters including the use of lossy materials. On the other hand, it is interesting to contemplate what will happen if an ECR instability does occur within the beam. It is tempting to think that we might be able to have an EBIS and ECR source operating simultaneously in in one device.

The RBE appears advantageous insofar as the highly spinning beam in the drift tube tends to increase the collision probability per unit length within the device. This, in turn, could lead to an increase in the yield of heavily stripped ions. Or the trap region could be made shorter in the device. Shortening the interaction region has numerous advantages including a less expensive superconducting main magnet, easier alignment, and likely a quieter and more efficient beam.

Further, it is anticipated that the RBE will be able to be operated over a wide range of voltages and currents. The reason for this is the relatively high magnetic field in the beam formation region of the MIG which tends to control the shape of the beam. As a final point, there is a high degree of experimental flexibility in the proposed scheme of this paper. For example, coil currents in the TPPFR can be arranged so that there is no magnetic field reversal which will provide an essentially rotation-free beam in the interaction region of the EBIS.

REFERENCES

1. Courtesy of Robert W. Schmieder, EBIS TN-016 (1986), unpublished.
2. R. True, IEDM Tech. Dig., 436 (1983).
3. R. True, Fifteenth IEEE Intnl. Conf. on Plasma Science, (1988).
4. R. True, IEDM Tech. Dig., 257 (1975).
5. R. True, IEDM Tech. Dig., 173 (1978).
6. G. Jespersen, Litton Rep. L-5-63 (1963), unpublished.
7. G.P. Scheitrum and R. True, IEDM Tech. Dig., 332 (1981).

TABLE I

PRELIMINARY SPECIFICATIONS
(July, 1985)

Cathode voltage	50-300	kV
Cathode current density	10-20	A/cm2
Beam current	0.2-1.0	A
Electrostatic compression	1-3	x
Magnetic compression	30-100	x
Axial magnetic field	5-6	T
Magnet bore	12-15	cm
Trap length	80-100	cm
Vacuum in trap	10(-12)	Torr

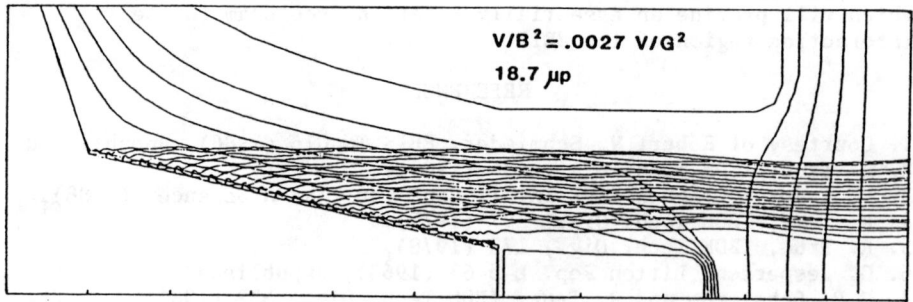

Fig. 1. Computer simulation of standard magnetron injection gun.

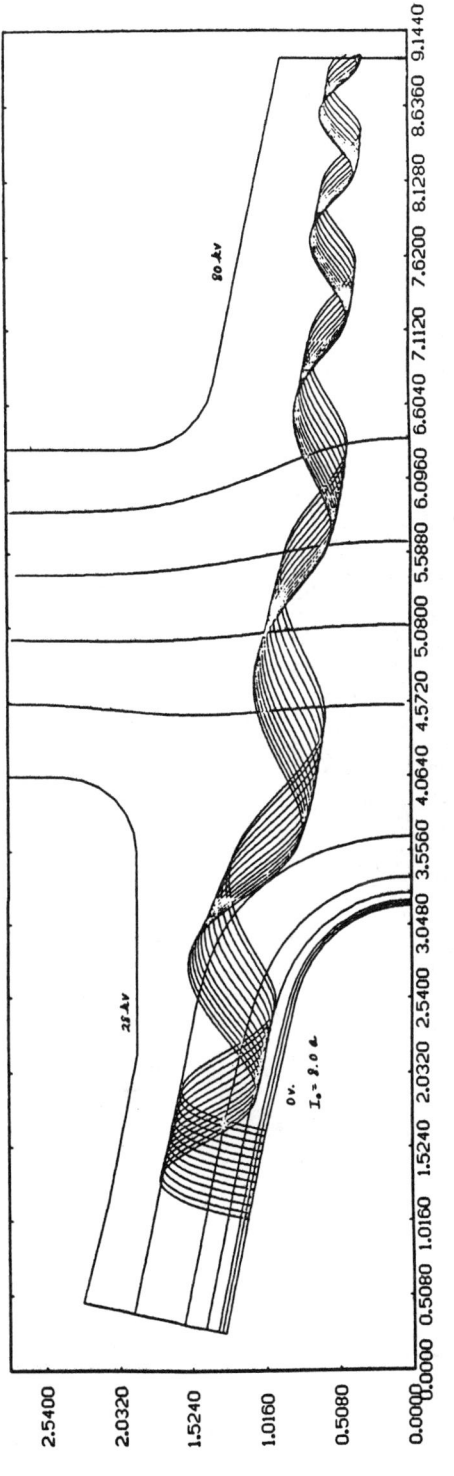

Fig. 2. Gyrotron gun simulation using deformable mesh code DEMEOS ($J=1.82$ a/cm^2 (temperature limited); $B_c=921.7$ G at $z=1.3335$ cm; $B=6279$ G at 9.2075 cm).

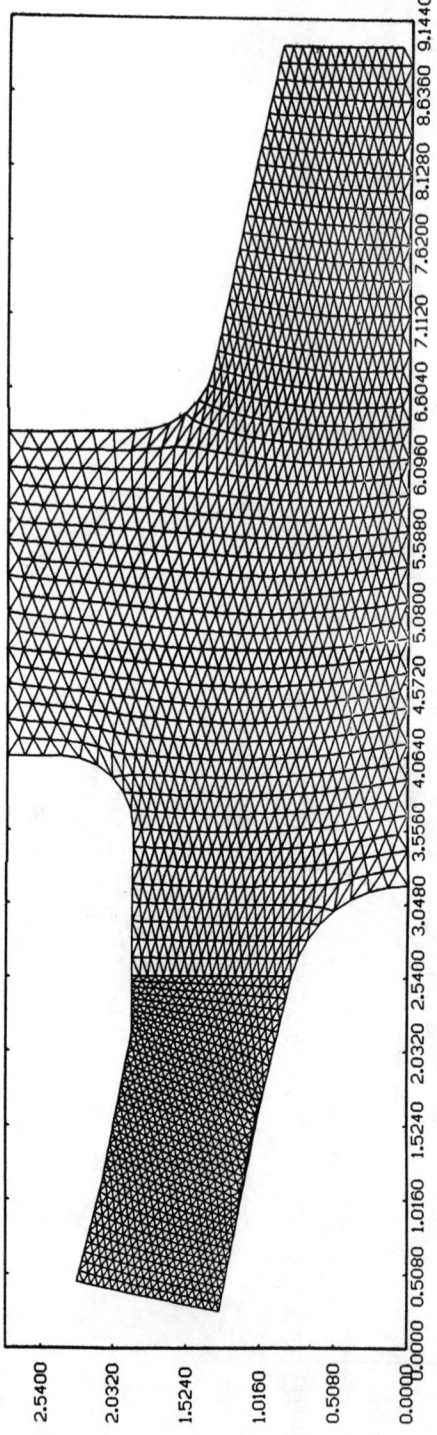

Fig. 3. Relaxation mesh used in DEMEOS for gyrotron gun problem. Problem rotated 90 degrees in logical space (cathode vertical and to the left). Computer rotates boundaries to real space position and sets up deformed mesh within them.

```
$ RUN TRMIG

TRMIG: MAGNETRON INJECTION GUN DESIGN
01-14-87, VERS 3, FORTRAN 77, MICROVAX II, R. TRUE
ENTER 1.0 FOR CM; 2.54 FOR INCHES
1
NOTE SMALL Z, SMALL Y BELOW IN CM OR INCHES

ENTER MICROPERVEANCE AND ANODE VOLTAGE
10, 2154
CATHODE: ENTER LENGTH, REAR DIA, ANGLE(DEG)
.508, .381, 10
FRONT DIA= 0.2018518
CATHODE CURRENT(A)= 0.9996973
CATHODE LOADING (A/CMSQ)=   2.149447

Z=  8.001626
N3=   6.270642

BHC(GAUSS)=   572.8898
Y=  7.958808
SMALL Y= 0.5052816
BEAM I.D.= 0.2032126
BEAM O.D.= 0.3808308
ENTER 1 FOR NEW MPERV, 0 TO CONT, 2 TO STOP
0

ENTER TUNNEL BEAM O.D.
.04076
BDT(GAUSS)=   50011.15
TUNNEL BEAM I.D.= 0.2174968E-01
AREA COMPRESSION=   498.3278
NORMALIZED POTENTIAL PHI=   28.36637

ALPHA=+9999 FOR NEW PHI; -9999 TO STOP
TYPICAL RANGE: -.1 LE ALPHA LE .75
ENTER ALPHA
-9999
FORTRAN STOP
```

Fig. 4. Sample printout from MIG design code.

TABLE II

MAGNETRON INJECTION GUN DESIGN SUMMARY

($P\mu = 10$, $I = 1$ AMP)

ANODE VOLTAGE (V)		2154	2154	4308	4308
CATHODE:					
	LENGTH (cm)	.508	.762	.508	.762
	d_c (rear) (cm)	.381	.381	.381	.381
	ANGLE (°)	10	6.7	10	6.7
	$L/\overline{D}$	1.74	2.61	1.74	2.61
	J_c (a/cm^2)	2.15	1.43	6.08	4.05
	BHC (G)	572.9	565.3	810.2	799.5
	V/B^2 (V/G)	.0066	.0067	.0066	.0067
	J_{B1} (a/cm^2)	12.3	12.2	34.7	34.4
	J_{B1}/J_c	5.71	8.51	5.71	8.49
DRIFT TUBE:					
	BDT (G)	50011	49995	49995	49991
	d_o (cm)	.0408	.0406	.0485	.0483
	d_i (cm)	.0217	.0215	.0259	.0256
	d_c/d_o	9.35	9.39	7.86	7.90
	J_B (a/cm^2)	1071.1	1076.3	2141.6	2152.3

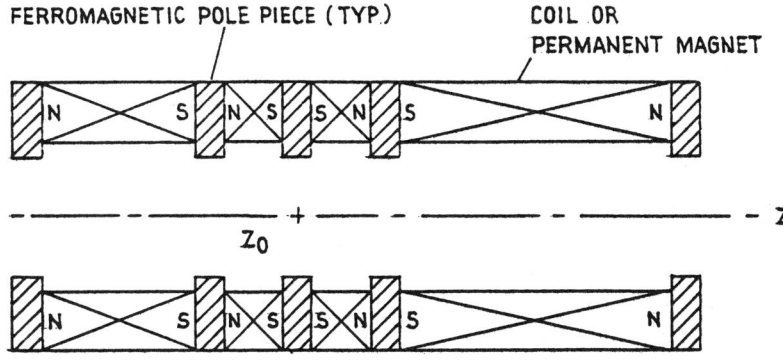

(a) TRIPLE POLE PIECE MAGNETIC FIELD REVERSAL ELEMENT

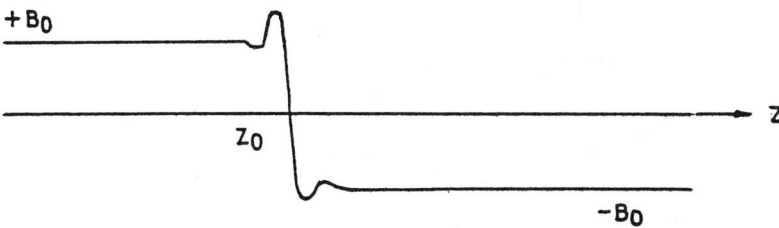

(b) MAGNETIC FLUX DENSITY THROUGH FIELD REVERSAL

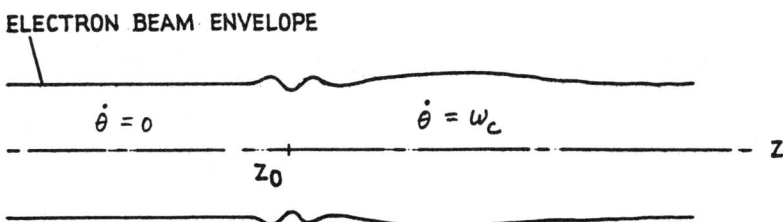

(c) BEHAVIOR OF BEAM PASSING THROUGH REVERSAL

Fig. 5. Behavior of beam in passing through triple pole piece magnetic field reversal.

FIG. 6. REVERSAL CALCULATIONS (NONRELATIVISTIC).

REVERSAL NEAR GUN $r_o = .1904$ cm $B_o = .0573$ T

$\omega_o = \eta B_o = 1.008 \; E + 10$ r/sec

$v_\perp = r_L \omega_o = r_o \omega_o = 1.919 \; E + 7$ m/sec

ENERGY (ev) $V = \dfrac{v^2}{2\eta}$ $V = V_\perp + V_z$

$V_\perp = 1.05$ kv

$V_z = 113.32$ kv

$V = 114.37$ kv   (APPLIED VOLTAGE)

$J_B = 1071.1$ a/cm^2 $r_o = .0204$ cm $B_o = 5T$

$\omega_o = 8.794 \; E + 11$ r/sec

$v_\perp = 1.794 \; E + 8$ m/sec

$v = 2.006 \; E + 8$ m/sec

$v_z = (v^2 - v_\perp^2)^{1/2} = 8.970 \; E + 7$ m/sec

$V_\perp = 91.49$ kv

$V_z = 22.87$ kv

$V = 114.37$ kv

$\alpha = \dfrac{v_\perp}{v_z} = 2$

BEAM SPINS UP IN ADIABATIC COMPRESSION REGION

FIG. 6. REVERSAL CALCULATIONS (CONTINUED).

J_B = 2141.6 a/cm^2 r_o = .0242 cm B_o = 5 T

$v_\perp$ = 2.128 E + 8 m/sec

LET α = 2 = $\dfrac{v_\perp}{v_z}$ v_z = $\dfrac{v_\perp}{2}$

$v^2 = v_\perp^2 + (\dfrac{v_\perp}{2})^2 = \dfrac{5}{4} v_\perp^2$

v = 2.379 E + 8 m/sec

v_z = 1.064 E + 8 m/sec

ENERGY BALANCE

$V_\perp$ = 128.75 kv

V_z = 32.19 kv

V = 160.94 kv

FIG. 7. REVERSAL CALCULATIONS (RELATIVISTIC).

J_B = 2141.6 a/cm^2 r_o = .0242 cm B_o = 5 T

V = 160.94 kv

$v = c \{ 1 - (\dfrac{E_o}{V+E_o})^2 \}^{\frac{1}{2}}$ = 1.947 E + 8 m/sec

E_o = 511.004 kv = REST MASS OF ELECTRON

$\beta = \dfrac{v}{c}$ = .649

$\gamma = (1 - \beta^2)^{-\frac{1}{2}}$ = 1.315

$\omega_o = \dfrac{\eta B_o}{\gamma}$ = 6.688 E + 11 r/sec

$v_\perp = r_o\, \omega_o$ = 1.618 E + 8 m/sec

$v_z = (v^2 - v_\perp^2)^{\frac{1}{2}}$ = 1.082 E + 8 m/sec

$\alpha = \dfrac{v_\perp}{v_z}$ = 1.496

DISCUSSION

<u>Faure</u>: How do you extract the ions from this beam?
<u>True</u>: Well, the ion extraction system, I presume, would be similar to what is being used now. In other words, the beam goes down into the drift region. You have your potentials and then you drop your well. The advantage to this device is that you can take the ions out of the collector or the gun because it is a hollow beam.

EXTRACTION OF A STEADY STATE ELECTRON BEAM FROM HCD PLASMAS FOR EBIS APPLICATIONS*

A. Hershcovitch, V. Kovarik, K. Prelec
AGS Department, Brookhaven National Laboratory
Associated Universities, Inc., Upton, NY 11973 USA

ABSTRACT

Experiments to extract high brightness electron beams from hollow cathode discharge plasmas are now in progress. A unique feature of these plasmas, which in principle can facilitate the extraction of large current low emittance electron beams, is the existence of a relatively high energy electron population with a very narrow energy spread. This electron population was identified in a self-extraction experiment, which yielded a 35 eV, 600 mA electron beam with parallel energy spread of less than 0.5 eV. Application of a very modest extraction voltage yielded a steady state extracted electron beam current of 6.5 A of which 5.7 A had a preacceleration parallel energy spread of no more than 0.25 eV. The end result of this endeavor would be an electron beam current of 6 A even though, preliminary results strongly suggest that much larger electron beam currents can be produced.

INTRODUCTION

Tentative long-range plans for the BNL heavy ion program call for the development of an electron beam ion source (EBIS) with an electron beam current of 6 A. Present day EBIS devices operate with electron beam current of less than 1 A. These beams operate steady state (or with very long pulses), and they have low emittance. Because of the later requirement, no serious considerations were given to electrons extracted from plasmas. Plasma cathodes can easily yield multi-amperes of electrons, however, at a very high emittance. Therefore, this type of electron beams have very limited use (mostly used to excite molecules in powerful gas lasers). For almost all other applications, electron beams are injected from guns utilizing either thermionic cathodes, or photocathodes. These surface emitted electrons have energy spreads of about 0.5 eV or less (as compared to plasma electrons with energy spreads of a few eV at least). But, space charge problems pose severe limitations on the total steady state (or long pulse) electron current that can be extracted from the various surface emitters. Other surface emitters, e.g., semi-conductor photoemitters have an inherent limitation to very short pulses only at high current outputs.

*Work performed under the auspices of the U.S. Department of Energy.

Hollow cathode discharge (HCD) plasmas have two electron populations: the bulk electrons with a density of up to (and/or slightly above) 10^{14} cm^{-3} with a temperature of several eV, and fast primary electrons having a density of about 10^{11} cm^{-3} or less with an energy corresponding to the cathode potential and a thermal spread which is close to the cathode temperature of 0.17 eV. (In order to avoid confusion in terminology, it is important to note that most European researchers refer to this type of discharges as hollow cathode arcs HCA). Existence of fast electrons in HCD plasmas was either postulated[1] to explain properties of external plasmas in HCD's, or accepted as a possibility[2] that a minute portion (2%) of electrons emitted from a hollow cathode surface could survive their "trip" to the anode without loss of energy. Some theoretical models[3] even attempt to predict HCD operation based on full thermalization of the surface emitted electrons.

To date, there is no theory which can fully explain HCD operation. As a result, there is no theory which could predict whether the electron population is a Maxwellian with a long superthermal tail, a pure Maxwellian, or whether there are two distinct electron populations.

We gambled on the existence of two distinct groups of electrons. Initial results indicate that at higher operating pressures the electron population is most likely thermal with a long superthermal tail, while at lower operating background pressures there are two electron populations, one of which has a narrow energy spread. Self extraction of this population yielded a beam current of 600 mA with a parallel energy spread of less than 0.5 eV. Application of an extraction voltage of only 100 V yielded a total electron current of 6.5 A of which 5.7 A were surface emitted electrons.

EXPERIMENTAL SET-UP AND RESULTS

In principle, this scheme is based on extracting the energetic electrons only. Schematically, the experimental configuration[4] relied on repelling the bulk (thermal) electrons with a negatively biased grid and further selection was to be done with a magnetic mirror in case the discharge would not operate properly without a magnetic field. Fortunately, we obtained very encouraging results with unmagnetized discharges. After numerous experiments and changes, our experimental configuration evolved to that which is shown in Figure 1. The plasma is generated in a 3mm Ta hollow cathode.

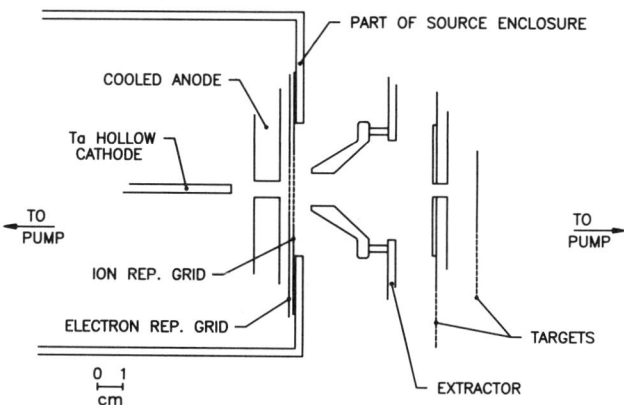

Fig. 1. Schematic of the electron selection and extraction system.

The length of the discharge, i.e., the cathode-to-anode distance, was 9mm. This distance should be made as short as practically possible to minimize velocity space relaxation due to multiple small angle scattering of the fast electrons by the plasma particles. The hollow anode is followed by a negatively biased grid designed to repell the bulk electrons. Next, we have a positively biased ion repelling grid, which proved necessary in self-extraction experiments for preventing ions from reaching the target. The whole source is enclosed in a differentially pumped cylinder which can be negatively biased. The extractor and target are at ground potential.

A large number of measurements failed to show the existence of a distinct fast electron population with a narrow energy spread for "normal" modes of operation[5] with a background pressure in the 10^{-4} Torr range. If this population existed, the electron current on target would have shown a very sharp drop or rise as the electron repelling grid bias was increased or decreased. This sharp transition was expected at a grid bias close to the arc voltage (the cathode potential). With the background pressure at about

10^{-4} Torr (or even as low as 7.2×10^{-5} Torr), target-current repelling-grid-voltage characteristics showed a gradual response. This was indicative of either a larger than expected thermal spread, or of the existence of a velocity distribution with a very long tail. Finally, when the background pressure in the discharge was reduced to 1.8×10^{-5} Torr, the desired behavior was observed. Figure 2 is a plot (on an x-y recorder) of the target current as a function of the bias on the electron repelling grid. The very sharp rise extends for 600 mA (off scale in Fig. 2). The resolution is about 0.5 V, hence the parallel energy spread is about or less than 0.5 eV. Furthermore, this sharp transition did occur at an electron-repelling grid bias of -33 V which roughly equalled the arc voltage. The plasma current was 18 A and the bias on the ion repelling grid was +42 V.

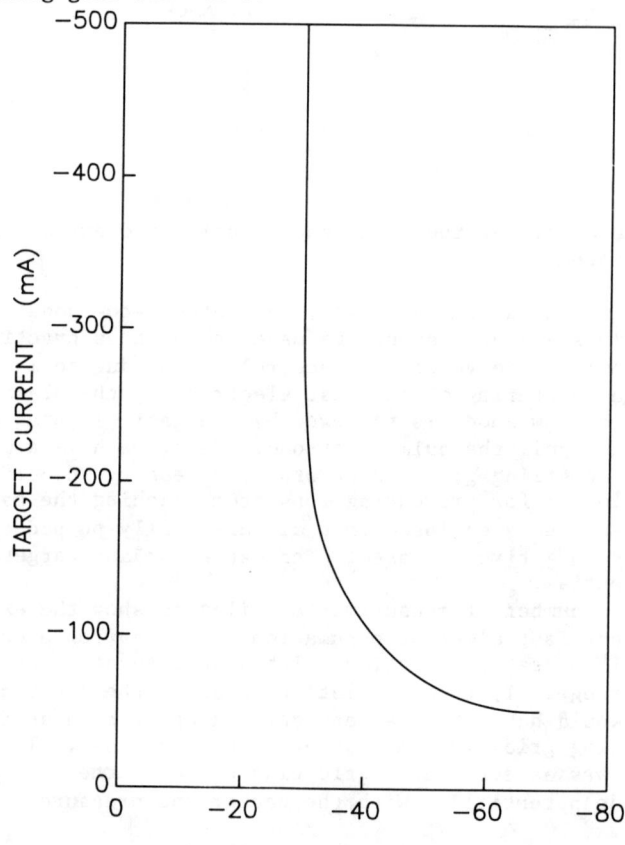

Fig. 2. Self-extracted electron current versus bias on the electron repelling grid.

Initial high voltage extraction results were performed in an experimental configuration identical to that shown in Figure 1 except for the addition of a grounding screen at the end of the source enclosure to prevent arcing between the source (floating at a negative voltage) and the pump (at ground potential). The inclusion of the screen reduced pumping which forced operation at twice the optimum pressure, for which the results from Figure 2 were obtained. Up to 1.2 A of electrons were extracted at 2.5 kV. However, the dependence of the extracted electron current on the electron repelling grid bias did not exhibit as sharp an increase/decrease as shown in Figure 2 (most likely due to the operation at non-optimal pressure).

To rectify the situation (without having to acquire expensive pumps), a slight modification of the source was made to allow for HCD operation at lower pressure. This new configuration is shown in Figure 3. The only substantial difference is that the hollow cathode is now inserted inside the hollow anode. This change facilitated HCD operation with 1.1×10^{-5} Torr pressure in discharge chamber. Best results to date are shown in Figure 4. Up to 6.5 A of total electron current was extracted with the source enclosure floating at only -100 volts, total gas flow was 0.22 ℓ/min at STP. HCD operation was in argon with an arc voltage of 36.8 V. From Figure 4, the fast electron population (the sharp increase) seems to account for 5.7 A of the extracted electron current. Based on our visual monitoring of the gauges (Voltmeter, and Amperemeter) measuring the electron current, the energy spread

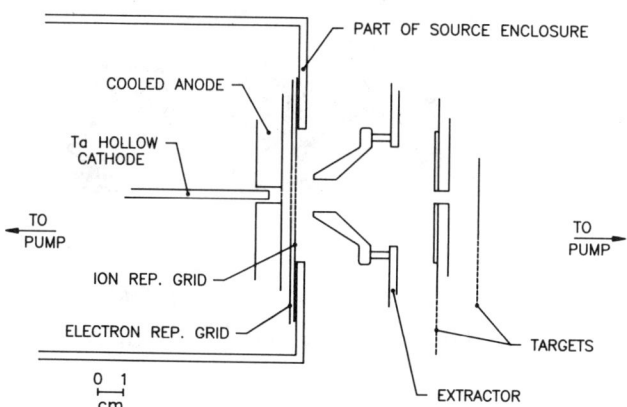

Fig. 3. Modified experimental apparatus.

of the fast electron population is 0.23 eV, or lower. The most spectacular result, however, is that the plasma current in this case was only 7 A. And, as it was lowered, so was the total extracted current (roughly proportionally).

Results shown in Figure 4 coupled with the fact that the total arc current was only 7 A strongly suggest the existence of a fast electron population with a very narrow energy spread (at least in the parallel direction). Since this group of electrons accounts for 5.7 A, which is over 80% of the total arc current, the results support theories and experiments asserting that the bulk external plasma existing in a hollow cathode discharge consists of almost fully thermalized ions and electrons drifting slowly away from the cathode. Schram and his group[6] have shown in a series of experiments that ions do indeed drift against the electric field towards the anode; a further enhancement of the assertion that the cathode emitted electron carry most of the arc current.

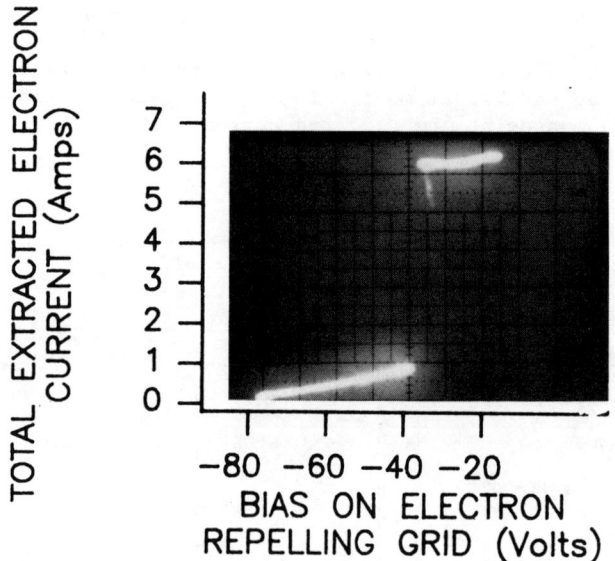

Fig. 4. Total extracted electron current vs bias on electron repelling grid. Bias voltage was swept manually.

FUTURE PLANS

Initial results identifying a fast electron population with a narrow parallel energy spread, which accounts for the bulk of the arc current is extremely encouraging. Since arc currents of up to 100 A have been obtained with 3 mm Ta HCD's (and up to 1 kA with larger cathodes), generation of several 10's of Amperes of steady state electron beams with low emittance may be possible.

Our next set of experiments (with smaller cathodes and lower current discharges) is designed to measure carefully the parallel energy distribution of the electrons. Following these measurements, an E x B filter to separate fast electrons from the Maxwelliam tail will be built. With the filter in place, we plan to measure emittance of an electron beam consisting of the extracted fast electron population. If results meet requirements for an EBIS, experiments with long life cathodes will be performed.

REFERENCES

1. L.M. Lidsky, S.D. Rothleder, D.J. Rose, S. Yoshikawa, C. Michelson, and J. Mackin, J. Appl. Phys., 33, 2490 (1962).
2. J-L Delcroix and A.R. Trindade, Advances in Electronic and Electron Physics, 35, 87 (1974); Academic Press, New York, L. Martin, editor.
3. A. Lorente-Arcas, Plasma Phys., 14, 1651 (1972).
4. A. Hershcovitch, V.J. Kovarik and K. Prelec, Bull. APS, 32, 1788 (1987).
5. Our group at BNL has had years of experience using hollow cathode discharges for a number of purposes, i.e., Rev. Sci. Instr. 52, 1459 (1981), Rev. Sci. Insr. 53, 819 (1982), Rev. Sci. Instr. 54, 328 (1983), Rev. Sci. Instr. 55, 8 (1984), Rev. Sci. Instr. 55, 1744 (1984), and Rev. Sci. Instr. 57, 827 (1986).
6. J.M.M.J. Vogels, J.C.M. deHaas, D.C. Schram, and A. Lunk, J. App. Phys. 59, 71 (1986).

RELATED TECHNOLOGIES;

PRIMARY IONS

MAGNETIC PRECISION ALIGNMENT OF A LONG HORIZONTAL ULTRA-STRAIGHT SOLENOID

Martin P. Stockli, C.L. Cocke, J.A. Good† and P. Wilkins†
J.R. Macdonald Laboratory, Department of Physics
Kansas State University, Manhattan, KS 66506, USA

† Cryogenic Consultants Ltd, Metrostore Building
231 The Vale, London W3 7QS, Great Britain

ABSTRACT

We present a magnetic alignment method for long horizontal ultra-straight solenoids based on the rotating Hall probe technique. We demonstrate how the sag can be eliminated, which otherwise would cause serious alignment errors. Only a flat Hall probe with a 4½ digit controller and two standard ball-bearings are needed, which makes our method very inexpensive. It requires a position accuracy which can be easily achieved with a ruler and an angular precision which can be eyeballed. We give detailed instructions for our simple, straightforward method which allows one to align a solenoid very accurately in less than one hour. Our analysis and our results indicate that our solenoid is aligned within two-hundredths of a millimeter.

INTRODUCTION

There are many good reasons to strive for a perfectly straight solenoid in an Electron Beam Ion Source.[1,2] A hypothetical straight solenoid has one field line which is perfectly straight, - the magnetic axis. In an EBIS it is important that the magnetic axis is straight over the whole length, which includes the fringing fields, because this field line guides the electron beam through the solenoid starting out in the almost zero field on the cathode, transversing the high field inside the solenoid and ending in the collector, which is on the other side in a very low field level. We wound our solenoid, which is shown in Fig. 1, with a patented winding technique[3] which produces straight, uniform helical and coaxial wire layers and essentially eliminates the propagation of wire positioning errors, and therefore guarantees superior straightness. Straightness should not be confused with homogeneity. The advances of magnetic resonance imaging brought great improvements in the achieved homogeneity of magnetic fields. However, this high level homogeneity is mostly achieved by adjusting various correction windings, which improve the homogeneity in a very limited area, but in general degrade the homogeneity in the rest of the space. Naturally, such solenoids do not have very straight fringing fields and therefore may be a mixed blessing for an EBIS. Fig. 2 shows a typical solenoidal field. Since the field line B, which goes through the origin, is not perfectly aligned with the z-axis, we find a small transverse field component B_t, or its

© 1989 American Institute of Physics

carthesian components B_x and B_y. If we measure the field off axis on a radius r, we find in addition a radial component B_r, if the field is not homogeneous. The relation between the radial field component and the inhomogeneity can be deduced from the conservation of the magnetic flux: $-2 \cdot B_r/r = dB/dz$. The angle between the reference axis z and the magnetic field line crossing the z axis at the position z, $B_t/B_z(z)$, is a local quantity, which occasionally is called straightness. However, the most common and reasonable definition of straightness is the maximum absolute angle max$<B_t/B_z(z)>$ encountered in a certain range on the axis. This definition is a useful and meaningful measure of the "straightness" quality inside the solenoid, but becomes questionable in the fringing fields, because the axial component B_z decreases rapidly. It seems to be more adequate to specify in the fringing fields the maximum transverse component. From the point of view of an electron, it would be reasonable to specify the maximum deviation of a field line from the optical axis,[4] but this quantity is rather cumbersome to evaluate.[5] However, field mapping or quality control of such a solenoid exceeds the scope of this paper and will be reported elsewhere.[5] We will restrict the discussion to the alignment of a straight solenoid to a given reference axis.

THE SOLENOID

Our 5 tesla superconducting solenoid is 1 meter long and has an 8 cm cold bore. A detailed description is given in another report;[6]

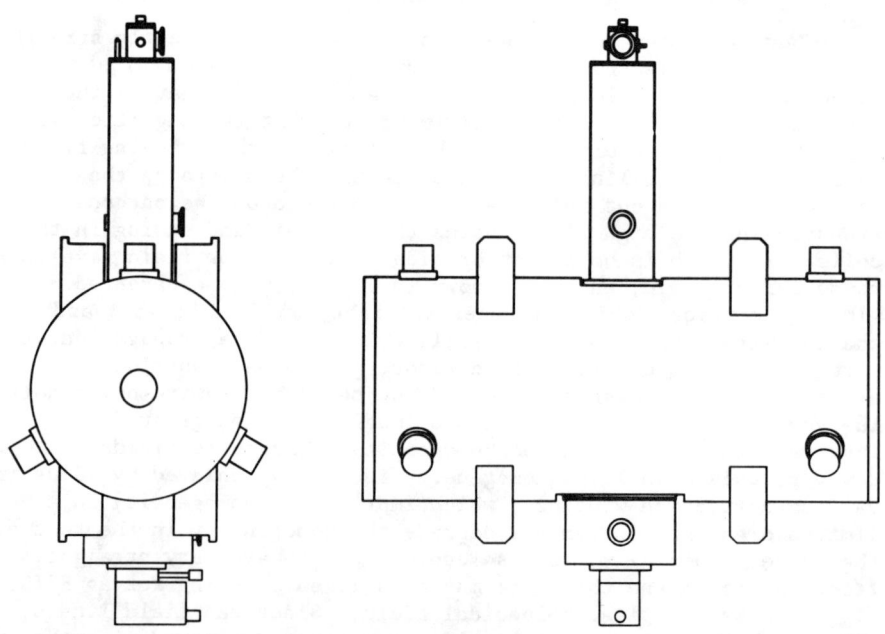

Figure 1: End view and side view of our 5 tesla ultra straight CRYEBISsolenoid. The three large knobs at each end allow one to align the solenoid inside its surrounding magnetic shield.

here we restrict the discussion to its alignment system. The
solenoid is mounted inside a vacuum vessel, shown in Fig. 1, which
is made from low carbon steel which also serves as a magnetic
shield. Therefore, we had to align the axis of the solenoid with
the axis of the shield. We made sure that the axis of the shield
goes through the center of the axial openings on both ends of the
shield,[4,7] which we used for alignment. The solenoid itself is
suspended from the shield with 6 axial and 6 radial titanium rods,
which have a high strength but a low thermal conductivity. The
radial rods are attached to the 3 large knobs at each end shown in
Fig. 1. They allow for moving the solenoid within the shield during
alignment. The radial rods are only 58 mm away from the end of the
solenoid in the axial direction, which gives approximately a
one-to-one translation scale between the position of the solenoid
end and the position of the rods. We added dials with 0.01 mm
divisions and found the position of the solenoid to be reproducible
within our test accuracy of one- to two-hundredths of a millimeter,
if all the rods are under sufficient tension. We would like to
explain briefly the simplicity of a 3 rod system, because of the
persisting discussions over the 3 versus 4 rod systems.[8] The main
argument for the 4 rod system is that one can move the system easily
in either a vertical or horizontal direction, without changing the
tension in the system, by moving the opposing rods the same amount.
However, having 3 rods makes very little difference. For vertical
displacement we move the vertical rod for the full amount, but the
two lower rods only for half the required amount (sin(30)=.5). For
horizontal movements we leave the vertical rod, but move both lower
rods in the same direction. The solenoid will move approximately
15% more than the rods were moved (1/cos(30)=1.15), and that is
quite easy to calculate for Americans, who are used to figuring the
tip every time they pay in a restaurant. And those, who do not
depend on horizontal or vertical orientations, can move easily
parallel or perpendicular to any of the three rods, which gives a
total of 6 axes. In addition, we should not forget the amount of
liquid helium one can save by omitting 2 unnecessary rods.

ROTATING HALL PROBE TECHNIQUE

The rotating Hall probe technique can be used to measure a
variety of axial symmetric fields,[9] however, we will discuss here
only the technique used to measure transverse components of a
solenoidal field. Fig. 3 shows the principle: A flat Hall probe is
rotated around the reference axis z and mounted normal to its radius
r. In this way the Hall probe will ignore the main component B_z in
the direction of the reference axis and only measure the field
component in the direction of the radius r. That component is
composed of a constant contribution from the radial component B_r and
a contribution from the transverse component, which varies with the
angle between the Hall probe and the transverse component B_t. For
simplicity, we assumed in Fig. 3 that the transverse component is
vertical, but it could be under any angle ϕ. When the Hall probe is
rotated we measure a sinusoidal Hall voltage U_H as a function of the
rotation angle ϕ, as shown at the bottom of Fig. 3. The phase and

the amplitude correspond to the orientation and size of the
transverse component B_t. The DC offset corresponds to the radial
component B_r if the probe is perfectly parallel to the axis of
rotation. A misalignment between the two will cause an additional
DC-offset proportional to the misalignment and B_z. Hence, even if
one mounts the Hall probe directly on the axis (r=0), one is still
likely to find a DC offset. For certain tasks it can be an
advantage to mount the probe off axis, because it would allow one to
determine the radial component B_r and hence the inhomogeneity dB/dz
without a differential measurement or a differential probe, although
a differential axial probe would be simpler. If one mounts the Hall
probe far away from the axis, the large radial component B_r can cause
a DC offset which is large enough to reduce the sensitivity for the
small variation caused by the transverse component. We compromised
and mounted the flat Hall probe 8 mm off the axis, which left enough
space to mount an axial probe on the axis, which surveyed the axial
component simultaneously. But since in this paper we are only
interested in the transverse component which is fully determined by
the amplitude and phase of the Hall voltage, we do not have to worry
about the various DC offsets, as long as they are not too large. On
the other hand, it is important that the probe is kept in a very
well defined axial position and is rotated around a well defined
axis. Otherwise, if the probe is rotated in an inhomogenous field
region without adequate constraints, the measured radial field
component varies depending on the actual axial and radial position,
which then could be falsely interpreted as a part of the transverse
field component. If the rotating Hall probe is used in high fields,
one has to consider the transverse Hall effect because it is

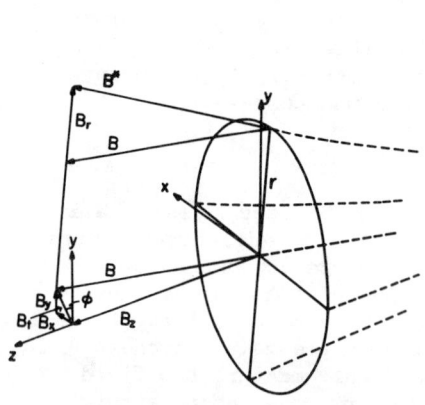
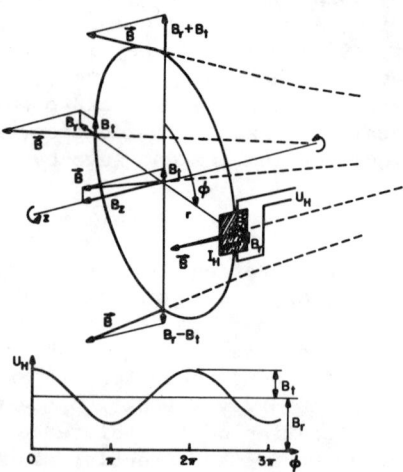

Figure 2: The slightly diverging solenoidal field (dashed field lines) is misaligned in the system of reference x,y,z or r,φ,z, giving rise to the transverse component B_t. B_r is the radial component which one finds off-axis in an inhomogeneous region.

Figure 3: A flat Hall probe rotated around the axis of reference z detects a transverse field component, caused by a misalignment between magnetic axis and z. The bottom part shows the Hall voltage as a function of the angle of rotation.

proportional to the square of the field component in the plane of the Hall probe.[10] We will align the solenoid until the Hall voltage is independent of the angle of rotation, which eliminates the transverse component. In this final position the angle between the Hall current and field component in the plane of the Hall probe is independent of the angle of rotation of the probe, and therefore the transverse Hall effect is reduced to a DC-component, which we can ignore.[11]

THE HALL PROBE SETUP

Fig. 4 shows a schematic cross section of our setup of the two Hall probes. A thermally insulated tube with a 38 mm inner diameter was mounted through the cold bore of the solenoid to provide atmospheric access for the probes during the operation of the solenoid. On each side a double flange provided a vacuum tight seal between the warm bore tube and the end shields. A smaller tube with an outer and inner diameter of 31.8 and 25.4 mm, respectively, was chosen to serve as a revolving tube. Its outer diameter at each end was reduced to mount ball-bearings with a 30 mm inner diameter, one end being press-fitted to provide a precise axial location, the other having a less tight fit for installation. Each bearing was then mounted inside a ring flange which was bolted onto the sealing flange. That arrangement allowed the inner tube to spin freely inside the warm bore and to provide revolving support for the probe carrier which could be moved inside the tube to any axial position. The revolving tube extruded only 37 mm from the right bearing but 234 mm from the left bearing, which allowed us to map the whole length of the solenoid in a single pass of the probe carrier without turning it around. The probe carrier was centered inside the revolving tube with an O-ring at each end, 246.4 mm apart. The O-ring grooves were designed to compress the O-ring cross sections by 10% in the radial direction to give a well-centering seat without having excessive friction when the probe carrier was pushed through the tube to adjust the axial position. Electrical tape pressed the flat Hall probe[12] against a flat reference surface on the side of the probe carrier, 7.4 mm away from the axis. An undersized groove and electrical tape provided cable stress relief. The first measurements showed that the flat Hall probe was tilted inside its thin epoxy matrix by 1.1 degrees. Therefore, we had to shim the reference

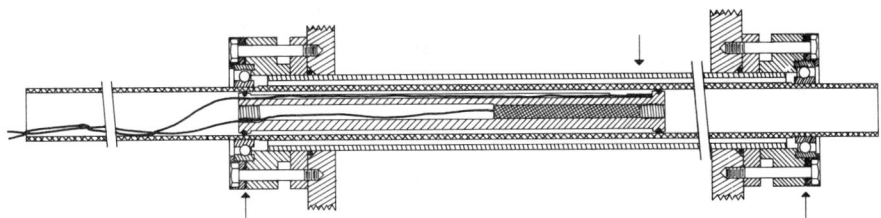

Figure 4: Cross section through the warm bore, the revolving tube (crosshatched) and the probe carrier. The sensor location of the axial (crosshatched) and the flat (black) Hall probe are indicated with the arrow above. The black filled circles represent O-rings. The lower arrows indicate the location bearings, which support the revolving tube.

surface on the probe carrier by that amount to reduce the DC component in the homogenous field region. The probe carrier had an 8 mm center bore to accommodate an axial Hall probe,[12] which had its sensor right at the tip. Several nylon screws provided cable stress relief and kept the probe tightly in the same axial position as the flat probe. The carrier was pushed in the proper axial position with a thin-wall aluminum tube, which had inch marks for reference. This tube was removed completely for every single measurement to minimize disturbance of any kind. At the beginning we used special precision bearings,[13] which were centered and retained with four set screws. Special precision bearings have typically only 0.01 mm radial and 0.1 mm axial play and so should provide a well defined center of rotation and axial position. However, the first measurements were reproducible only if one adhered to very strict and complex measuring procedures and the results did not always look like a harmonic function, as one can see in the upper row of Fig. 5. After some experiments it became clear that the tight bearings torqued the revolving tube in certain orientations. A special precision bearing of this size has typically only 0.16 degrees of angular play. If the mount of the inner- or the outer- race or the

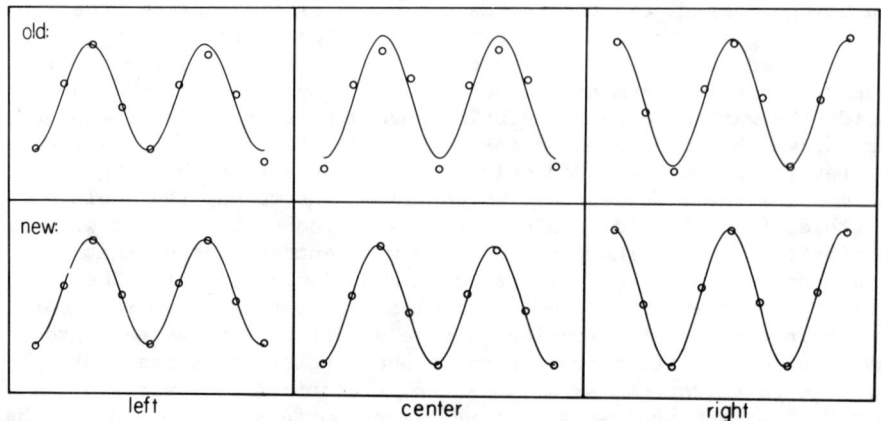

Figure 5: A sinus function was fitted to the rotating Hall probe measurements, which were taken 4 mm outside from the "left" end of the solenoid, in the "center" of the solenoid and 40 mm inside from the "right" end of the solenoid. All tranverse components were caused by the sagging axis of rotation. One notes in the upper row strong deviations from the fit, especially in the center, which is most sensitive to a torque created by a tight misaligned bearing. The data in the lower row were taken with the new torque-free bearings and show essentially a perfect agreement with the fitted function.

Figure 6: A torque-free bearing was made by cutting away the side segments and lifting the top segment of the outer race, with both remaining segments mounted in the dashed groove.

straightness of the revolving tube are not within this limit, the
bearing will in certain orientations show friction and produce a
torque on the revolving tube, which then bends and dislocates the
probe. After realizing that one needs a support, which is very
tight at the bottom but loose on the top, to allow for angular
misalignments, we considered several solutions. We were not sure
whether the standard precision bearing[14] with a typical radial play
of 0.02 mm and a typical angular play of 0.25 degrees would be
sufficient. Self-aligning bearings are designed for angular
misalignments, but their restoring force would still produce a small
torque. Finally, to save time and money we produced a simple
torque-free bearing mount as shown in Figures 4 and 6. We took two
off-the-shelf standard precision bearings[15] and cut the outer race in
four equal segments. We designed new ring flanges with grooves to
mount the segments of the outer race. The outer radius of this
groove was 0.3 mm larger than the outer bearing radius, but the
center was 0.3 mm above the flange center, so that the center of the
bearing remained in the center of the flange. The groove was 0.1 mm
wider than the outer race so that the segments fit in the groove
without increasing their original radius. Only the top and the
bottom segments of the race were mounted. Now when the revolving
tube was rotated, only the lowest two or three balls were in contact
with the outer race. The tight tolerances of the inner race, the
balls and the lower outer race segment,[14] still at its original
radius, provided a very accurate axial position and center of
rotation. If one moved the probe carrier inside the revolving tube,
it moved upwards, until the few upper balls hit the top quarter of
the outer race, which then provided a torque-free resistance to the
thrust load. After such a positioning, it was indicated we should
turn the revolving tube one turn back and forth, so that the balls
had a chance to settle in the center position of the outer race.
The substantial improvements can be seen in the lower row of Fig. 5.
We wish to point out that the optical axis is defined by the center
of the inner race of the two bearings and therefore the accuracy of
the optical axis is determined by the jitter of the center of the
inner race of the bearings, which normally cannot exceed the radial
play.

THE SAG AND ITS ELIMINATION

One of the inevitable problems with horizontal earthbound
structures is the sag caused by the gravity of our mother planet.
However, with the proper choice of materials and dimensions, the sag
can be kept to acceptable limits. For example the maximum
deflection in the middle of the cold bore of our solenoid is
approximately 0.02 mm, which should be acceptable for an EBIS.[1] The
real problem comes here, as often, with the probes because of the
limitations of the access space. In our case the rather small
revolving tube has only two locations of support which are 1511 mm
apart, which causes a sustantial sag. It is important to realize
that the revolving tube rotates around its sagging axis, and not
around the optical axis. But sag is well known and engineering type
formulas are readily available.[16] All the deflections of beams with

uniform cross-sections, which is an acceptable assumption in our case, are indirectly proportional to the modulus of elasticity and the moment of inertia. The moment of inertia can be easily and accurately calculated from the geometrical dimensions.[16] The modulus of elasticity is approximately 70 GPa for aluminum, but is not accurately known, because it can depend on the alloy, temper, and treatment.[17] Therefore the revolving tube was set up on a bench, loaded with different weights and the changes in deflection in the center were monitored with a dial gauge. This simple measurement determined the modulus of elasticity to be within 1% of 73.1 GPa. With these two numbers, one can predict all the deflections quite accurately. For example the dash-dotted curve in Fig. 7 shows the actual sag of the revolving tube by itself, which deflects up to 0.23 mm below the optical axis. The shape of the revolving tube

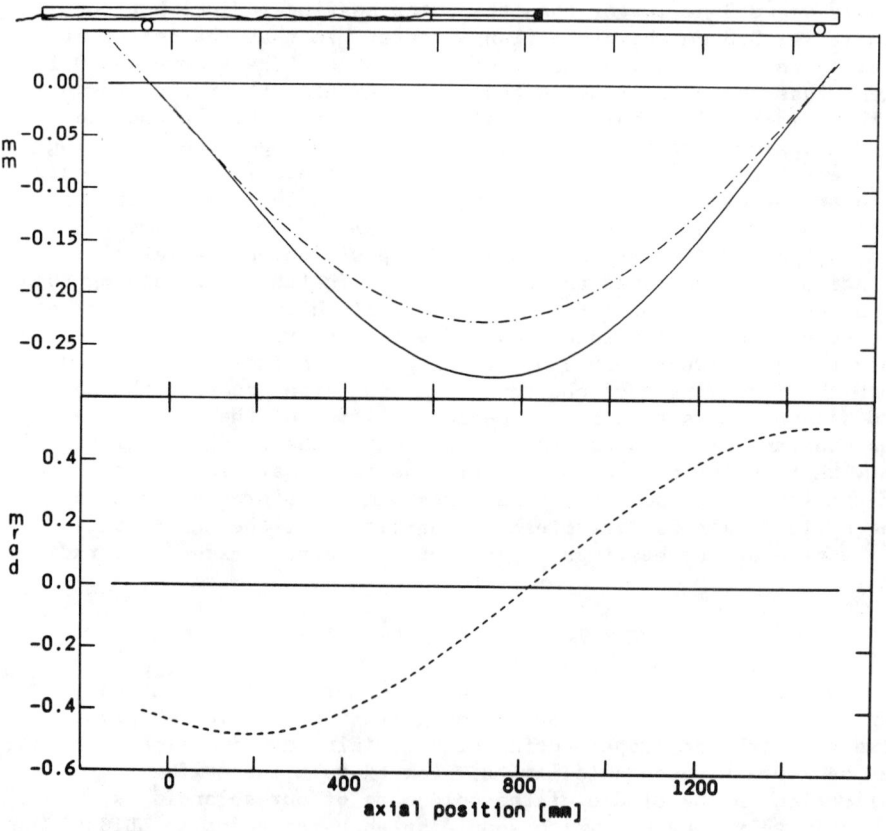

Figure 7: The dash-dotted curve shows the sag of the revolving tube by itself. The full curve shows the dislocation of the center of rotatation and the dashed curve the angle between the axis of rotation and the optical axis encountered by our rotating Hall probe for the different axial positions. The schematic on top of the figure indicates the locations of the revolving tube supported by the two bearings (open circles), the probe carrier, indicated by the the centering O-rings, and the sensor location (filled circle). The indicated position is used in the alignment procedure to put the solenoid parallel to the optical axis.

loaded with the probe carrier in a certain location can be easily calculated by using the sum of a few sag formulas, but that is not what the full and dash curves in the Figures 7-9 show, although it would look quite similar. We are more interested in the sag effects encountered by the rotating probe. The probe carrier is very stiff and short, and has a maximum sag of 0.8 μ in the center between the two centering O-rings, hence we assumed the probe carrier to be straight. With this assumption the location of the center of rotation and the slope of the axis of rotation can be evaluated easily from the deflections of the revolving tube at the locations of the two centering O-rings. But for each axial location of the probe carrier, one has to consider the corresponding position of its weight (82.1 g in both O-ring locations, which includes the weight of the probes) and the changing length of the cable (0.04 g/mm for both), which all have to be supported by the revolving tube. Therefore, we developed a computer code, which also allows us to take into account the unequal overhang and the reduced outer diameter at the end of the revolving tube. The code allows us in addition to add point loads or uniformly distributed loads. This code was used to produce Figures 7-9, in which the full curves show the displacement of the center of rotation and the dashed curves show the angle between the axis of rotation and the optical axis, which are encountered by the rotating probe in the various axial locations. One can see in Fig. 7 that the probe carrier adds only 23% to the maximum sag and much less at the ends when the probe carrier is close to one of the supports. The figure also shows that the axis of rotation is tilted up to 0.5 mrad, except when the probe is located at 821.4 mm where the axis of rotation is perfectly parallel to, but 0.274 mm below the optical axis. This position is

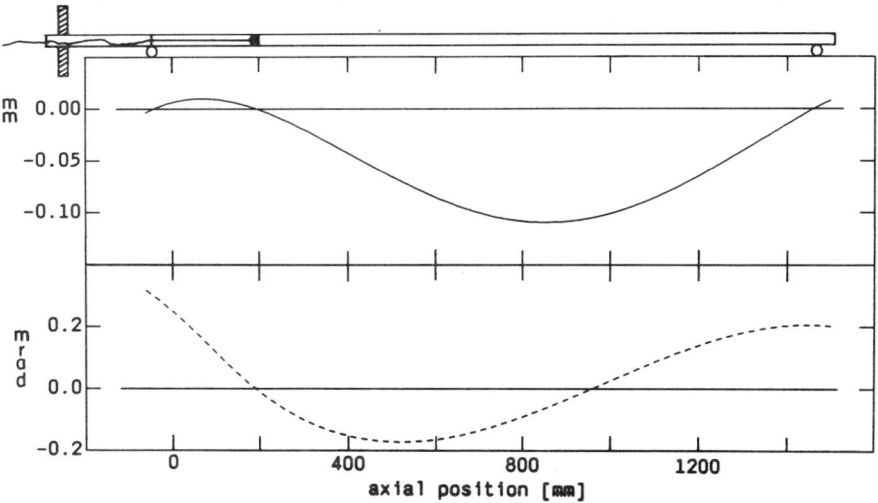

Figure 8: As in figure 7 except that a weight was added on the left end of the revolving tube, which compensates for the sag of the rotating probe in the left position, as indicated in the schematic on top. This position was used to align the left end of the solenoid.

indicated in the schematic on top of the figure. In case we
misplaced the probe carrier by one full millimeter, the angle
between the optical axis and the axis of rotation would be a
negligible amount of 1.2 μrad. With a little trick, one can
eliminate in selected locations the sag and the slope
simultaneously. If one locates one of the centering O-rings inside
one of the supporting bearings, and one uses a counter weight to
compensate for the sag in the location of the other centering
O-ring, the probe carrier will be located perfectly on and rotated
perfectly around the optical axis. In this position the probe is
190 mm inside of the shield and 16 mm outside of the solenoid. We
produced brass disks with a 30 mm center bore to be mounted on the
long left end of the revolving tube and weighted them accurately.
Using these values we calculated the adequate position for the
counter weights to eliminate the sag at the location of the second
centering O-ring. Fig. 8 shows that we can eliminate all sag
effects in the described location on the left end of the solenoid,
if we mount 1311 g 252.6 mm to the left of the shield, just as
indicated in the schematic on top of the figure. If we mount 2175 g
246.1 mm to the left of the shield, and we insert the probe carrier
from the right, all sag effects will be eliminated with the probe 16
mm outside of the solenoid, just as shown in Fig. 9. In either
case, if the counter weight is misplaced by one full millimeter, the
center of rotation moves only 0.6 μ off the optical axis which gives
rise to a slope of only 2.5 μrad between the axis of rotation and
the optical axis. Or if the probe carrier is axially misplaced by
one full millimeter, the center of rotation is still within 0.15 μ
of the optical axis and the resulting angle between the two axes is
less than 1.2 μrad. That shows that neither the center of rotation
nor the axis of rotation are very sensitive to the precise position
of all components, which makes this method very reliable.

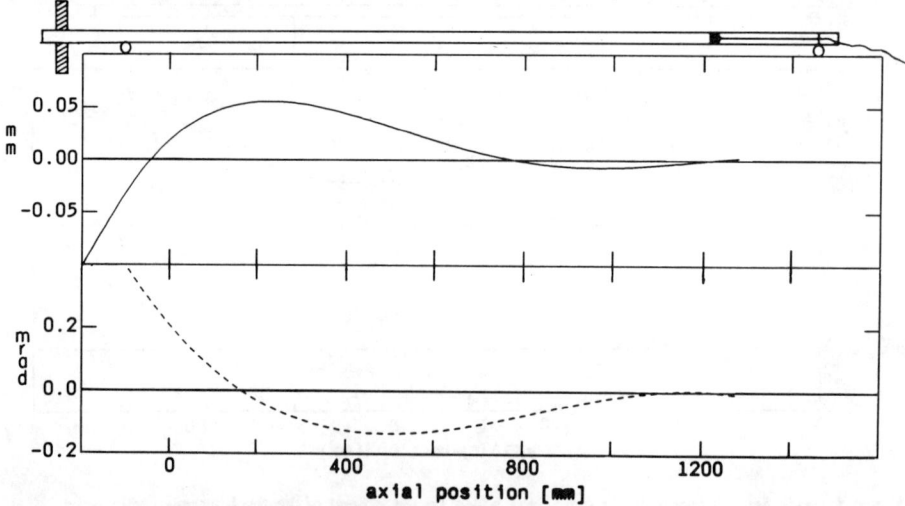

Figure 9: As in figures 7 and 8, except that the heavier weight compensates for the sag of the rotating probe in the position on the right as indicated in the schematic on top. This position was used to align the right end of the solenoid.

291

THE ALIGNMENT PROCEDURE

We discussed above that we can measure the transverse field component with the rotating Hall probe and adjust the solenoid until the transverse field component disappears, or the Hall voltage is independent of the angle of rotation. The transverse component can be zeroed within a very few steps if one puts the Hall probe perpendicular to each of the three adjustment rods and then measures the component of the transverse field in the direction of each rod. The difference between each reading and their mean value will tell directly how much each rod has to be moved.[18] However zeroing the transverse component at both ends does not necessarily align the solenoid. For example, if we first insert the Hall probe and eliminate the sag on the left side, and move the solenoid on the left side until the transverse component is zero and then in step 2 move the Hall probe to the right, adjust the elimination of the sag, the Hall probe would not show a tranverse component, despite the

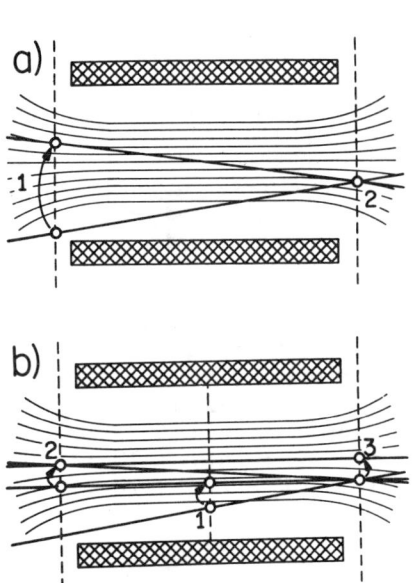

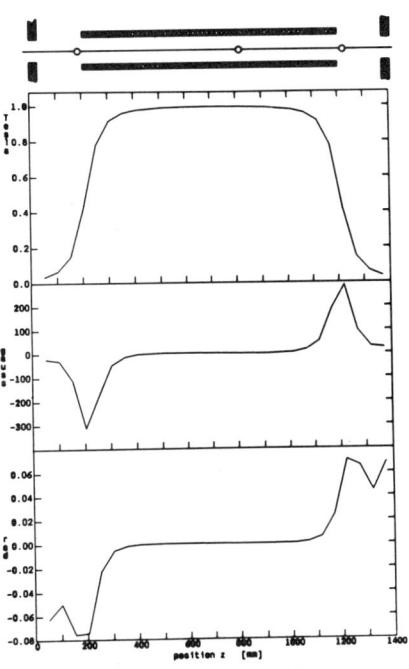

Figure 10: In this figure we moved the optical axis instead of the solenoid to demonstrate the incorrect (a) and correct (b) alignment procedures. The Hall probe (open circle) is moved to the different locations for the subsequent steps (numbers) in the alignment procedure.

Figure 11: The axial field component B_z (top), the radial component B_r 8 mm away from the magnetic axis (center) and their ratio B_r/B_z (bottom) are plotted as a function of the axial position z. B_r/B_z shows the divergence of the field, which is proportional to the relative inhomogeneity. The schematic above the frame indicates the positions of the end shields (shaded), the solenoid (crosshatched) and the probe during alignment (open circles).

fact that we adjusted only one side of the solenoid. Fig. 10a shows how we end up with two different field lines, which show for reasons of symmetry no transverse components on both sides. No transverse field component means only that the field line is aligned in the tangential sense, but not that the field line is straight. One would have to map the field to tell whether or not a field line is straight, and that is a time consuming and cumbersome procedure. The method described above ends most likely with a solenoid which is tilted around its axial center position as shown in Fig. 10a. We wish to point out that the center of the solenoid is the best location to define or measure the orientation of the magnetic axis, but not its location, because all field lines are parallel in this homogeneous region. In the fringing fields of a straight solenoid, the magnetic axis is the only field line which has the same orientation as in the center, and hence serves well to find the location of the axis. These thoughts lead us to the proper procedure for aligning the solenoid, as indicated in figure 10b: First, one places the probe close to the center where the axis of rotation is parallel to the optical axis, and moves the solenoid on either side to zero the transverse component, which will put the solenoid parallel to the optical axis. In step 2, one mounts the Hall probe on the left, eliminates its sag and reduces the transverse component by half of its measured value. Eliminating only one half of the transverse component avoids the symmetric misalignment shown in Fig. 10a, and theoretically brings the magnetic axis right on the axis of rotation in the location of the probe. In step 3 one switches the Hall probe to the right, adjusts the elimination of its sag and zeroes the transverse component. This whole procedure can be repeated for fine tuning and double checks. As a matter of fact, a repetition of step 1 and 2 would be sufficient for alignment, but step 3 is a useful and sensitive check.

We showed above that it is sufficient to adjust the positions of the counter weight and the probe carrier with a simple ruler or yard stick, which has divisions of the size of a millimeter. In addition we would like to point out that the angle of rotation of the probe can be eye-balled without a scale, because at least after the first cycle of adjustments the transverse component will be very small and hence will lose its sensitivity to the angle of rotation.

DISCUSSION

If one would use the proper alignment procedure but would neglect the sag by not using counterweights or any other corrections, one would have located the solenoid too low. The center of rotation would have been approximately 0.12 mm below the optical axis, and 0.05 mm would have been added by the 0.45 mrad tilted axis of rotation, which would have shown no transverse component with a field line tilted by 0.45 mrad, which would have added up to an almost parallel misalignment of 0.17 mm.

We made some approximations in the course of this paper which were not discussed. For example, the probe carrier is put close to the center of the solenoid, where it is parallel to the optical axis, which sets the sensors 0.274 mm below the optical axis and 114.4 mm away from the center, where the field is not perfectly homogeneous. However one can see in figure 11 that the inhomogeneity is very small[19] in this area and yields a misalignment of less than 3 μrad. A more serious fact is that the stiff probe carrier sags and its axis of rotation is tilted compared to the straight line approximation by 10 μrad at the location of the sensor. We neglected this error because it yields only 0.01 mm over the length of the 1 m long solenoid. However, these errors could have been easily corrected by moving the probe 11.1 mm closer to the center, to the axial position at 810.3 mm, where the slope of the revolving tube would cancel both effects discussed above. Naturally the same problem occurs in the two sag compensated positions at both ends of the solenoid. However, there the field has such a strong inhomogeneity[20] that the 10 μrad slope results in a position error of only 1 μ, which can be neglected. This analysis futher justifies the chosen probe positions for the alignment, as we showed that the center position is not very sensitive to the sag of the probe carrier and the high relative inhomogenetiy in the fringing field reduces the sensitivity of the probe to an angular error.

Naturally one also has to wonder about the straightness of the revolving tube. We mentioned before that only the axis of rotation matters, which is the load dependent sagging axis of rotation of the revolving tube. If the revolving tube or the probe carrier are not perfectly straight, the axis of rotation will remain, and only the distance of the probe from the axis of rotation and the angle between the probe and the axis of rotation could be slightly different than our expectations, which both are unimportant for the alignment. It is more complex if the revolving tube is tilted at the bearing locations, or the inner races are mounted with a tilt. The nonparallel component of the misaligned bearings would cause the balls to deviate from the center position in the race, depending on the angle of rotation. That would cause the inner race to enhance the up and down movements and at the same time reduce the amount of the normally dominant right and left movements, which all should stay within the radial play of the bearing. Therefore we can neglect this effect because it lies within the accuracy of the definition of the optical axis. The parallel component of a bearing misalignment causes the sensor to change its axial position as a function of the angle of rotation and hence falsely indicates a transverse component, which one would compensate with a real transverse component created by tilting the solenoid. If both bearings are tilted by 1 degree in a parallel manner, the probe would move ± 0.37 mm around its mean position. In the center alignment position, 114.4 mm away from the center, the radial component B_r changes very slowly[21] which would cause a misalignment of only 0.6 μrad. If in addition the rotating probe is misaligned in respect to the axis of rotation, one would encounter an additional 0.13 μrad per degree of misalignment. This problem is more serious in the alignment positions in the fringing fields; the misplacement of the solenoid, caused by a parallel misalignment of the bearings, is proportional to the change in inhomogeneity d^2B_z/dz^2 or dB_r/dz. Hence, one would not misplace the solenoid if the probe would be located approximately 10 mm inside of the solenoid, where the radial component B_r has a maximum. The misplacement increases with the distance from the solenoid and yields in our location[22] approximately 0.02 mm per degree of misalignment. If in addition the probe is misaligned with respect to the axis of rotation, we have to add in our case 0.013 mm for each degree of misalignment. The last component is proportional to the inhomogenity and hence would be at worst where the radial component has a maximum. Non-straightness and misalignment of 1 degree would be easily detected and therefore we can neglect the discussed errors. However we would like to point out the importance of placing the radial probe in a reasonable axial position and also the importance of minimizing the angle between the probe and the axis of rotation. The discussion above also shows that 0.1 mm axial play is acceptable.

We should also evaluate the requirements for a gaussmeter. If one wants to align the solenoid with a main field of 1 Tesla in the center position to 10 μrad, one has to be able to measure 0.1 gauss, which can be done with a 3½ digit instrument, if the DC component is less than 200 Gauss. That requires that the Hall probe is parallel to the axis of rotation within 1.1 degrees, or a fraction of a degree for alignments at higher main fields. If one wants to position the solenoid at the ends within 0.01 mm, one has to be able to measure a transverse component of 0.01%,[20] or approximately 0.3 Gauss for a 1 Tesla main field. That could be measured with a 3½ digit instrument, if the main field is reduced to keep the radial field in the end alignment positions below 200 Gauss. The numbers above explain why we recommend a 4½ digit instrument, such as ours,[12] which has always sufficient resolution for our alignment method.

The previous discussion brings some light back onto the bearing problem. Our torque-free bearing worked quite accurately and measurements with a dial gauge showed that the radial jitter was below 0.02 mm. But if the revolving tube is rotated in a careless and insensible way, one could encounter an increased radial jitter far above the normal radial play. This risk could be reduced with a standard precision or a self-aligning bearing. Such bearings might be adequate, if sensibly mounted, and could be easily tested, by fitting a sinus function to the measurements of a small transverse component. The center of the solenoid (z=707mm) is most sensitive to a torque problem, and the 0.13 mrad between the axis of rotation and the optical axis provide the transverse component, as shown in Fig. 5. The bearing is the key component which determines the accuracy of alignment.

Our solenoid was aligned under considerable time pressure and we were not able to identify the bearing problem in time. It became clear when we were mapping the solenoid for quality control. Hence we had to make some minor final adjustments according to the evaluated field lines. We do not recommend basing the alignment on the mapped field lines because it is very time consuming and requires very careful data analysis.[5] At a later time the alignment was confirmed with a laser beam which passed two 1 mm apertures mounted inside the cold bore and hit a transparent target disk mounted in the axial openings in the end shields. However the precision of this test was not better than 0.1 mm. Finally, we are happy to report that on the first try the electron beam passed right through the solenoid and the transmission could not be improved by changing the alignment of the solenoid.

SUMMARY

We showed that one can align a straight solenoid with the precision of a traditional Swiss watchmaker. Our method is more reliable and significantly more accurate than other methods. One can expect an accuracy of a very few hundredths of a millimeter, which is mostly limited by the characteristics of the bearings. The accuracy of our method was proven by evaluating all effects which could cause alignment errors. Our method requires only a flat Hall probe with a 4½ digit controller, but an additional axial Hall probe is helpful for diagnostics. No accurate positioning devices or position and angle sensors are used to reach the stated accuracy. It can be reached without demanding any excessive precision from the personnel, but it does require thoughtful and careful work. Our alignment procedure is straightforward and centers the solenoid on the optical axis with very few steps. A simple computer code to calculate the sag and the slope for the various arrangements is useful but not a necessity. The demonstrated procedure is highly efficient compared to the time consuming trial-and-error or the cumbersome field mapping. However, only field mapping can demonstrate the straightness of our solenoid, which will be reported elsewhere.[5]

Many people from Kansas State University provided very valuable assistance to this work: Dave Hill who remachined the reference surface with a tilt of 1.1 degrees, Dionisia Stockli, Dea Richard and Chris Koci Swiler who corrected and finalized the manuscript, Andrea Canelos and Marie Dawes who drafted the figures and Paul Gibson, who measured the modulus of elasticity of the revolving tube, to name a few.

This project was funded by the Division of Chemical Sciences, U. S. Department of Energy.

References

1. R. Becker, H. Klein and M. Kleinod, "All Union Accelerator Conference", Dubna, USSR, (1980).
2. C. Goldstein, H. Laurent and M. Malard, Proceedings of the "Third International EBIS Workshop, Ithaca, NY, USA (1985) p. 1.
3. J.A. Good, Patent 1456197, The Patent Office, London, (1976).
4. M.P. Stockli, C.L. Cocke and P. Richard, "Specifications of Superconducting Solenoid" (1985) unpublished.
5. M.P. Stockli, to be published.

6. Martin P. Stockli, C.L. Cocke and P. Richard, Proceedings of the 4th International Symposium on EBIS and their Applications, Upton, NY, USA (1988), to be published.
7. Martin P. Stockli, K. Carnes, C.L. Cocke, B. Curnutte, J.S. Eck, T.J. Gray, J.C. Legg, P. Richard and J. Arianer, Nucl. Instr. and Meth. B<u>10/11</u>, 763 (1985).
8. R.W. Schmieder, K. Battleson, D. Buchenauer, A.R. Van Hook, J. Vitko, J. Weeks, L. Hansen, R. Wolgast, V.O. Kostroun and R. Becker, Technical Specifications (1986) unpublished.
9. *e.g.* V. Jung, Nucl. Instr. and Meth. 101, 225 (1970).
10. M. Turin, Nucl. Instr. and Meth. 91, 621 (1971).
11. O. Runolfsson, Proceedings 6th Int. Conf. on Magnet Technology, Bratislava, 1977, (ALFA, Bratislava, 1978), p. 802.
12. RFL Industries Inc., Boonton, NJ, USA, the model 912039 flat probe and the model 912312 axial probe were operated with one model 912 gaussmeter and a model 913 differential adapter. A second 912 gaussmeter would have substantially simplified the calibration and operation, but Santa did not bring us one.
13. grade ABEC 3 (Annular Bearing Engineers Committee).
14. grade ABEC 1.
15. MRC Bearings Inc., Jamestown, NY, USA, model 106KS.
16. *e.g.* E. Oberg, F.D. Jones and H.L. Horton, Machinery's Handbook 22nd ed. (Industrial Press Inc., New York, 1984), p. 292.
17. *e.g.* H.E. Boyer and T.L Gall, Metals Handbook Desk Edition (American Society for Metals, Metals Park, Ohio, 1984) p. 6-58.
18. A component B_{tx} of the transverse field B_t in direction of rod x requires a correction of δx for rod x:
$\delta x = 2 \cdot (B_z \cdot dz/dB_z) \cdot B_{tx}/B_z$; the expression in parenthesis is the inverse relative inhomogeneity, which yields in our case approximately $\delta x = 105$ mm$\cdot B_{tx}/B_z$.
19. relative inhomogeneity $dB_z/dz/B_z = 20$ ppm/mm.
20. $dB_z/dz/B_z = 2\%/$mm.
21. relative slope of inhomogeneity $d^2B_z/dz^2/B_z = 0.4$ ppm/mm^2.
22. $d^2B_z/dz^2/B_z \sim 0.01\%/$mm^2.

DISCUSSION

Levine: What is your structure made of?

Good: It is a rather unusual alloy. It is English. It is not very far from something we call HE30. I do not know the U.S. number off-hand. It is partially precipitation hardened, then stretched by 4% after its final heat treatment, basically to align all the grains and align the stress of the billet. Then they machine it from there. We receive it in the pre-stretched condition. I think it is actually made for aircraft undercarriages which also have to be absolutely dead straight.

Levine: Do you heliarc weld any of it?

Good: Yes we do. We weld the outside of it. But we only run a very small ring weld round the end and, in order to enclose it, we use a flexible structure of soft aluminum.

Levine: And it is sealed by heliarc welding?

Good: Yes. Well, sometimes it is done by electron beam welding but that comes back with so many holes in it, we go over it. This happened to the last one that was done.

Levine: And that picture we saw, roughly what was the tension that he was putting on the conductor?

Good: Oh, it would be round one-half to 1 kilo.

Amboss: Do I understand you that you wind a layer, you put epoxy on it, you machine the epoxy, you take the winding out to the end, and then you somehow ....

Good: You cut if off.

Amboss: You cut if off? Wow!

Good: And then you put the next layer on. It is very tedious business. You join all these ends up, and you also have to make the ends rotate round. If you do not do that, you will find that if you put all the ends together, you get a nice bump in the field at the end. So each layer is a turn and 3/20 of a turn, or something like this. It keeps going round. When you have gone up your 20 layers, you come right back to where you started. And, of course, it is always an even number of layers.

Amboss: Do you ever wind pancakes?

Good: We have wound pancakes and we have wound trapezoidal-shaped pancake coils for the sort of things nuclear physicists like when they want a magnetized sample but they would really rather not have any magnet there because they would like to look in all directions. So very grudgingly they give you 10 degrees up here and 10 degrees down there. Then they say, 5 Tesla please, which means at least 10 Tesla on the windings. But, no. We built some 2-1/2 Tesla ones for SIN, for Dr. Mango, who is an old sparring partner. Each magnet is more complicated than the last one. When we think at last we have gotten them sorted out, he comes up with a more difficult design.

Hershcovitch: Was that a superconducting sextupole at SIN, by any chance?

Good: No, the ones at SIN are all dipoles, basically.

Good: When you do the room temperature measurement, not only is the magnet much stiffer when you rotate it than when you rotate the Hall probe but, of course, to first order it does not matter if it sags because you get the same sag in all directions so it is still symmetric.

Stockli: That is right.

Good: So you do very well on that sort of measurement.

Amboss: You have a winding which is basically a bobbin and then you put it into a warm bore and support it so that it is very highly insulated so that you do not have too many places where you support it. To what accuracy do you align the warm bore with the axis of the solenoid?

Stockli: Maybe Jeremy Good would like to answer this question.

Good: The warm bore is mounted inside the iron shield and is not particularly accurate. But the rotating tube is centered with a bearing mounted on the iron shield and can be positioned within mils.

Amboss: Do you provide an indication as to where that has to be?

Good: The solenoid is supported inside the iron shield at each end with three titanium rods which have high strength and low thermal conductivity. These rods are attached to large knobs on the outside which allow one to center the solenoid within the iron shield.

Amboss: Why don't you simply put on saddle coils and buck out the transverse field?

Stockli: That is a problem because a saddle coil can improve the field only in a certain limited location and is very likely to degrade the field in the rest of the space, especially in the fringing fields. Many coils would be needed to improve the straightness of the solenoid over the whole length including the fringing fields. However, if a solenoid shows a dominant defect in a very limited area, saddle or Helmholtz coils can provide the simplest correction, as shown on CRYEBIS II.

Amboss: Basically, we do that in the microwave industry, by adding little pieces of iron, which we call shunts.

Good: They do that also in the NMR industry.

Stockli: It is also done on EBISs.

Becker: One modifies the return flux.

Stockli: Shims or shunts are commonly used in EBISs, however it is inefficient for us to use them because our solenoid is shielded.

Good: You cannot correct the middle without messing up the ends. If you have a problem at one end, it is easy to add a transverse field, which straightens the field locally.

Amboss: It is better to use a pair, so that you can bring your beam truly back on the axis.

Good: If one starts adding coils over the whole length, it gets out of hand.

Amboss: I am not so sure because the transverse fields are not so localized. You might want to put in saddle field deflection coils.
Good: You could add a series of surface coils and calculate the required corrections to straighten out the measured field.
Stockli: One could also computerize all corrections coils, but that might take forever. Another problem was mentioned before: the coaxiality of the magnetic field and the cold bore. We mounted apertures inside the bore, cooled the system down and shot a laser beam through the solenoid, which showed that the magnetic field and the bore are coaxial within the 0.2 mm accuracy of the measurement.
Good: This is the accuracy of the aluminum bore.
Stockli: That is right, and some small contributions from the position of the different wire layers.
Amboss: Could I just make one comment. I do not want to monopolize all this. I use a transverse probe which I had made up by the Bell Gaussmeter people except I support the probe at two ends with the Hall element in the center. I have a screw at one end which I tighten so that the probe never touches anything as I rotate the magnet. If the probe touches the magnet bore you get an erroneous reading.
Stockli: Our system is more difficult because of the 1.5 meter length.
Amboss: If you can get it tightened, then you are alright. Then you have a carriage and you move the probe through on a carriage.
Stockli: But your carriage can also sag. There is very little space in our 1.5 inch diameter warm bore tube. Val Kostroun also used a system in which he reduced the sag with tension, but it only reduces the sag; it does not necessarily eliminate the problem depending on the weight and the length.
Becker: Is it correct that the transverse field you measured is greater than the transverse field measured in the lathe?
Stockli: Yes. This is most likely due to the fact the solenoid buckles under the tension of the longitudinal tie rods.
Becker: Did you measure the largest transverse field at the end where the iron shield is?
Stockli: Well, we measured the largest angle between the field and the axis in this location. But the measurements in this area are unrealistic because there was a 3.6 inch diameter hole in the shield. We will mount magnetic diaphragms with much smaller holes into the large shield openings which will dramatically increase the axial field and substantially reduce the nonstraightness.
Becker: I would like to point out that this is the most critical location for the electron transmission.

Stockli: That is correct. We thought about that a couple of weeks ago when we were not able to launch a beam through the system, which was most likely caused by a trivial problem. Our setup with the warm bore tube does not allow us to measure the field with the magnetic diaphragms mounted because they block the access. However, we are planning a different setup to measure the field just inside the magnetic diaphragm. Shims and diaphragms can be a source of trouble.

Marrs: I would like to know how stable the alignment is in regard to the hysteresis of the iron shield?

Stockli: We did check that and made many other tests to make sure that the alignment was stable. It cannot be double-checked easily after we remove the warm bore tube. We measured the remanent field after a 5 Tesla field run and a 20° K warmup to quench any remaining flux. The largest transverse field we measured was 0.04 G, which shows that the alignment is not sensitive to the hysteresis of the shield. We also cycled and measured the field at 1 Tesla, 2 Tesla and 5 Tesla. We warmed the magnet to room temperature and cooled back down. We moved the magnet in a certain direction and then moved it back. A comparison of all the data shows that the magnetic field lines are always reproducible within approximately one mil, but it is important that all the titanium rods are under tension.

Marrs: What if I take a strong permanent magnet from an ion pump, stick it on your ion shield, then take it away. Does that change the transverse field so much that you have to remap?

Stockli: No. We used such magnets to keep jigs in position. The shield is 6 cm thick and not saturated. Hence, such an external field does not affect the field on the axis. It might be a problem in the gun and collector regions which are outside of the shield. However, the physical access is quite limited which makes me believe that the remanent field would not cause serious problems even in this area. Such a magnet can help with diagnostics, but I was not allowed to perform such diagnostics.

Hershcovitch: I would like to make one comment. People who work on tokamaks, even the main line expensive machines, go through a lot of effort to map the fields and make them accurate. Some tokamaks operate at a magnetic field of over 10 Tesla. And yet, in spite of all the efforts that went into the main field coils, what makes them finally work is to flimsily wind number one wires on the tokamak. In my doctoral thesis, I worked on a gated electron trap which is exactly like an EBIS except that you trap the electron beam. This system was 4.2 meters long. To align the electron beam I used a phosphorus screen on the last gate, after the structure was first aligned optically with a laser. I literally took wire and wound some horizontal and vertical coils on the vacuum chamber for its full length. These wound coils were connected to two small power supplies, which were subsequently set to levels that centered the electron beam with respect to the last electrode for each setting of the main solenoidal field. The electron beam was aligned so well that after 50,000 transits, when dumped back to the cathode, there was little discoloration around the 1 mm hole in the first (extracting) electrode; i.e., the electron beam was still well centered. The point that I am trying to make is that the addition of some simple external coils can compliment the other methods used in the alignment process.

Becker: I want to point out that there is a very good way to correct magnets as long as no iron is involved. Jeremy has proven this with elliptical windings on our DR magnet. We have programs where we put in any transverse field in one plane, break it up into two perpendicular planes, and then you can just define, in an interactive fashion on a computer, elliptical windings of any slope in order to buck out the transverse fields. And this has worked in our magnet to 10^{-4} and better. Elliptical windings are wonderful with superconducting wire on the surface of a cylinder.

Amboss: In which plane? Perpendicular to the axis?

Becker: Yes.

Good: You carve little ellipses with 50 to 100 turns in them, then, if you have a horizontal solenoid, you put the ellipse around it. Then you can create vertical and axial fields. But the combination that Reinard has described was meant to take away the axial field and just leave you with a vertical field. The alternative is to make some sort of saddle coil arrangement. It is quite difficult to make precisely, although it can be done, as long as you keep to only a few turns.

Faure: We made this type of measurement of CRYEBIS I when it came back from Orsay because in Orsay they use, what I call, wild shimming. That means a large piece of iron to try to have the correct transmission of the electron beam. So we found that the axis was not so bad. I mean, maybe 1 milli-radian or something like that in effect. We corrected the end off of the shielding and did not need any more shimming. We have not made axial measurements on DIONE because they take a lot of time. There is a strong incentive to start up the source as quickly as possible because there are a lot of things to be done.

Stockli: The final proof will be the electron transmission and the ion yields.

BRIEF COMMENTS ON TEST RESULTS OF BOTH METAL AND ELASTOMER VACUUM SEALS*

Kimo M. Welch, Gary T. McIntyre,
Joseph E. Tuozzolo, David J. Pate
Brookhaven National Laboratory, Upton, NY

INTRODUCTION

A brief over-view of the subject of sealography was given. Those results will be published elsewhere. However, results of metal vacuum seal tests, conducted at Brookhaven National Laboratory will be herein summarized. The paper then summarizes results of work done by the first author, at the Stanford Linear Acceleration Center, on the extended outgassing properties of elastomers including Viton® and Buna N®.

CONSIDERATIONS IN SELECTION OF A VACUUM SEAL

When selecting a particular seal joint design for use in ultra-high vacuum (UHV) application we must make a number of compromises. Considerations which must be entertained in selecting a joint design include the: 1) reliability; 2) material outgassing rates; 3) seal permeation rate; 4) bakeability; 5) mating surface materials; 6) ease of installation; 7) required sealing force; 8) flange sizes in typical use; 9) provision for cleaning of parts; 10) initial costs; 11) operating (i.e., replacement) costs; 12) seal availability for replacement; 13) degree of radiation "hardness"; 14) shelf life; 15) handling and storage fragility; and 16) safety. Most of these 16 variables are coupled. Each was briefly discussed in the talk, but such discussion is beyond the scope of this paper.

METAL SEALS

All-metal seals are self-destructive in that they can only be reused, at most, two to three times. Common varieties of all-metal seals include the: 1) ConFlat® (i.e., knife-edge); 2) Wheeler® (special wire seal);[1] 3) Foil (Al, In, Cu); 4) Wire (Al, Au, Ag, Cu, In, etc.); 5) C-Rings, coated with soft metals; 6) C-Rings, either coated or uncoated, and reinforced w/spring materials; 7) C-Rings, with "diamond" edges, and spring reinforced (i.e., the Delta® Seal); and 8)

---

*Work performed under the aupices of the U.S. Department of Energy.

"diamond" seals. Advantages and disadvantages of several seal configurations were discussed in the talk. Only the results of recent metal seal tests at Brookhaven will be reported. The seal configurations tested included:

- A. Inconel® C-Rings, coated with In,
- B. Spring reinforced, Inconel® C-Rings, jacketed with Al,
- C. Spring reinforced, Inconel® C-Rings, coated with Pb,
- D. Spring reinforced, Inconel® C-Rings, jacketed with an Al sheath on which "diamond" edges are machined (i.e., the Delta® seal),
- E. Conventional Al "diamond" seals.

Applications at the Alternating Gradient Synchrotron (AGS) require that leak tight, all-metal vacuum seals be made with flange materials including aluminum, stainless steel and enamel coated stainless steel. For the greater part, most of the metal seals in the AGS were Type-A of the above listing. However, problems were encountered from time to time with these Inconel® C-Rings spontaneously cracking. Past studies indicated that such problems resulted from H_2 embrittlement, due to some process in the fabrication of the seals. Also, In coated seals had very poor shelf life. Oxides of In would form on the sealing surfaces, even though the seals were stored in sealed bags. Rather than "study" these problems again, we faced up to the fact that the seals were marginal, at best, and launched a study program to find a suitable alternative.

Seals of the above listed configuration were tested. The test consisted of squeezing the seal between a stainless steel mandrel, and sealing plates made of the three different materials. The mandrel was positioned in a precision hydraulic ram. The joints were leak checked by pumping between the two plates through a hole in the bottom of the mandrel. A leak detector with a sensitivity of $\sim 10^{-10}$ Torr-ℓ/sec He was used. Any indication of a leak constituted a seal failure.

Results of these tests are given in Tables 1-4. The numbers listed vertically in the columns in the first three tables represent the number of times the seal was cycled. Stress-strain measurements, and yield data were recorded. Each time a successful seal was achieved, seal deflection was measured. Then the hydraulic ram was "released", and the seal height again measured. The plates were not parted and their respective indexing with the seal changed between each measurement. Because of this, one may not interpret the existence of successive data in any one column as indicating that the seal is reusable. However, the absence of additional data in any one column implies that the seal would not reseal on the next application of pressure. The results for the In coated C-Rings was very poor. This is because they were used "as is", on removal from the sealed bags in which they had been stored for many months. New, freshly coated C-Rings would have yielded far more reliable results. But, this is just a restatement of the initial shelf life problem.

Because of these results, the Delta® Seal was adopted as the standard seal for future use in the AGS. During the 1988 summer shutdown, approximately 300 Delta® Seals were used to replace the In coated, 21.6 cm diam. C-Rings used throughout the AGS. Problems of installation technique or inexplicable failures occurred in ~ 3.0% of these seals at the time they were installed. However, after successful installation, there have been no failures (i.e., leaks) in these seals.

ELASTOMER SEAL TEST RESULTS

The most widely used elastomer seal material is Viton A®. However, where large quantities of seals (i.e., O-rings) are needed, cost considerations sometimes dictate use of another elastomer, such as Buna N®. If cost is no object, Polyimide has the best vacuum characteristics of all the elastomers. However, this material has some disadvantages. Peacock provides an excellent review paper on the properties and uses of elastomers.[2]

In tests at Stanford, elastomer gasket materials were exposed to high vacuum for up to three years in duration. The purpose of these tests was to determine if there were irreversible hardness changes in the materials as a consequence of this exposure. Approximately 100 specimen O-ring segments, measuring 0.51 cm ϕ × 2.5 cm long, were individually suspended on wire frames mounted within Cu pinch-off tubes. These, in turn, were welded to a large manifold pumped by a 500 L/sec sputter-ion pump. Specimens were pinched off this manifold over a three year period. The weight and hardness of each specimen was measured before and after extended immersion in high vacuum. Some samples were baked at moderate temperatures (i.e., ~ 200 C°). Numerous control samples, not exposed to vacuum, were also tested for hardness and weight changes throughout the duration of the experiment. Results of the vacuum tests were as follows:

1) All Buna N® O-rings showed significant irreversible weight loss as a consequence of exposure to vacuum. This lost material is probably not exclusively water vapor, but rather plasticizers and polymers which would contaminate the vacuum system (e.g., plug sieve materials; poison cathodes). Irreversible weight losses in unbaked Buna-N® O-rings amounted to approximately 10 to 50 Atm./cm^3 of O-ring.

2) Unbaked Viton-A® O-rings showed no significant irreversible weight loss or hardness change. Weight losses in Viton® ranged from 3 to 7 Atm./cm^3 of O-ring. Hardness changes in unbaked Viton-A® were completely reversible. With baked Viton-A® O-rings, the higher the initial Shore-A durometer hardness reading, the less the irreversible weight loss due to baking. This suggests the presence of less plasticizers and unreacted polymers in the initially harder compounds.

REFERENCES

1) Roth, A., J. Vac. Sci. Technol. **9**, No.1, 14 (1972).
2) Peacock, R., N., J. Vac. Sci. Technol. **17**, No. 1, 330 (1980).
3) Wahl, H., *et. al.*, CERN Tech. Note SPS/81-8, December, 1981.
4) The authors are indebted to Alain Poncet of CERN PS. Dr. Poncet provided the authors with the data which led to this study. He has extensive experience in UHV sealography, including the successful use of "diamond" seals throughout the CERN Proton Synchrotron.

Table 1. Sealing Force For Metal Seals of Different Configuration When Sealing Against Two Stainless Steel Surfaces

Sealing Force in Lbs./Inch	C-Ring In Coated			C-Ring Pb Coated			W/Spring "C" Pb Coated			Delta Seal @			Al Diamond			Al C-Ring W/Spring			Al C-Ring W/Spring
Sample No.->	1	2	3	1	2	3	1	2	3	1	2	3	1	2	3	1	2	3
101 to 150																		
151 to 200											1	1						
201 to 250							1			1	2							
251 to 300							2			2	3							
301 to 350							3			3			1					
351 to 400							4			4								
401 to 450	1,2			1,2						5								
451 to 500													2	1,2				
501 to 550				3						6	2			3				
551 to 600											3				6			
601 to 650				4	1													
651 to 700											4				7			
701 to 750				5	2					7								
751 to 800				6														
801 to 850					3					8					8			
851 to 900																		
901 to 950													3		9			
951 to 999																		
>1000																		
Wouldn't Seal	X	X	X			X							X			X	1	1

Table 2. Sealing Force For Metal Seals of Different Configuration When Sealing Against An Aluminum and Stainless Steel Surface

Sealing Force in Lbs./Inch	In Coated C-Ring			Pb Coated C-Ring W/Spring "C"			Pb Coated C-Ring W/Spring "C"			Delta Seal®			Al Diamond			Al C-Ring W/Spring			Al C-Ring W/Spring		
Sample No.–>	1	2	3	1	2	3	1	2	3	1	2	3	1	2	3	1	2	3	1	2	3
101 to 150																					
151 to 200											1	1									
201 to 250											2	2									
251 to 300											3	3		1							
301 to 350										3		4									
351 to 400																					
401 to 450				1	1,2	1,2				4		5		2							
451 to 500				2										3							
501 to 550					1,2	3					4		1								
551 to 600																					
601 to 650				3	3	4								4							
651 to 700				4		5				5	5		2		1						
701 to 750				4	4	5					6										
751 to 800				5	5	6								5	2						
801 to 850																					
851 to 900																					
901 to 950														6							
951 to 999																					
>1000										3											
Wouldn't Seal	X	X	X													X	X	X			

Table 3. Sealing Force For Metal Seals of Different Configuration When Sealing Against A Stainless Steel & Enamel Coated Surface

Sealing Force in Lbs./Inch	In Coated C-Ring			Pb Coated C-Ring w/Spring "C"			Delta Seal @			Al Diamond			Al C-Ring w/Spring			W C-Ring w/Spring		
Sample No.→	1	2	3	1	2	3	1	2	3	1	2	3	1	2	3	1	2	3
101 to 150																		
151 to 200									1									
201 to 250							1	1	2									
251 to 300								2	3									
301 to 350							2	3	3									
351 to 400								3	4	1								
401 to 450							3		5									
451 to 500							4	4		2								
501 to 550				1			5	5	6									
551 to 600																		
601 to 650				2			6											
651 to 700																		
701 to 750																		
751 to 800				3							1							
801 to 850																		
851 to 900																		
901 to 950											2							
951 to 999																		
>1000																		
Wouldn't Seal	X	X	X		X	X						X	X	X	X	X	X	X

309

Table 4. Qualitative and Quantitative Considerations In The Selection of a Vacuum Seal Configuration.

Vacuum Seal Configuration	Sealing Force Lbs./Inch	Bakeable (yes, no)	Bakeout Temp. C°	Reuseable (yes, no)	Maximum Flange Diam. Inches	Cost (1=High)	Reliability (1=High)	Shelf Life	Fragility (1=High)
ConFlat® Gasket	1350[1]	YES	625	NO	10	3	1	GOOD	5
Wheeler® Gasket	>1500[2]	YES	450	NO	30	1	2	GOOD	4
Indium Coated Inconel C-Ring	<400	NO	—	?	>30	2	?	POOR	4
Al Jacket C-Ring, Spring Reinforced	1000	NO	—	NO	>30	4	?	GOOD	4
Pb Coated C-Ring Spring Reinforced	400–600	?	?	?	>30	2	3	POOR	3
Al Delta® C-Ring Spring Reinforced	100–250	YES[3]	200[3]	YES	40	4	2	GOOD	1
Diamond, Al	200–800	NO	—	NO	>40	5[4]	3	GOOD	2

DISCUSSION

Knapp: I have a few questions about c-rings. Are either the Inconel™ "c" or the spring material used in it magnetic? Does it have a high magnetic permeability?

Welch: No to all three questions. In fact, the reasons we use Inconel™ for the beam chambers in the alternating gradient magnets of the AGS include the following: 1) Inconel™ has a fairly high resistivity, and 2) Inconel™ is nonmagnetic (i.e., a very low permeability).

Knapp: What about the spring material?

Welch: No, it too is nonmagnetic. It is made of Inconel™.

Herrlander: Concerning the Helicoflex Delta Seals®, I think we have one in Stockholm which was purchased from a French company, that we hopefully will be able to bake to 300°C.

Welch: I would be very interested in that result. What is the major diameter of the gasket?

Herrlander: It is not circular. It is rectangular.

Welch: Yes; well, good luck. We are investigating the use of oval Delta Seals® for some of our magnet chambers.

Herrlander: Did I understand you correctly that the circular (Delta Seals® or diamond?) are not bakeable?

Welch: Alain Poncet, at CERN/PS, has had a great deal of experience with diamond seals. He feels that they are not bakeable. I have baked diamond seals to roughly 150°C with very limited success. In the "old days", Welch used to supply their turbo pumps with diamond seal flanges. In order to do base pressure measurements of pumps, the pressure dome and flange (pump) required baking. We had lots of problems. I, too, do not feel diamond seals are bakeable. Hartmut Whal, at CERN/SPS introduced me to the Delta Seal®. He has conducted bakeout tests of these seals and reports that they may be successfully baked to temperatures of 150 - 200°C.

Herrlander: We are also working with cryopumps that have been modified so as to be bakeable.

Welch: I would also be interested in that result. Is it a Gifford-McMahon type refrigerator?

Herrlander: It is a "Tri-Star", or something. I can not really be sure.

Welch: Is it a gaseous He cryopump with a compressor?

Herrlander: Yes.

Welch: What do you mean by "bakeout"?

Herrlander: 150°C.

Welch: 150°C. That is very interesting. I would be very careful about excessive bakeout temperatures. Bakelite (i.e., phenolic) and Teflon materials are used in the expander of the refrigerator. I would be concerned about creep in the seal material.

Good: That should not matter, actually within those systems, I call it toponol, you call it something else, the plastic parts you have to bake at 100°C anyway because it is full of water otherwise. You have got to get rid of all the water, so ...
Welch: To condition it.
Good: Yes, to condition it.
Welch: The phenolic cylinders (with materials used as first and second stage regenerator beds) are baked to 100°C prior to installation of the seals and insertion in the expander cylinders. I am concerned about creep of the He seals at elevated temperatures.
Good: So the material is a very good quality and it should take bakeouts up to 150°C, I recall.
Welch: I have baked them out (cryopumps) to 100°C, on a CERN dome, so that I could do speed measurements at very low pressures. But I have never exceeded this temperature. I would be very interested in that result, too.
Stockli: You mentioned that some metal seals are reusable. It is possible to remake ConFlat® flanges. It is possible to re-seal both diamond and Delta® seals if one does not part the flanges on releasing the sealing force, and then reapply the pressure. If the indexing of the seal, with respect to the flanges, is changed, they are not going to work.
Welch: I have heard of people using ConFlat® gaskets several times. The method involved using successively thinner shim stock between the mating flange faces. Each time the flange is resealed more shim stock is removed so that the knife-edge "bites" deeper into the Cu gasket. This seems awfully labor-intensive for a relatively inexpensive Cu gasket. Secondly, it was stressed in my talk that we did not change the index of the sealing surfaces with respect to the gasket in our tests. Our reason for cycling the seals at successively higher loading was to enable us to measure elastic strain vs. stress of the respective seals as well as permanent seal deformation. Whal, at CERN/SPS reports that the Delta® seal is reusable up to three times. We have not yet attempted to verify this finding.
Stockli: I do not reuse any seals because the risk of leaks and the resulting additional labor is not worth the cost of a seal. However, our machines do not have to be as reliable as in your case on the RHIC. But I would like to add silver wire to your list of sealing materials. Silver works very well, can be welded easily to form rings, and is much less expensive than gold. Again, that is not a solution for RHIC because you have to be able to reassemble in a minimum of time.
Welch: Roth does reference the use of Ag wire seals in his book.

Becker: Martin, ten years ago we bought Hall flanges, as they were called. They seal the identical Cu gasket of the ConFlat® flanges. We have found them to be very reliable and you really could not scratch them. It was wonderful because you could not damage these seals.

Welch: When a company, formerly called Ultek, made their flanges the same diameter as the Varian ConFlat®, the Ultek flanges, called Curvac Seals, would work with the ConFlat® flange and gasket. The Ultek flanges had rounded, i.e., half toroidal, edges that coined the gasket, rather than a knife-edge which cuts into the gasket.

Becker: No, that is a completely different design. That is Ultek, Perkin-Elmer. I am wondering where this seal has gone. It has not been established in the market.

Welch: I do not know. Actually, Hall was the founder of Ultek, an offshoot of Varian. The seal to which Dr. Becker referred was first described by Reynolds. Peacock offers a very good, one-day course on sealing technology. It is sponsored by the American Vacuum Society. This is not a "plug". It is only a very good course for those interested in the subject.

PRIMARY ION SOURCES FOR EBIS DEVICES

Roderich Keller
Lawrence Berkeley Laboratory, Berkeley, CA 94720, USA.

ABSTRACT

The ion-optical conditions for primary ion sources that could be installed in an EBIS injector are derived, assuming a realistic set of fixed parameters to be imposed by the EBIS. It is shown how these requirements may be met, and that beam currents of up to 2 mA can be generated with the postulated emittance. This derivation, even though carried out for one specific case, gives general guide lines how to proceede for other conditions as well. In the second part, different types of ion sources are presented that are likely candidates for EBIS injector sources. Beam current examples are given and the basic features of the sources discussed. The emphasis of this paper is put on the reliable production of ion beams, rather than attempting to furnish a representative cross section of the existing ion source varieties.

INTRODUCTION

The following discussion of primary ion sources that may be used to inject lowly charged ions into EBIS devices will be founded on a specific set of parameters a primary ion source has to fulfil. After outlining a few basic principles of the ion beam formation process the achievable beam parameter limits will be determined for the given conditions. On this basis, individual ion sources are presented that may be used for the primary beam production, stating their general characteristics and performance data. Even though this paper actually refers to one specific case of injector source requirements only it is intended to furnish a useful guidance how to judge the performance of injector source candidates for other cases as well.

Before starting this discussion, the distinction between the ion generator and the beam formation system has to be stressed; both together form the entire ion source. This distinction is rarely being pointed out in publications on ion sources, but obviously a good ion generator can be used with many different extraction systems, and a given extraction system, essentially the electrode configuration and aperture shapes, can be attached to many ion generator types, as long as the generator yields ions with matching current density and sufficient uniformity across the extraction aperture zone. Especially high-current plasma ion sources are well suited for this kind of system interchanges, and in the following this source type only will be further dealt with.

A wide range of delivered ion species, together with operational reliability, are the dominant criteria for the choice of a plasma generator. This aspect of injector sources for EBIS devices will be treated in the final section of this paper.

---

* This work was supported by the Office of Energy Research, Office of Basic Energy Sciences, Department of Energy under Contract No. DE-AC03-76SF00098

BEAM PARAMETER SPECIFICATIONS

The following set of desired beam parameters would suit the requirements of the DIONÉ EBIS:[1]

Extraction voltage	10 kV
Normalized emittance	$10^{-8}\ \pi$ m = 0.01 π mm mrad (area of the emittance pattern)
Absolute emittance	13.7 π mm mrad (derived for 10 keV from normalized value)
Beam current	15 µA for argon; possibly much higher; metal ions desired, too
Pulse length	300 µs
Repetition pattern	5 pulses with 150 ms distance.

The primary ion beam has to be decelerated upon injection into the EBIS. The design of such a deceleration stage cannot be treated in detail here, but it is not particularly difficult if one makes use of the existing particle transport simulation codes. One must respect, however, the danger of deteriorating the effective beam emittance by inducing distortions due to optical aberrations. Else, 10 keV are already a comfortable energy for beam transport to the EBIS over a typical distance of 1 m.

The absolute emittance limit of 13.7 π mm mrad practically limits the choice of the extraction system to a single-aperture type. This is not at all a serious restriction, in view of the required current value, and rather allows to also consider generator types with narrow plasma columns, such as the duoplasmatron.[2,3] Pulse lengths of 300 µs are rather short for plasma sources, but there is no principal difficulty involved when a generator with thermionic cathode is used; the best way is to insert a pulsed switch in the main discharge circuit that can be triggered according to the desired time pattern.

PRINCIPLES OF BEAM FORMATION

A detailed discussion of the beam formation process is given in a recent publication.[4] In essence, ion beam currents generated by a given extraction system with one round aperture scale as:

$$I_{tr} = P^* \sqrt{\zeta/A}\ U^{3/2} \frac{S^2}{1 + aS^2} \quad (1)$$

where I_{tr} is the beam current (mA) transported within a give angle; ζ is the ion charge-state; A is the atomic mass; U is the extraction voltage (kV); S = r/d is the aspect ratio of aperture radius over extraction gap width; and a is a dimensionless factor representing the effect of optical aberrations in the extraction system. Formula (1) was derived from the Child-Langmuir law[5,6] to fit results obtained with a duopigatron ion source.[7] The value of a was found to be a ≈ 3 in the original study[7] and a = 1.7 for a more sophisticated extraction system[8] attached to the multi-cusp/reflex discharge ion source CHORDIS.[9] The proportionality factor P^* in (1) depends on the allowed acceptance angle of the transport system; $P^* = 1.9$ mA/(kV)$^{3/2}$ was found for ± 20 mrad half-angle with the extraction system described in Ref. 8 and can be taken as a reference value. The rightmost factor in formula (1) indicates S ≈ 1 as a practical limit for the aspect ratio above which the gain in beam current is quite low.

The divergence half-angle is not a fixed beam parameter but can be adjusted for a given extraction voltage by varying the ion current density of the plasma generator. This matching effect is demonstrated in Fig. 1. for two density values. The extraction system for which the emittance diagrams of Fig. 1. are calculated is shown in Fig. 2. High-brightness extraction systems like this one are essential for producing primary ion beams with quality requirements as listed above. They consist of outlet-, ground-, and screening electrode because electrons that are created by the beam itself must be kept from being accelerated back into the source. These triode structures are commonly called accel/decel systems; their aperture contours are in most cases optimized by computer simulations of the beam formation process.

In order to maximize the beam brightness one will always cut off some part of the beam halo and try to transport as much of the beam core as possible. If only a smaller divergence angle is allowed for the extracted beam, compared to any given condition, the beam current has substantially to be decreased. A reduction factor of 2.83 applies between the cases illustrated in Figs. 1.a and 1.b, to fit the beam core into a divergence half-angle of 5 mrad. Still, the beam brightness of 1.b is higher than that of 1.a. The ultimate minimum angle, below which no substantial gain in brightness can anymore be achieved, depends on the ion temperature of the source plasma; for a given extraction aperture radius this results in a lower limit for the beam emittance:[10]

$$\varepsilon_{n,4rms} = 0.0653 \, r \, \sqrt{kT/A} \quad [\pi \text{ mm mrad}] \quad (2)$$

where $\varepsilon_{n,4rms}$ is the normalized, 4rms, emittance which encompasses about 89 % of a beam with realistic density profile; r is the aperture radius (mm); kT is the ion temperature (eV); and A is the atomic mass number.

DETERMINATION OF BEAM EMITTANCES

The finite divergence of the beam core in Fig. 1.b, for example, is caused by an ion temperature of 0.1 eV assumed in the calculation. Application of Eq. (2) with A=1 leads to $\varepsilon_{n,4rms} = 0.10 \, \pi$ mm mrad; whereas the size of the displayed emittance pattern that represents the beam core alone is $\varepsilon = \pi * 1.5$ mm $* 5$ mrad $= 7.5 \, \pi$ mm mrad, and normalization to 50 keV beam energy results in $\varepsilon_n = 0.0774 \, \pi$ mm mrad. This must be called a good agreement between two independent methods of determining an emittance, considering also that the two emittances are differently defined. Ion temperatures have rarely been measured in high-current sources, but as an example, the value of 0.2 eV was found for a multi-cusp source operated with helium,[11] whereas 0.1 eV best fits the CHORDIS results.[12,13]

Assuming now 5 mrad as minimum achievable half-angle and 0.1 eV ion temperature, one can immediately deduce from Eq. (2) that the maximum allowed radius of the outlet aperture amounts to 3 mm if the normalized emittance of an argon beam is to be kept below 0.01 π mm mrad. According to Eq. (1), the 100 mA proton ion current value on which the simulated emittance pattern of Fig. 1.b bases, reduces to 15.8 mA in the case of argon. And scaling back from 50 kV / 5 mm radius / 6 mm gap width to 10 kV / 3 mm / 3 mm results in an expected current value of 2.0 mA.

On the other hand, it is well confirmed by many different simulations as well as measured emittance values that a high-brightness ion beam is compressed near the extraction aperture to 1/2 the outlet aperture width or even less. This immediately leads to another emittance formula that does not explicitly depend on the ion temperature and represents the beam core alone:

$$\varepsilon = 2.5 \, r \quad [\pi \text{ mm mrad}] \tag{3}$$

where ε is the absolute emittance, and r is the aperture radius (mm). For singly charged argon ions with 10 keV beam energy, Eq. (3) is converted to normalized emittance values as follows:

$$\varepsilon_n(10 \text{ kV}) = 0.001825 \, r \quad [\pi \text{ mm mrad}] \tag{4}$$

allowing r = 5.48 mm as maximum tolerable radius for the case of regard. Now, however, the condition S = 1 imposes a larger gap width, completely offsetting the gain in current that would result from the larger outlet aperture area, see Eq. (1). The only benefit then lies in a reduction of the necessary current density, 0.54 mA/cm^2 compared to 1.8 mA/cm^2 needed for the same current to be extracted out of an aperture with 3 mm radius. In any case, these values are both much lower than the maximum current densities that are produced by CHORDIS[9] when generating high-current argon beams, namely 30 mA/cm^2.

In consequence, the beam quality requirements introduced above should be easily fulfilled by many ion sources equipped with a customized extraction system. The reason why high-current sources are preferred even though the control of the beam emittance, rather than the generation of high current, is essential for the outlined goal, bases on the fact that the principles of beam extraction just explained work only if the source plasma is quiet (free of oscillations), cold (less than 0.2 eV ion energy), uniform within a few %, and free of magnetic fields in the outlet plane. These conditions are best fulfilled by high-current ion sources.

The extraction system shown in Fig. 2. can serve as a master design for any other system of different size. When varying details, one should generally try to scale all dimensions, but minor changes are not at all critical, as many simulations and tests performed by the author have proved sufficiently well.

PLASMA GENERATORS

It is not possible here to mention all high-current ion sources described in the literature that could in principle be employed as plasma generators with the qualities outlined above. Multi-cusp sources[14] certainly offer the best prospects in terms of plasma density variation, but at the densities needed for EBIS injector sources this is not a critical issue. The CHORDIS system[15] bases on a multi-cusp reflex discharge and includes source versions for gases and vapors, with internal or external ovens, or with a sputtering electrode for the generation of refractory metal ion beams.[16] The sputtering version of CHORDIS is shown in Fig. 3. Operation with argon as auxiliary gas and a molybdenum sputter target exhibited a very stable molybdenum ion beam, comparable to pure noble gas beams. The transported molybdenum ion current amounted to 1.1 mA, using a 0.8-cm^2 single-aperture extraction system at 20 kV with a 4.2-mm main gap that could have been operated up to 40 kV. By making use of the sputtering technique ions of elements virtually covering the entire periodic table can be generated.

With CHORDIS, the share of metal ions exceeded 10 % of the entire beam current in tests with molybdenum, tantalum, and titanium, without yet reaching saturation. It is worthwhile mentioning that titanium beam production with the sputtering technique proved to be much simpler and much less trouble-affected than when using a chemical compound ($TiCl_4$), instead.[16] The metal share of the entire beam was comparable in both cases. The main difficulty found with the chloride is the deposition of solid material on the first extractor (screening) electrode, leading to frequent high-voltage sparking, especially after an interruption of the beam extraction process. Furthermore, a source that was operated with halogenides needs a diligent servicing procedure after operation.

One should be aware that multi-component plasmas always demand the installation of a beam analysis system between injector source and EBIS. Pure metal plasmas, on the other hand, have been generated with the oven version of CHORDIS,[15] and 37 mA bismuth ion beam current was obtained at 36 kV, using 7 outlet apertures with 2 cm^2 total area. Temperature control of the oven, however, and the servicing procedure for the source are rather elaborate, and will sometimes outweigh the advantage of the single ion species. In addition, the oven cannot possibly be pulsed with a time pattern as the one given above. This means that solid material will be continuously deposited on the extractor electrodes, drastically reducing the operational reliability of the high-voltage system.

Pure multi-charged metal ion beams are also delivered by the pulsed metal arc ion source MEVVA.[17,18] Its discharge pulses of about 1 ms are long enough for the time pattern required from an EBIS injector source, and recently the poor pulse-to-pulse repeatability reported with uranium and titanium beams[19] has been improved.[20]

Another high-current ion source that would suit the needs of an EBIS injector is the duopigatron.[21] This source type, like the duoplasmatron from which it was derived, utilizes a two-stage discharge constricted by an axial magnetic field and an intermediate electrode; the main discharge stage is enclosed by two reflector electrodes and a hollow, cylindrical anode. The axial magnetic field expands towards the outlet reflector electrode and does not generally impede the formation of bright ion beams. A duopigatron for high-current accelerators and industrial applications[22] has set a standard for high-current, high-brightness beams, yielding 98 mA of xenon ions at 35 kV from 7 outlet apertures with 1.4 cm^2 total area. Two-stage discharges allow to apply the two-gas technique[23] where a protective gas (argon) is fed into the first chamber, impeding the fast consumption of the hot cathode filaments, and a reactive gas is added to the second chamber where no more hot metal electrodes are present. This method is convenient for aggressive vapors like chlorides and facilitates the production of ions from low vapor-pressure elements that form volatile compounds. Even oxygen ion beams have been generated in the duopigatron just mentioned, using pure oxygen in addition to argon as feeding gases.[24]

For the purpose of creating a high-brightness ion beam with as little current as 2 mA or less, the effort with the high-current ion sources described so far may seem to be somewhat high. But scaling down the size of the plasma generator would lead to an intolerable loss of ionization efficiency, due to the reduction of the primary electron mean free path in the discharge chamber. On the other hand, an aperture of only a few mm width needs only a narrow plasma column, and this is the reason why the duoplasmatron source[2,3] is a serious option for an EBIS injector source. Historically, the first high-current ion source type was a duoplasmatron, equipped with an expansion cup[25] that causes the dense plasma of the anode column to cool down and assume a uniform density profile.

A duoplasmatron for industrial applications[26] is shown in Fig. 4. This source yields the same ion currents as the CHORDIS sources, up to outlet aperture diameters of 10 mm; a standard argon beam current value is 6 mA, obtained at 30 kV from a single aperture with 0.5 cm^2 area. In comparison to duoplasmatrons optimized for multi-charged ion production,[23] this high-current version is much less subjected to plasma oscillations and shows constant currents for singly-charged noble gas ion beams over many hours.

The generation of ions of non-gaseous elements by a duoplasmatron can be accomplished with the same methods as outlined for duopigatrons. In addition, a sputtering duoplasmatron version has been built[27] that is capable of producing metal ion beams for over 100 hrs.[28] For this purpose a hollow, cylindrical sputtering target is inserted between intermediate electrode and anode and biased to about -250 V below cathode potential. The beam stability with bismuth is better than ± 10 % over a few hours, not quite as good as with the CHORDIS sputter version.[16] To assure stable metal plasma production the duoplasmatron anode must run hot near its aperture, thereby inhibiting the condensation of metal atoms that can lead to droplet formation. A detailed design is shown in Fig. 5.

The only draw back inherent in the application of the sputtering technique (or the use of chemical compounds) is that necessarily the ion beam current has to be shared among some different constituents. 10 % for the useful species is a reasonably pessimistic assumption whenever accurate data are not known. Thus, the 2 mA limit derived above for the entire beam from the emittance restriction means that in reality about 200 µA of metal ion beam current can reasonably be expected. Because this value still exceeds the desired beam current specified at the beginning, it appears preferable to avoid the difficulties involved with the production of pure metal plasmas using evaporator ovens.

REFERENCES

1. A. Hershcowitch, Brookhaven Nat. Lab. and J. Faure, Laboratoire National SATURNE, private communication (1988).
2. M. von Ardenne, Tabellen zur Angewandten Physik (VEB Verlag der Wissenschaften, Berlin, 1962) vol. I, p. 653.
3. H. Fröhlich, Nukleonik 1, 183 (1959).
4. R. Keller, Ion Extraction, in: I.G. Brown, ed., The Physics and Technology of Ion Sources (J. Wiley, New York, in press)
5. C.D. Child, Phys. Rev. (Ser. 1) 32, 492 (1911).
6. I. Langmuir and K.T. Compton, Rev. Mod. Phys. 3, 251 (1931).
7. J.R. Coupland, T.S. Green, D.P. Hammond, and A.C. Riviere, Rev. Sci. Instr. 44, 1258 (1973).
8. R. Keller, P. Spädtke, and K. Hofmann, Springer Ser. in Electrophys. 11, 69 (1983).
9. R. Keller, F. Nöhmayer, P. Spädtke, and M.-H. Schönenberg, Vacuum 34, 31 (1984).
10. P. Allison, J.D. Sherman, and H.V. Smith, Report LA-8808, Los Alamos Nat. Lab. (1981).
11. A.J.T. Holmes and M. Inman, Proc. 1979 Linac Conf., BNL-51134, Brookhaven Nat. Lab., p. 424 (1979).
12. R. Keller, P. Spädtke, and H. Emig, Vacuum 36, 833 (1986).
13. R. Keller, 1987 IEEE Particle Accelerator Conf. 87CH2387-9, p. 382 (1987).
14. K.W. Ehlers and K.-N. Leung, Rev. Sci. Instr. 54, 1296 (1983).
15. R. Keller, P. Spädtke, and F. Nöhmayer, Proc. Int. Ion Engineering Congr., Kyoto, p. 25 (1983).
16. R. Keller, B.R. Nielsen, and B. Torp, Proc. 7th Int. Conf. Ion Implantation Techniques, Kyoto, June 7-10, 1988 (in press).
17. I.G. Brown, J.E. Galvin, and R.A. MacGill, Appl. Phys. Lett. 47, 358 (1985).
18. I.G. Brown, this Workshop Proc.
19. I.G. Brown, J.E. Galvin, R. Keller, P. Spädtke, R.W. Müller, and J. Bolle, Nucl. Instr. Meth. in Phys. Research A245, 217 (1986).
20. I.G. Brown, Lawrence Berkeley Laboratory, private communication (1988).
21. R.A. Demirchanov, H. Fröhlich, U.V. Kursanov, and T.I. Gutkin, BNL-767 (C-36), Brookhaven Nat. Lab., p. 224 (1962).
22. M.R. Shubaly, Inst. Phys. Conf. Series 54, 333 (1980).
23. R. Keller, Radiation Effects 44, 201 (1979).
24. M.R. Shubaly, R.G. Maggs, and A.E. Weeden, IEEE Trans. Nucl. Sci. NS-32, 1751 (1985).
25. O.B. Morgan, G.G. Kelley, and R.C. Davis, Rev. Sci. Instr. 38, 467 (1967).
26. R. Keller, ROKION Darmstadt, unpublished material (1984).
27. R. Keller and M. Müller, Inst. Phys. Conf. Series 38, 40 (1978).
28. N. Angert, R. Keller, and M. Müller, Proc. Int. Ion Engineering Congr., Kyoto, p. 225 (1983)

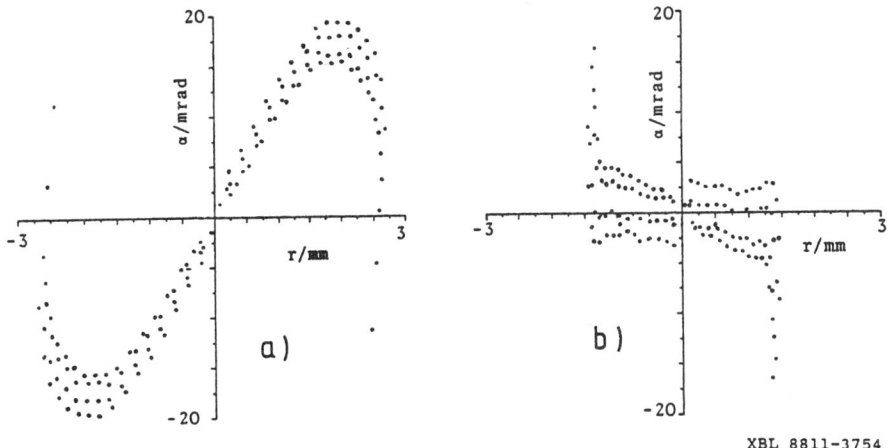

Figure 1.
Simulated emittance patterns of the beams generated by a single-aperture extraction system shown in Fig. 2. The patterns are calculated for a plane 2 mm downstream of the ground electrode. a) For system with contours shown by solid lines, with ion current density chosen so as to yield a maximum transported ion current within 20 mrad half-angle, amounting to 283 mA for protons. b) For system with modified screening and ground electrode contours, as shown by the broken lines. The ion current density in this case was chosen so as to yield a maximum brightness value for the beam core, totally neglecting the tails of the emittance pattern (beam halo). The beam core then contains 100 mA proton current within 5 mrad divergence half-angle. After Ref. 12.

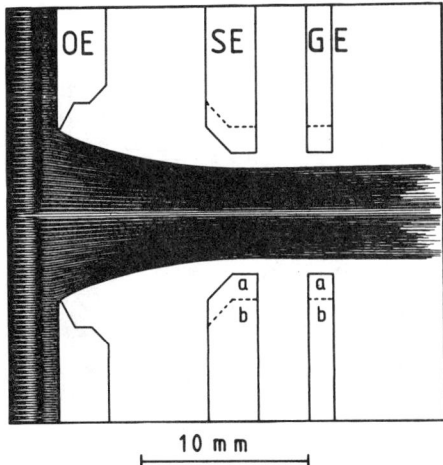

Figure 2.
Round, single-aperture, accel/decel extraction system.[12] The broken lines mark a geometry modification that leads to higher beam brightness, at the expense of beam current, see Fig. 1. OE, outlet electrode, 50 kV. SE, screening electrode (electron trap), -4 kV. GE, ground electrode, 0 kV.

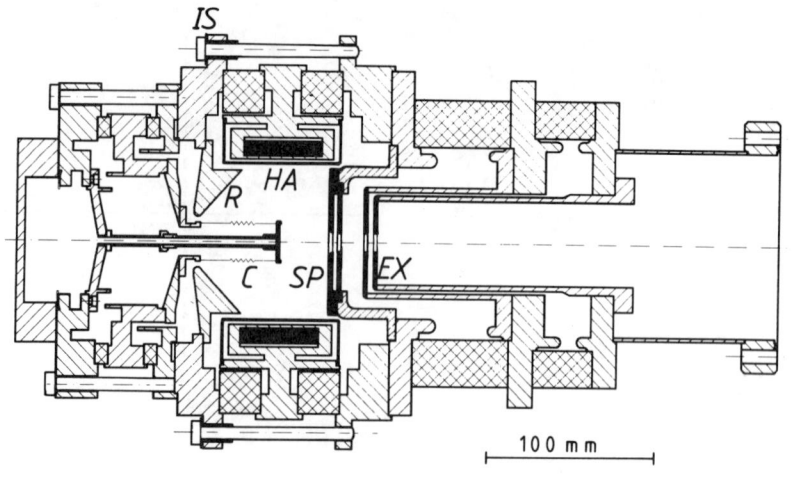

Figure 3.
Sputtering version of the ion source system CHORDIS.16 C, cathode. SP, sputter electrode placed on the inside of the outlet reflector electrode. EX, accel/decel extraction system. HA, hot anode. R, reflector on cathode side. IS, insulator introduced to electrically separate the two reflector electrodes. 18 rare-earth/cobalt magnets are placed around the cylindrical anode.

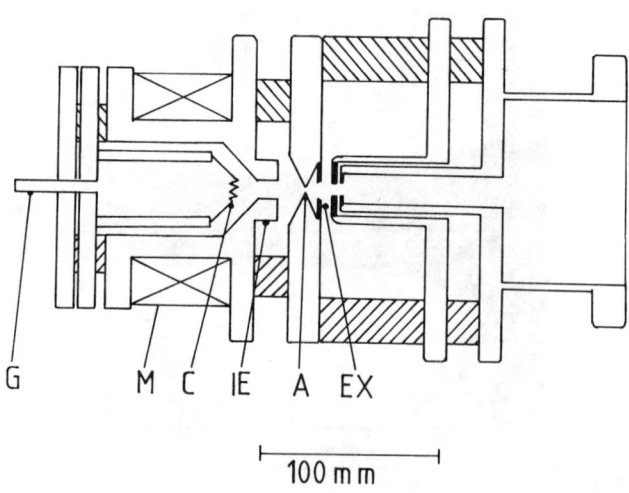

Figure 4.
Duoplasmatron with high-current extraction system.[26] G, gas inlet. M, magnet coil. C, cathode filament. IE, intermediate electrode. A, anode. EX, accel/decel extraction system. The expansion cup is the chamber between anode and the leftmost of the three extraction electrodes.

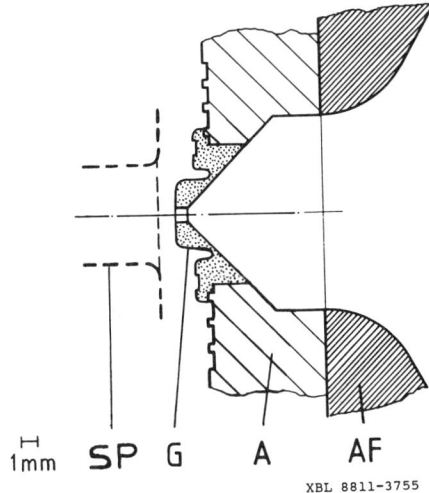

Figure 5.
Details of the anode region of a sputter duoplasmatron for multi-charged ions.[28] SP, sputter target. G, graphite insert. A, non-magnetic anode. AF, magnetic anode flange. This design already features a plasma expansion cup, but for high-quality ion beam production the cup should be terminated on its right side by an outlet electrode with shaped aperture contour, see Fig. 2.

DISCUSSION

<u>Levine</u>: I would not characterize MEVVA as extremely reliable. We can get along with it because the variation of intensity is not serious for our particular experiment. However, I would like to say that we have a very stringent requirement on injection temperature because we compress our ions by a factor of 900 between the entrance aperture and the trap. This is so ions will be cool enough to stay in the electron beam. To do this requires a temperature of .01 eV at the aperture. It is done by starting with a high current beam and using a very small part of the beam. The principle advantage of the MEVVA is its simplicity and the large variety of metallic ions it can inject. MEVVA only works about 95% of the time. It works well with gold, but with some of the harder metals it is less reliable.

PROPOSED EBIS BASED SYSTEMS

THE RELATIVISTIC HEAVY ION COLLIDER (RHIC)
PROJECT AT BROOKHAVEN*

T. W. Ludlam
Brookhaven National Laboratory, Upton, NY 11973

I. THE RHIC CONCEPT

High energy and nuclear physics have undergone extraordinary changes during the past fifteen years. The existence of quarks as the elementary constituents of hadronic matter has been confirmed, and the field of nuclear physics has embraced a wide-ranging program of research to study the structure of nuclear matter in terms of the quark degrees of freedom. As a result the characteristic energy of nuclear collision experiments is no longer confined to the few MeV range of traditional nuclear structure experiments, exploring a system of neutrons and protons, but extends into the many-GeV range which once was the exclusive domain of particle physics. This has led to a new generation of large accelerator facilities for nuclear physics. One example of these is the CEBAF 4 GeV electron beam facility now under construction in New Port News, Virginia.

The most spectacular of the new high energy approaches to nuclear physics is in the field of relativistic heavy ion collisions, for which it is anticipated that construction of the RHIC facility will begin soon. Here the goal is to subject large volumes of nuclear matter to such extreme conditions of temperature and pressure that a new form of matter is produced in which the recognizable components are not the familiar neutrons and protons, but are quarks.

The thermodynamic conditions required to bring about this phase transition from ordinary nuclear matter to a plasma of quarks and gluons can be estimated from the observed properties of quarks in high energy scattering experiments, and detailed theoretical studies can be carried out via quantum chromodynamics. The interesting temperatures are of the order of 100-200 MeV, and the corresponding densities about 10 times that of ordinary nuclear matter. These conditions are characteristic of the expanding universe a few microseconds after the Big Bang. The relevant densities might be attained in the cores of neutron stars. The possibility of exploring, in the laboratory, nuclear matter under such extreme conditions has sparked the recent widespread interest in colliding heavy nuclei at very high energies.

During the past two and a half years the first steps have been taken to carry out this kind of research. Ion beams have been accelerated in the Brookhaven AGS (14.5 GeV/u) and the CERN

---

*Work performed under the auspices of the U.S. Department of Energy

SPS (200 GeV/u). These beams, impinging on fixed targets, provide c.m. collision energies which are about an order of magnitude below the values which are ultimately desired. To date, the accelerated ion beams have been limited to nuclei of relatively low mass ($A \lesssim 32$). Nonetheless, important new data have resulted:[1] These experiments have established that states of compressed nuclear matter can be created and studied under laboratory conditions in which extreme values of energy density are achieved. The general characteristics of the observed events bear out the theoretical expectations for large thermal energy deposition in violent nuclear collisions, and they point the way for extending further the range of thermodynamic conditions over which to study new forms of dense hadronic matter. The most exciting results are the several indications that, indeed, when these extreme conditions are reached, there are new physical phenomena to be explored.

The planning for future facilities in both the U.S. and Europe call for a continued pushing back of the frontiers of this new field. Within a few years it will be possible to accelerate the heaviest nuclei (e.g., gold, lead, uranium) at both the SPS at CERN and the AGS at Brookhaven. The next step is the realization of RHIC in which high energy colliding beams will be available in a dedicated facility for heavy ion research.

The Relativistic Heavy Ion Collider at Brookhaven will extend the present heavy ion capabilities of the AGS into an energy domain not available at any other laboratory within the foreseeable future. The Brookhaven site map in Fig. 1 shows the accelerators and connecting beam tunnels involved in the heavy ion program, i.e., the Tandem Van de Graaff, the Heavy Ion Transfer Line, the AGS, the Booster Synchrotron presently under construction, and the existing ring tunnel for the proposed collider. Operation of the AGS for heavy ion experiments started in October 1986 with the delivery of O^{8+} beams. Subsequently, the mass range was extended with the AGS delivering typically 2×10^8 Si^{14+} ions/pulse at an energy of 14.5 GeV/u. Completion of the AGS Booster Synchrotron in 1991 will extend the mass range to the heaviest ions, typically ^{179}Au, with ^{238}U a definite possibility.

The collider[2] consists of two rings of superconducting magnets for accelerating and storing circulating beams of ions. The facility will have the flexibility of using the full range of ion species from protons to gold, and will cover the entire range of collision energies from those available at the Brookhaven Alternating Gradient Synchrotron (AGS) up to the top collider energy (100 GeV/AMU for gold beams). Average luminosities (particle fluxes) of $10^{26} - 10^{27}$ $cm^{-2} sec^{-1}$ are predicted for gold-on-gold collisions at full energy, depending on the details of the experimental layout. Proton energies up to 250 GeV in each beam will be available at luminosities of about 10^{31} $cm^{-2} sec^{-1}$. The capability for collisions between different species will also be provided. The RHIC ring lattice has six crossing regions where the beams collide and experiments can be performed. Four of these experimental areas will be utilized initially, with the remaining two available for future expansion of the research program. RHIC will use the existing AGS and Tandem Van de Graaff accelerator complex at Brookhaven as an injector. The new accelerator will be built in the existing CBA tunnel (3.8 km circumference), and will utilize the experimental halls, support building and liquid helium refrigerator from the partially completed CBA project. The RHIC project is proposed to begin construction in 1991, with the first collider experiments beginning in 1996.

II. INJECTORS AND MACHINE PERFORMANCE

The RHIC facility is designed to incorporate the flexibility to study collisions of all nuclei, from the lightest to the heaviest, and allow experiments to be carried out over the full range of energies, from a few GeV/u in the c.m. (AGS fixed target) up to the top collider energy, with no inaccessible gaps, and with adequate intensity for sensitive experiments. A plot of luminosity vs. energy for various ion species, with corresponding beam lifetime is shown in Fig. 2.

As discussed above, the AGS is presently capable of accelerating ions up to silicon with an energy of 14.5 GeV/u, using one of the two BNL Tandem Van de Graaff accelerators as injector. By the end of 1991, using the AGS Booster Synchrotron, which is presently under construction, the range of ion masses which will be available from the AGS will be extended up to Gold. The Tandem-Booster-AGS complex will then not only be available to provide beams for fixed-target heavy ion experiments, but will serve as injector for the RHIC collider.

The design specifications for RHIC call for 1.0×10^9 particles per bunch of fully stripped ^{197}Au ions, with normalized emittance of 10π mm·mrad and a bunch area of 0.3 eV·sec. The AGS will deliver a single bunch to RHIC, where nominally 57 bunches are accumulated in boxcar fashion.

The acceleration of heavy ions will be started using one of two existing Tandem Van de Graaff accelerators. The second tandem will be available as an alternate injector, providing an additional degree of operational reliability. A Tandem Van de Graaff accelerator is an attractive choice for starting the injection of heavy ions into a synchrotron. The ion source is conveniently located close to ground potential, and a wide variety of negative ions are available. Other important advantages are related to the excellent beam quality and stability. The transverse emittance at the exit of the accelerator is better than 1π mm·mrad and the energy stability is one part in 10^4. Of particular importance for RHIC operation is the available ion-source output currents prior to stripping and acceleration in the Tandem. A pulsed mode of operating the source has been developed at Brookhaven, demonstrating that ten 300 μsec long pulses per second could be injected at maximum intensity without adverse effects on the performance or stability of the accelerator. Up to 240 μA of useable ^{197}Au are reliably available from a sputter source. Larger currents are available for lighter ions, with up to 500 µA for ^{12}C. The design for injection to RHIC assumes a source current of 200 µA and a pulse length of 110 µS. The Tandem then delivers 1.2×10^{10} gold ions per pulse to the booster, in a charge state +14, with kinetic energy equal to 1.07 MeV.

The Booster Synchrotron, with a vacuum of 10^{-10} torr will be capable of accelerating the partially stripped ions, without significant beam loss, for final stripping before injection into the AGS. The final energy of Au^{+14} ions after acceleration in the Booster is 65.5 MeV/u. After extraction from the Booster, and on their way to the AGS, the ions pass through a stripping target which, for gold ions, consists of a copper foil 35 mg/cm^2 thick. The following table shows the number of ions per bunch and the normalized emittance for various ion species after acceleration in the Booster and stripping prior to injection into the AGS:

Species	^{16}O	^{127}I	^{197}Au	^{238}U
Charge State	8	53	77	90
Ions/bunch ($\times 10^9$)	8.3	2.3	2.0	0.6
Normal emittance ($\pi\cdot$mm$\cdot$mrad)	6.1	3.6	4.0	2.3

Inside the AGS all ions with $A \geq 127$ will be fully stripped. Heavier ions such as ^{197}Au (77^+) and ^{238}U (90^+) will be accelerated with 2 electrons in a filled K-shell.

After acceleration in the AGS, the buckets are transferred one at a time into RHIC. For the heaviest ions, it is necessary to pass through a final stripping foil to remove the last two electrons in the K-shell. At the top AGS energies (10.7 GeV/u kinetic energy for ^{197}Au), a design beta function value of 50, and a rms scattering angle of .046 mrad for a 100 mg/cm^2 copper foil, would correspond to an emittance growth of .033π mm$\cdot$mrad. Thus an insignificant increase in beam emittance is expected at these energies. At these energies, negligible losses in particle numbers per beam bunch are also expected during the final stripping process.

Single bunches of ions are injected 57 times into each ring in boxcar fashion. Filling time per ring will be about one minute. For gold, as an example, there will be ~ 1.1×1^{-9} ions/bunch, or 6×10^{10} ions in 57 bunches in each ring. For the lightest ions, hydrogen and deuterium, approximately 10^{11} ions/bunch can be stored in the machine. Acceleration will take approximately 60 seconds. Bending and focussing of the ion beams is achieved with superconducting magnets. Given that the machine will be built in the existing CBA tunnel, a cost optimization is achieved by filling the circumference with relatively low field magnets. The maximum energy of 100 GeV/amu for gold ions (250 GeV for protons) is reached with a magnetic field of 3.4 Tesla. Maximum operational flexibility is obtained with the magnets of each ring in separate vacuum vessels, with the beams in the arcs separated by 90 cm. The layout of the storage rings is shown in Fig. 3a.

The six beam crossing regions are designed to accommodate a range of detector configurations to fulfill the needs of experiments. The free space available for experimental equipment in each crossing region is 9 meters on either side of the intersecting point. For head-on collisions with gold ion beams at top energy, a luminosity of 2×10^{26} cm^{-2} sec^{-1} averaged over a 10 hour beam lifetime is expected. For protons the expected luminosity is 1×10^{31} cm^{-2} sec^{-1}. The r.m.s. length of luminous interaction region where the beams cross is $\sim$ 20 cm. Collisions of unequal species, e.g., protons in one beam and gold in the other, will be possible as well. The Accelerator Physics group has considered posssible future upgrades of the machine performance, and these were discussed in detail in a recent workshop on the ultimate performance of such a collider.[3]

III. PRESENT STATUS; OUTLOOK FOR EXPERIMENTS

As noted above, a large fraction of the RHIC facility already exists. For the injector complex, the Tandem Van de Graaff, AGS, and Heavy Ion Transfer Line are already operational; the Booster Synchrotron is under construction. Most of the conventional construction for the collider is complete, including the ring tunnel, main service building and experimental halls for four of the six intersection regions. In addition, the liquid helium refrigerator, capable of cooling all of the superconducting magnets in the collider has been completed (as part of the CBA project) and successfully tested. The refrigerator has a capacity of 25 kilowatts at a temperature of 4.3K. The estimated heat load for RHIC is $\sim$ 10 kilowatts at 4.6K.

One of the most important elements of the RHIC proposal is the design of the superconducting magnets for the accelerator rings. These magnets are the largest component of the cost of the machine, and their fabrication and installation is the major determinant of the construction schedule. The design of these magnets is based on the cosine theta coil structure developed at Brookhaven for the Isabelle/CBA magnets, which has since been adopted for the Tevatron, HERA and SSC magnets as well. The RHIC dipole magnets are designed to operate at a relatively low field (3.5 T), and thus the coil can be wound in a single layer of superconductor. The dipole magnet cross section is shown in Fig. 4. Figure 5 shows a magnet assembly, consisting of a dipole, quadrupole and corrector coils, mounted in a cryostat. These magnets are fully designed, and have been the major component of an intensive RHIC R&D program which has been funded by the Department of Energy beginning in fiscal year 1987. Full-size, "machine quality" prototypes of dipoles, quadrupoles and sextupoles have been built and successfully tested. In the case of the dipoles, the largest component of the magnet system, three of the prototypes built thus far have been assembled by an industrial manufacturer. All of the prototype dipole magnets (5 have been tested so far) have reached fields of approximately 4.6 T, or 35% higher than the operating field of RHIC with virtually no training. The RHIC magnets have been designed to be amenable to fabrication in quantity in industrial facilities, and it is planned that a significant fraction of the magnets for RHIC will be built by commercial manufacturers.

The magnet R&D program is continuing, with work now in progress to prepare a full cell of the machine magnet lattice (see Fig. 3) consisting of two dipoles, two quadrupoles, sextupoles and lumped corrector package, which will be installed and tested as a system prior to the production of final magnets.

The project has been reviewed and validated by the U.S. Department of Energy, and construction will begin in fiscal year 1991 if funds are made available. The accelerator construction cost is roughly 215 million dollars, with an additional 74 million dollars budgeted for detectors (these figures are in FY 1989 dollars).

Of the six crossing regions built into the RHIC rings, those at the 2, 4, 6 and 8 o'clock positions (see Fig. 1) have completed experimental halls, including support buildings and (except in the 4 o'clock "open area") crane coverage. The RHIC plan calls for mounting experiments initially in these four areas, leaving the remaining two unfinished until some later time.

The recording of events at RHIC, with sensitivity corresponding to the expected signals for new phenomena in nuclear matter under extreme thermodynamic conditions, will require the extension of known techniques for particle detection beyond the present ranges of application in elementary particle and nuclear physics. With beam energies up to 100+100 GeV/nucleon and ion masses up to ≥200, the total energy in each collision can reach up to 40 TeV in the center-of-mass: a range far beyond that of any present accelerator or any existing detector system. Unlike the proposed SSC collider, which would accelerate elementary particles to such an energy and produce hundreds of very high energy particles in the final state, the most interesting events at the RHIC collider are expected to produce tens of thousands of final-state particles in each collision, with proportionately less energy carried away by each particle. Thus, while the basic detector technology will have much in common with the detector systems developed for colliding beams of elementary particles, the design of detector systems for the heavy ion collider must address a different set of problems.

In a series of workshops[4-6] held since 1985, designs for large detector systems have been examined to study their technical feasibility and their impact on the design and modes of operation of the collider. These workshop studies also showed that the most useful improvements to the machine performance - from the point of view of extending the physics reach after, say, a first generation of experiments - would be those which increased the luminosity of the machine, rather than extending the energy range. Subsequent accelerator physics studies have explored possible future upgrades of the machine which could increase the luminosity from all ion species by about a factor of ten beyond the current design values.[3]

REFERENCES

1. For summaries of these results see P. Braun-Munzinger and S. Nagamiya (eds.), Proc. Int. Conf. on Ultra-Relativistic Nucleus-Nucleus Collisions, Lenox, MA, Sept. 1988 (to be published Nucl. Phys. A).

2. Conceptual Design of the Relativistic Heavy Ion Collider RHIC, BNL Report 51932 (1986). An updated edition is being prepared, and will be available in May, 1989.

3. F. Khiari et al., eds., "Workshop on the RHIC Performance", BNL 41604 (1988).

4. P. Haustein and C. Woody, eds., Proc. of the Workshop on Experiments and Detectors for a Relativistic Heavy Ion Collider, Brookhaven National Laboratory Report BNL 51921 (1985).

5. H. G. Ritter and A. Shor, eds., Proc. of the Second Workshop on Experiments and Detectors for RHIC, LBL-24604 (1988).

6. B. Shivakumar and P. Vincent, eds., Proc. of the Third Workshop on Relativistic Heavy Ions, BNL 52185 (1988).

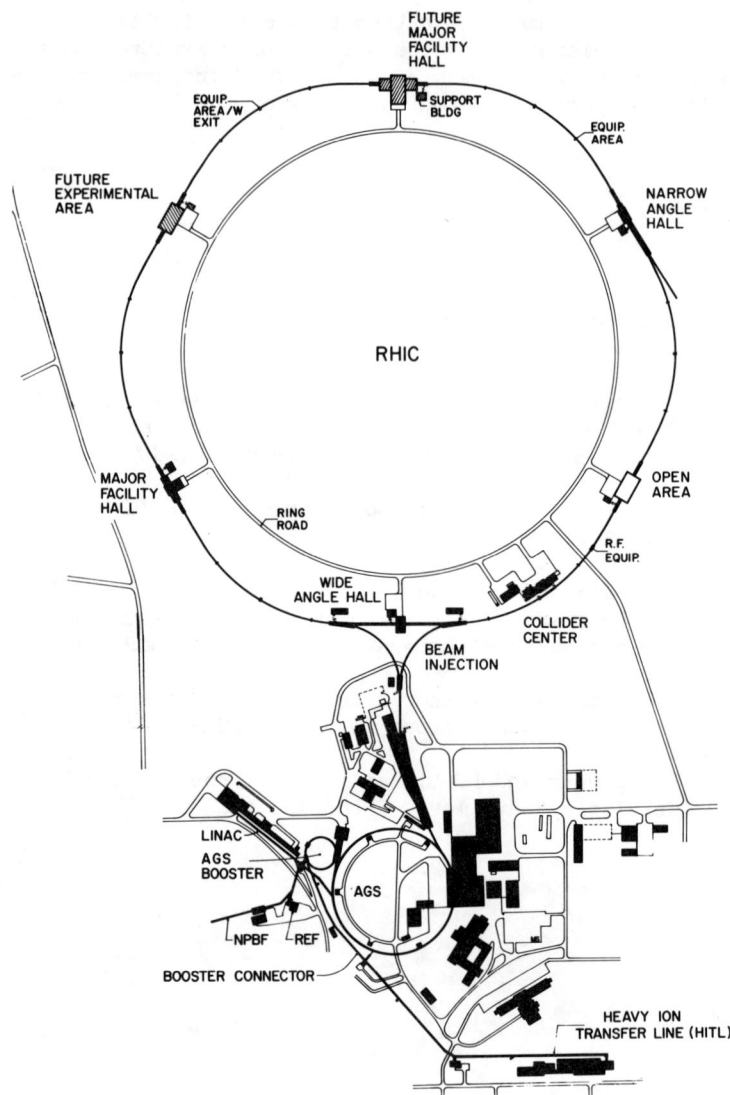

Fig. 1 Site map of present and proposed accelerators at Brookhaven. The Tandem Van de Graaff and the AGS with its linac injector are existing machines. The Booster Synchrotron for pre-injector to the AGS is currently under construction. The RHIC colliding beams accelerator to the north of the AGS complex is a proposed construction project.

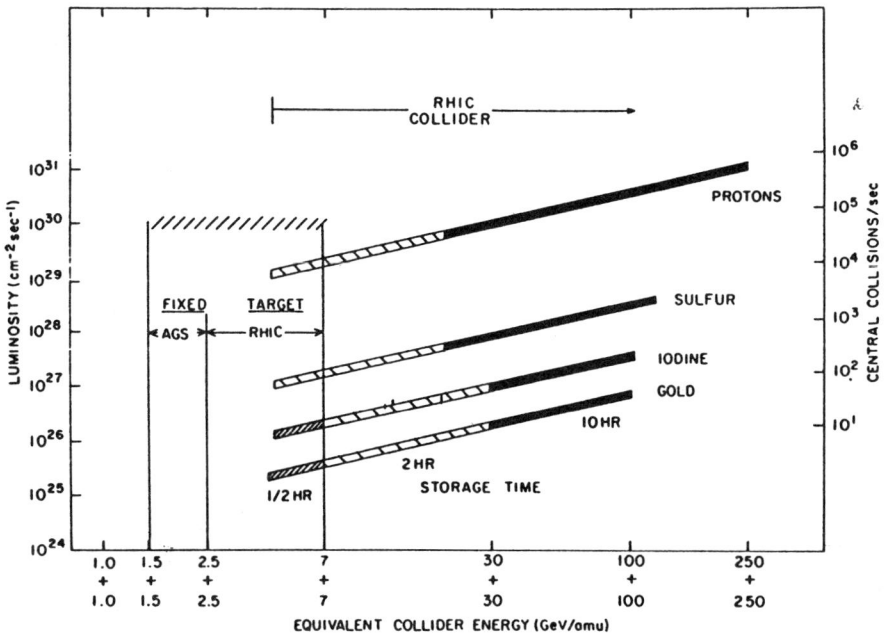

Fig. 2 The design luminosity, for various ion masses, as a function of collision energy over the full range accessible with AGS and RHIC. On the left-hand scale, central collisions correspond to impact parameter less than 1 fermi.

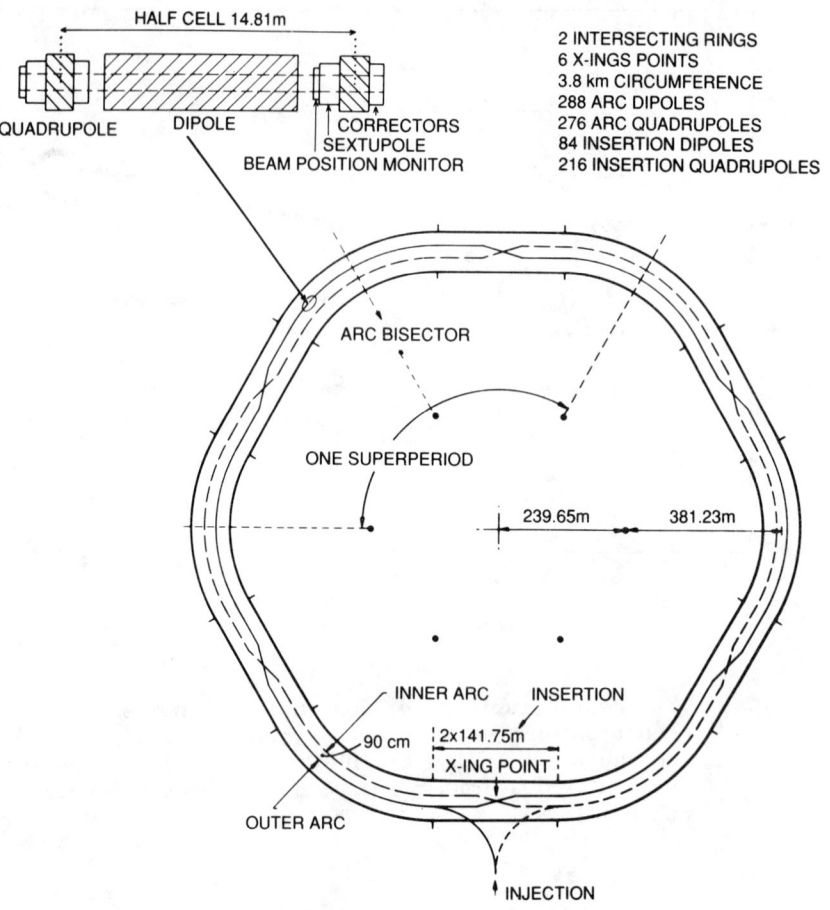

Fig. 3 Layout of the storage rings for the RHIC collider.

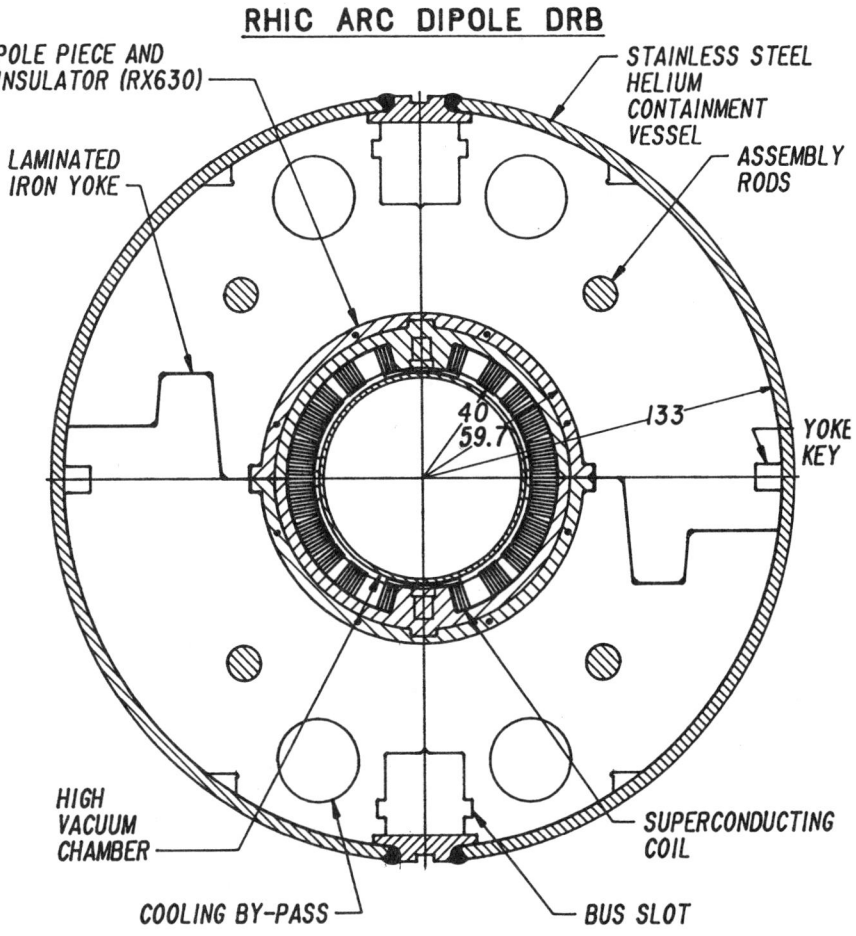

Fig. 4   Cross section of RHIC dipole magnet.

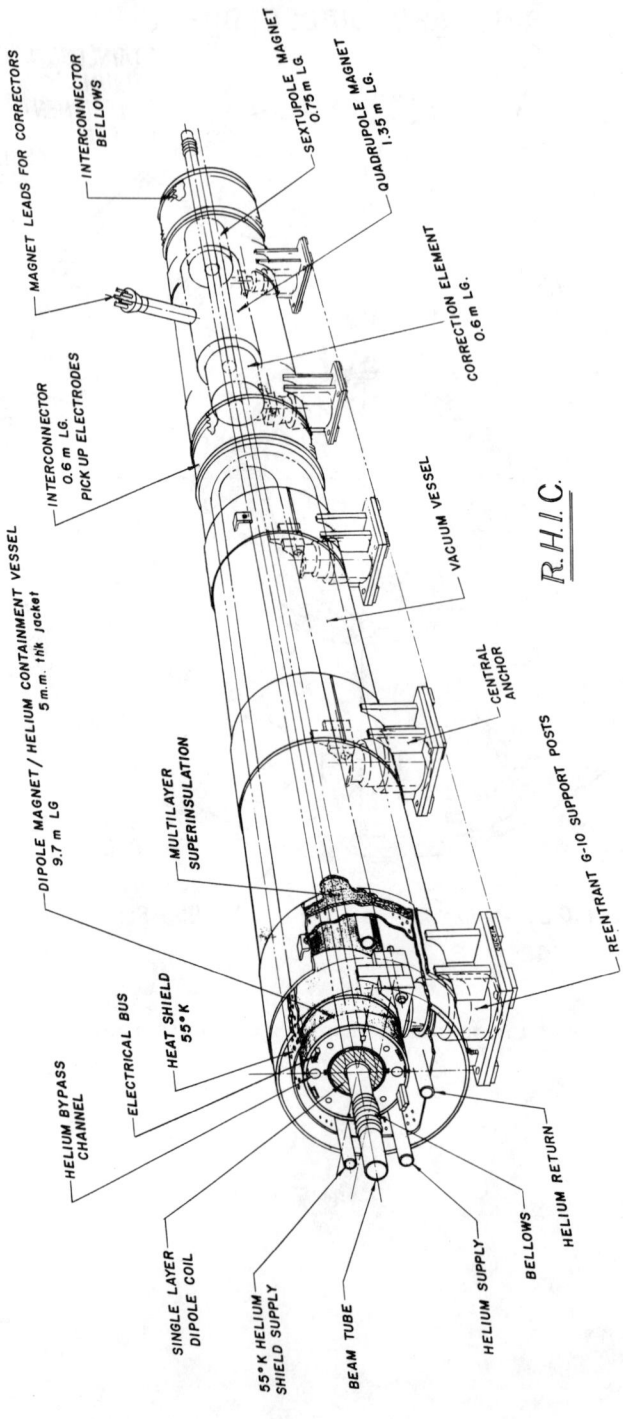

Fig. 5 RHIC magnet assembly: The drawing shows a half-cell of the arc magnet lattice, including a dipole, corrector package, quadrupole and sextupole magnets enclosed in their cryostat.

DISCUSSION

Staples: Tom, what are the ultimate limitations and the intensity that you can get into this machine? Is it IVS or you are talking now about upgrading a 2×10^9 per bunch? What are the mechanisms that eventually will ......

Ludlam: There are a number of things which ultimately limit how far you can go just by the brute force technique of stuffing more particles in and, for example, there are limitations on the beam beam tune chips.

Staples: What is the beam beam tune chip, at 2×10^9 bunch? Do you know?

Ludlam: Maybe one of my colleagues here has.

Hahn: .02; a total of 6, if I remember correctly. But that is consistent. I mean that is the design number which we were given from the SPS as the acceptable limit. We have the same number, again for all three, not per insertion but for all. We have 6. So per insertion, I mean it is less.

Ludlam: I am glad to introduce Harold Hahn, who is the accelerator physicist in charge of these.

Hahn: Around the machine S achieved at the SPS and recently at the Fermi Lab Tevatron in the pp experiment.

Staples: Tom, what are the ultimate limitations and the intensity that you can get into this machine? Is it IVS or you are talking now about upgrading a 2×10^9 per bunch? What are the mechanisms that eventually will ......

Ludlam: There are a number of things which ultimately limit how far you can go just by the brute force technique of stuffing more particles in and, for example, there are limitations on the beam beam tune chips.

Staples: What is the beam beam tune chip, at 2×10^9 bunch? Do you know?

Ludlam: Maybe one of my colleagues here has.

Hahn: .02; a total of 6, if I remember correctly. But that is consistent. I mean that is the design number which we were given from the SPS as the acceptable limit. We have the same number, again for all three, not per insertion but for all. We have 6. So per insertion, I mean it is less.

Ludlam: I am glad to introduce Harold Hahn, who is the accelerator physicist in charge of these.

Hahn: Around the machine S achieved at the SPS and recently at the Fermi Lab Tevatron in the pp experiment.

SOURCE OPTIONS FOR RHIC*

K. Prelec
AGS Department, Brookhaven National Laboratory
Associated Universities, Inc., Upton, NY 11973 USA

INTRODUCTION

Conceptual designs of the RHIC facility are matched to the parameters of existing tandem Van de Graaff accelerators at Brookhaven National Laboratory. It has been shown that tandems could produce many ion species up to gold[1] with sufficient intensities for injection into the BNL booster and further acceleration and storage in AGS and RHIC sychrotrons. There have been, however, questions about the long-term performance of tandem accelerators, in view of their reliability, cost of maintenance, and expected requests for higher intensities than they could provide. A study[2] was done in 1986 to investigate the possibility of replacing the tandems with a more compact preinjector situated close to the booster. This report will review the options for such a preinjector.

RHIC CONCEPTUAL DESIGNS

At the time this symposium was held, there was only one published version of the RHIC Conceptual Design[1]; work has still been in progress on the modified version, with some preliminary results reported in BNL documents.[3] Assuming a negative ion current of 200 µA from the tandem source, stripping both, in the tandem terminal and after acceleration to the ground, booster and AGS beam intensities for a few selected ion species were estimated (Table I). It was thought that the acceleration of ions heavier than gold would not be possible because not enough of them could be produced in a fully stripped state, a condition, at that time, considered as necessary due to the vacuum in the AGS.

---

*Work performed under the auspices of the U.S. Department of Energy.

Table I

Species		C^{6+}	S^{14+}	Cu^{21+}	I^{29+}	Au^{33+}
	Booster					
	AGS	C^{6+}	S^{16+}	Cu^{29+}	I^{53+}	Au^{79+}
Output Booster Intensity*		14	6.7	4.7	3.2	2.2
Output AGS Intensity*		14	6.4	4.5	2.6	1.1

* $\times 10^9$, per bunch

As part of the review process of the conceptual design, a new scheme for injection was studied.[3] In this approach there would be no stripping foil between the tandem and booster and ions would be injected into the booster in a much lower charge state, resulting in a higher space charge limit. The final booster energy would still be high enough for fully stripping of all ion species except gold (and heavier), which would have to have the last two electrons remaining. Vacuum losses in the AGS for helium-like heavy ions would amount to only a few percent, as shown by subsequent analysis. Table II shows estimated intensities for this scheme.

Table II

Species		C^{5+}	S^{9+}	Cu^{11+}	I^{13+}	Au^{14+}
	Booster					
	AGS	C^{6+}	S^{16+}	Cu^{29+}	I^{53+}	Au^{77+}
Output Booster Intensity*		7.7	7.5	8.7	5.6	4.1
Output AGS Intensity*		7.7	7.5	8.7	2.3	2.0

* $\times 10^9$, per bunch

The values in Table II refer to the number of particles per bunch, which is relevant for RHIC injection (harmonic number was chosen h = 2 or h = 3). By comparing the tables one can conclude that a substantial gain in intensity would result for gold ions if the new scheme is used; another expected benefit of accelerating helium-like ions in the AGS is the extension of the range of available ions up to uranium.

EBIS OPTION

An EBIS offers several attractive features as a source of highly stripped heavy ions to be injected into a synchrotron: it is a pulsed device capable, in principle, to provide any charge state of any ion species by properly adjusting source parameters. The important design parameters are the electron beam energy eV, the product $j \times \tau$ of the electron beam current density j and interaction time τ and the product $I \times L$ of the electron beam current I and the trap length L. Assuming now that a device with design parameters similar to the DIONE is considered to be used in the RHIC preinjector, it is possible to estimate its performance as a source for RHIC. With an electron beam energy of eV = 20 keV, a beam current of I = 1A, a beam current density of 10^4 A/cm^2, and a length of L = 2m, the number of charges in the machine is:

$$Q_o = 10^{13} \, ILV^{-1/2} = 1.41 \times 10^{11}.$$

If the ion build-up is 75%, the trap could store

$$Q = 10^{11} \text{ charges.}$$

In order to fully utilize the electron beam energy of 20 keV, the maximum $j\tau$ should be about 10^3 As/cm^2 or 6×10^{21} charges/cm^2, which corresponds to an interaction time τ = 100 ms. An EBIS would best be followed by an RFQ and a short linac (equivalent voltage 10 - 20 MV) to raise the energy for booster injection. If properly designed, RFQ structures can capture and accelerate ions in a relatively wide range of the charge-to-mass ratios with an efficiency approaching 100%. Table III shows estimated intensities for charge states similar to those in Tables I and II; in order to achieve a higher abundance of the selected charge state ($\approx$ 50%), it was assumed that ions have been stripped to a full shell or subshell.

Table III

Source-booster	C^{6+}	S^{12+}	Cu^{11+}	Cu^{25+}	I^{17+}	I^{35+}	Au^{19+}	Au^{51+}
q/m	0.5	0.375	0.17	0.39	0.134	0.276	0.096	0.259
Beam Voltage V	5kV	5kV	10kV	10kV	10kV	10kV	10kV	20kV
$j\tau$†	10^{20}	2×10^{19}	$<10^{19}$	10^{21}	10^{19}	2×10^{20}	2×10^{19}	6×10^{21}
AGS	C^{6+}	S^{16+}	Cu^{29+}	Cu^{29+}	I^{53+}	I^{53+}	Au^{77+}	Au^{79+}
AGS Intensity*	16.6	8.2	7.0	2.8	2.1	2.1	1.8	0.8

* $\times 10^9$, per bunch
† cm^{-2}

It should be emphasized that the full electron beam energy eV and the highest beam current density j are necessary for the heaviest ions and highest charge states only; for lighter ions both requirements can be substantially relaxed (Table III; the storage capacity of the trap has been increased for voltages below 20 kV). A comparison of Tables II and III shows, that except for the heaviest elements, the performance of such an EBIS would be comparable to the tandem.

As shown in Ref. 2, another EBIS with a much higher electron beam current of I = 6A would not only increase the yield by a substantial factor compared to the tandem, but enable the production and acceleration of uranium ions as well. Its performance is summarized in Table IV, assuming again 50% abundancy and 75% ion build-up.

Table IV

Source-booster	C^{6+}	S^{12+}	Cu^{25+}	I^{35+}	Au^{51+}	U^{71+}
AGS	C^{6+}	S^{16+}	Cu^{29+}	I^{53+}	Au^{79+}	U^{92+}
AGS Intensity*	100	49	17	12	4.8	2.8

* $\times 10^9$, per bunch

A paper presented to this Symposium[4] describes the design of an RFQ matched to the EBIS. It would accept ions with $q/m \geqslant 0.23$ and deliver a beam with an energy of 300 keV/n. The selected design is very conservative, but it would not handle all the species considered in Table III. However, by a minor modification, its range should be extendable to all species but Au^{19+}. The question of the next state of acceleration, in a linac, has not been addressed in sufficient details.

On the basis of this analysis, we conclude that a two-stage approach in the EBIS design may be preferable. The state-of-the-art model, described first, could be built in five years. It would serve to gain the experience in the EBIS design, and it could replace the tandem for a part of the heavy ion program (operation with the fixed target; operation with light and medium heavy ions in the collider mode; tuning of the machines and beam lines with intensities below the full RHIC intensity). The other EBIS model would take five more years to build, but it should be capable to fully replace the tandem[2].

ECR OPTION

There is, at present, no ECR source capable of delivering ion beams with intensities as required by RHIC. ECR sources are inherently steady state devices, having a good output but with a rather wide charge state distribution. Although the extracted current is much higher than the time averaged current from an EBIS the number of particles in the desired charge state available during the short injection interval (e.g. 100 μs) may be orders of magnitude lower than for the tandem. The CERN Pb acceleration project[5] is based on an ECR source to be developed within a few years, delivering about 1 part.μA of lead ions in a charge state between 25+ to 30+. To match the expected tandem performance for gold ions, the source should deliver either 24 part.μA of Au^{14+} ions or 5 part.μA of Au^{33+} ions. It is unlikely that in the time frame specified before, one could develop a source with a performance about one order of magnitude above the state-of-the-art.

REFERENCES

1. Conceptual Design of the RHIC, BNL Rep. 51932 (May 1986)

2. Heavy Ion Preinjector Committee Report (December 1986; unpublished).

3. M.J. Rhoades-Brown, Private Communication.

4. J. Staples, paper presented at this Symposium.

5. R. Billinge, et. al., CERN/PS 88-67 (DL) (October 1988)

DISCUSSION

<u>Becker</u>: I would say that it could be much more reasonable not to operate with these current densities. Perhaps 100 A/cm^2 or even less could be enough and this relaxes all requirements on alignment, and so on.

<u>Prelec</u>: High current densities of the order of kA/cm^2 were used because for uranium in a high charge state you need more current density. That is the reason.

<u>Keller</u>: What did you get from these sputter sources, in what emittance, or would you prefer not to give any values?

<u>Benjamin</u>: Chuck Carlson, Peter Thieberger and I worked with Gerald Alton from ORNL and with Prof. Mori from KEK Laboratory in Japan on the Tandem. We saw about 3 mA directly downstream from the source. But when we magnetically analyzed and pre-accelerated to 200 kV we saw about 300 µA of gold. We are not sure if we had an alignment problem or an aperture problem. We thought there may have been space charge problems. We did not have a lot of emittance measuring equipment in our injector to understand the problems. With this particular source from KEK, we are having some trouble with the lanthanum hexaboride filaments. Where we were holding them, the contact was opening up. I guess it is something that other people had experienced and learned how to do. But in shipping it from Japan to us some slight misalignments of the filament posts must have occurred. We had a great lesson in changing filaments and not enough time to understand the intensity problems after the analyzing magnet. Now, of that 300 µA of gold entering the Tandem, about a milliampere of all charged states came out of the Tandem. This is about the same as we do with our present source. Typically the current at the high energy end of the Tandem is about 1 mA. We were hoping for an improvement factor from 2 to 10 on that. In the time we had we were unable to do much better.

RFQ for an EBIS-Based RHIC Injector

John Staples
Lawrence Berkeley Laboratory, Berkeley, Cal 94720[†]

Abstract

A Radio Frequency Quadrupole (RFQ)-based low energy accelerator integrates well with an electron beam ion source (EBIS) as the front end of a heavy ion synchrotron. A design example for an RFQ-based RHIC injector front-end is given which uses an EBIS. The characteristics of heavy ion RFQ's are outlined and several specific examples are given of operating machines.

Introduction

EB ion sources provide high charge-to-mass (q/A) ratios and pulse lengths suitable for injection into synchrotrons. The high q/A permits economical acceleration to high injection energy to overcome gas scattering and recombination, and closed orbit distortion due to residual fields in the synchrotron guide field magnets.

RFQ accelerators have been installed over the past few years as the front end of both proton and heavy ion synchrotron injectors in many laboratories. Recent advances in beam dynamics and mechanical design have made the RFQ a reliable and economical accelerator. The RFQ mates well with the EBIS to provide efficient heavy ion injectors.

In this paper, the characteristics of RFQ's will be briefly reviewed, the requirements for a RHIC injector will be summarized, and an RFQ design example will be discussed. The characteristics of several operating heavy ion RFQ's will be listed and compared with the RHIC injector design example.

RFQ Characteristics

The RFQ is a low-velocity ion accelerator characterized by relatively low injection energy and nearly complete capture and acceleration of the input beam. It is a compact, efficient structure with only one variable parameter: the input r.f. power.

The RFQ has been replacing the high-voltage Cockcroft-Walton d.c. preinjector on many proton and heavy ion synchrotrons, and is included in all newly proposed synchrotrons. The reasons are numerous: it is compact; it captures nearly 100% of the beam, which is essential for rarer ions such as those produced by high charge state sources; it is more reliable than high-voltage sets and it requires less capital and operating expense.

[†]This work was supported by the Director, Office of Energy Research, Office of High Energy and Nuclear Physics, Nuclear Science Division, U.S. Department of Energy under contract number DE-AC03-76SF00098.

Originally a Soviet invention[1], it was further developed by LANL[2] about 10 years ago, and subsequently by several other laboratories. LBL has produced four so far, two proton and two heavy ion machines[3]. There are now several dozen operating, under construction or under serious consideration world-wide.

The accelerating process in linear accelerators (linacs) by necessity causes a transverse defocusing of the beam (a three dimensional static potential well is excluded by Maxwell's equations in free space). Most linacs retrofit some form of strong (alternate gradient) focusing into the acceleration structure. Heavy ion linacs which use quadrupole magnet focusing are limited in available pole tip strengths, limiting the minimum length of the cell. This forces a high injection velocity requiring a high-voltage preaccelerator, or low frequency operation requiring a physically large structure. The RFQ, however, is derived from a strong focusing transport channel with acceleration added as a perturbation. This results in strong focusing along the entire structure and relaxes the input energy requirement.

The RFQ configuration, shown in Figure 1, creates an r.f. quadrupole field on the beam axis in a waveguide type structure, or with four parallel rods, which has the property of focusing the beam transversely as does an equivalent series of alternating quadrupoles spaced by $\beta\lambda/2$, where $\beta = v/c =$ the normalized ion velocity and λ is the free space wavelength of the r.f. excitation. For certain ranges of the quadrupole focusing field strength, ions are transported along the axis in stable orbits.

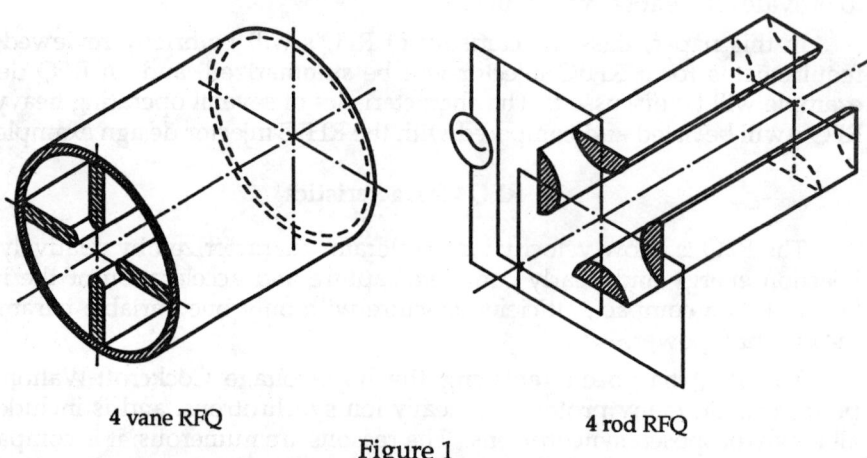

4 vane RFQ　　　　　　　　　　4 rod RFQ

Figure 1

To accelerate the ions, a longitudinal field is needed which has a traveling component synchronous with the ion velocity. This is generated by a particular pattern of ripples machined onto the quadrupole channel electrodes (vane tips). The wavelength of the ripples is $\beta\lambda$ and the strength of the longitudinal accelerating field is roughly proportional to the ripple amplitude.

If the longitudinal field amplitude increases slowly along the axis of the RFQ nearly all the beam is bunched adiabatically and accelerated to full energy. This adiabatic bunching characteristic is almost unique to RFQ linacs, where the longitudinal fields are determined during manufacture by the machined ripple amplitude. Since the focusing and acceleration forces are electrostatic, the RFQ characteristics, for low-intensity beams, are independent of the q/A of the beam, when the vane voltage V varies as $V \propto (q/A)^{-1}$.

The round input beam is matched into the RFQ channel by a *radial matcher* section. The beam is captured and the acceleration process begins in the *buncher* section, whose prescription depends on the specific application. For RFQ's where space charge forces are significant, one prescription is used, developed by LANL[2], and for low-intensity heavy ion beams, another developed at LBL[4] may be applied which shortens the structure. Finally, the *accelerator* section brings the fully bunched beam to the output energy. All of these sections are characterized by the amplitude of the ripple along the vanetips of the quadrupole structure.

The cell length $\beta\lambda/2$ and longitudinal field amplitude E_z are limited only by the ability to machine the ripples rather than by the length of the focusing quadrupoles found in other linacs. Therefore, the minimum cell length requirement is relaxed, and the RFQ can be injected at low velocity and operated at a high frequency resulting in a low voltage preinjector and a compact RFQ structure.

The amount of beam that can be betatron stacked into a synchrotron at low intensity is inversely proportional to the x-plane emittance of the injector. For RFQ's, the transmission is essentially 100%, and the transverse emittance is that of the ion source. The output longitudinal emittance, established by the injection energy and by the details of the longitudinal adiabatic capture and acceleration, is small and compatible with a subsequent Alvarez linac.

As a rule of thumb, the highest practical energy for an RFQ accelerator is about 2 MeV$\times q$, where q is the charge of the accelerated ion. The RFQ, being a Sloan-Lawrence type of accelerator, has an efficiency fall-off of β^{-1}, as well as having a practical length limit.

RHIC Injection Requirements

The proposed Relativistic Heavy Ion Collider (RHIC) at BNL will accelerate ions up to mass gold to energies of 100 GeV/n[5]. The heavy ion injector will include a negative ion source injecting a tandem accelerator, followed by a new booster synchrotron and the AGS ring. The beam will then be transfered from the AGS to the two 3.4 T superconducting rings in the existing CBA tunnel each with a circumference of 3834 meters. The average luminosity for gold-gold collisions with 1.1×10^9 ions in each of 57 bunches per ring will result in an average luminosity of 4.4×10^{26} cm^{-2}sec^{-1} with a lifetime of 10 hours.

The AGS is currently supplied heavy ions up to sulfur from existing Tandem van de Graaf accelerators[6]. A new transfer line connects the tandems to the AGS where energies up to 14.6 GeV/n are currently available. A new injector using an EBIS followed by an RFQ and DTL has also been considered[7] and is the subject of this paper.

For the purpose of deriving a specific design example, the following requirements were assumed for a RHIC injector:

- EBIS source for Au^{+51} to C^{+6}, for $q/A \geq 0.26$.
- Pulse length to 500 µseconds.
- Instantaneous intensity at least 1 emA in the macropulse.
- Low injection energy for simple EBIS extraction geometry and potential.
- An RFQ output energy of 300 keV/nucleon.
- 200 MHz frequency compatible with existing DTL r.f. technology.
- Transverse acceptance $\geq 0.05\pi$ cm-mrad, normalized.
- Length less than 230 cm, the length of the LBL NC mill.
- Vanetip dimensions machineable by a sufficiently large tool.

The vanetip machineability requirement is imposed by the desire of being able to use a sufficiently strong milling tool to cut the details of the ripples. This, in fact, is usually the critical parameter in the design of heavy ion RFQ's and determines other factors such as the injection energy and transverse phase space acceptance. The length is usually kept below about 2 r.f. free-space wavelengths to insure stability of the electric fields in the structure.

A Specific Design Example

A strawman design example is presented here for an RFQ-based accelerator compatible with an EB ion source and a 200 MHz Alvarez linac post-accelerator. This example will reflect current practice and will be a conservative design based on successfully operating RFQ's. A $q/A = 0.23$ design more than satisfies the 0.26 requirement for accelerating Au^{+51}, and allows gold charge states down to +45 to be accelerated for greater intensity.

The design process consists of a series of compromises, balancing parameters such as length and power requirement against others such as source extractor voltage and RFQ transverse acceptance. An acceptable accelerator is found usually after an extensive search of the controlling parameters while accepting the derived values of others, such as vanetip machineability. The controlling parameters for heavy ion accelerators are usually different than those for proton devices, as heavy ion accelerators are usually not influenced by space charge forces, but have a shorter cell size from the lower injection energy which makes the geometry more difficult to produce.

For this specific accelerator, the injection voltage and transverse acceptance are selected to allow the ripples in the vane tip to be machined with a

3/4 inch diameter ball end mill, which in our experience has adequate cutting life and strength to machine the steel vane material. The output energy of 300 keV/n (59 MeV gold) is compatible with the 96 inch length limit of the LBL numerically controlled milling machine.

The parameters selected for this strawman design are:

q/A	≥0.23	
Example Ion	Au^{+45}	
Injection Energy	11	keV/n
Output Energy	300	keV/n
EBIS extraction voltage	48	kV
Acceptance	0.09π	cm-mrad, normalized
Frequency	200	MHz
Length	227.6	cm
r_0	0.359	cm
Number of Cells	322	
Surface field	1.9	Kilpatrick
Vane Voltage	64.9	kV
Transmission	90%	
Space Charge Limit	10	emA
Output Phase Spread	±17	degrees
Output Energy Spread	±1.7	percent

Table 1. Strawman RFQ Parameters.

These design values are for a conservative RFQ design similar to two heavy ion injectors already produced at LBL, one at of the Bevatron Local Injector[8], the other at the CERN heavy ion injector[9].

The most critical manufacturing parameter, the minimum longitudinal radius of curvature of the ripples along the vane tip, is a rapidly varying function of the input energy. The selection of 11 keV/n implies a 48 kV acceleration potential from the EB ion source for Au^{+45}. For an 18% reduction of the input energy to 9 keV/n, the minimum longitudinal radius is reduced by 28%, which cannot be machined by an satisfactorily large tool.

The transverse focusing force due to the r.f. electric quadrupole field on axis is independent of the beam velocity. The r.f. acceleration causes a transverse defocusing, which is constant along most of the accelerator with this design, all the way to the 236 keV/n point. This particular design technique [4] shortens the RFQ by applying as much bunching voltage as possible consistent with radial stability. The beam size is $\sqrt{\beta \varepsilon_u}$ where this β is the betatron amplitude which is proportional to velocity, and ε_u is the unnormalized beam emittance, inversely proportional to velocity. Therefore, the beam size is constant along almost the entire length of the RFQ.

The input (output) beam is highly convergent (divergent), due to the short transverse betatron wavelength in the RFQ. The input beam is matched into the structure by an 8 cell radial matcher which transforms the d.c. round beam into a beam whose x-y axis ratio varies at the r.f. frequency along the rest of the RFQ accelerator. The bunched output beam has dissimilar divergences in the x and y planes when it emerges. The betatron parameters of the input and output beams are given in Table 2. The normalized emittance of the source is unchanged by the RFQ.

	β(cm)	α
Input beam	4.03	0.82
x-plane output beam	22.0	-2.7
y-plane output beam	19.4	1.9

Table 2. Input/Output Beam Parameters.

The input beam is highly convergent and requires a focusing lens located 10–15 cm upstream of the beginning of the vane. This lens may be either a solenoid or a quadrupole array. The lens must be located as close as possible to the RFQ while allowing a Faraday cup, a beam centroid monitor and a vacuum valve to be interposed. Pulling the lens further away from the RFQ results in an uncomfortably large beam radius in the lens.

The output beam is highly divergent and must be matched into the following drift tube linac (DTL). One of two approaches is usually taken: either the beam is immediately taken into the DTL with a minimal matcher or transported some distance by a system that includes longitudinal cavities to maintain the bunch structure. At LBL the RFQ is coupled immediately to the DTL with the ±17° output phase spread shearing to ±22°, within the phase acceptance of the DTL. The transverse match is accomplished by ramping the first six focusing quadrupole strengths in the DTL. At Saclay a different approach uses a long beam line transporting the beam from the RFQ to the entrance of the DTL[9].

Comparison with other Heavy Ion RFQ's

It is instructive to compare this RFQ with others that have been built and operated. These include the Si^{+4} injector at the LBL Bevalac[8] and the O^{+6} injector at CERN[9], both built by LBL, the Saclay heavy ion injector for Saturne II[10], the TALL RFQ at INS Tokyo[11] and the planned Cryring injector[12], built at IAP Frankfurt. All RFQ's are of the 4-vane variety except the Cryring RFQ, which is of the 4-rod variety, as shown in Figure 1. The 4-vane type is more suitable to high frequency operation. At low frequencies, the 4-vane geometry becomes inconveniently large and the 4-rod geometry is preferred. The lowest frequency RFQ built to date is a 6.9 MHz 4-rod fusion driver device built at ITEP, Moscow[13]. The TALL RFQ accelerates the beam to a higher energy than the others, and is longer. An RFQ similar to TALL will be used as the injector to the NIRS medical synchrotron facility at Chiba, Japan[14].

Table 3 lists major parameters for the RFQ's mentioned above and the proposed BNL RHIC injector RFQ.

RFQ	f_0(MHz)	q/A	Length(cm)	T_{in}†	T_{out}†	$\varepsilon_{N,A}$‡
BNL RHIC	200	0.23	228	11.0	300	0.09
Bevalac	199	0.14	225	8.5	200	0.05
CERN	201	0.38	86	5.6	139	0.09
Saclay	200	0.25	230	12.5	187	0.17
Tokyo	100	0.14	725	8.0	800	0.06
Cryring	108	0.25	154	10.0	300	?

†in keV/n. ‡normalized acceptance in π cm-mrad.
Table 3. Comparison of RFQ Parameters.

The proposed RHIC injector RFQ is similar to the others which are operating successfully. It uses a conservative design with an acceptance large compared to the expected emittance of the EB ion source.

Discussion

The RFQ accelerator is now a mature technology and many have been built and successfully operated. All new synchrotron injectors are designed with RFQ's and many older installations are retrofitting the high voltage d.c. preinjectors with RFQ's. The benefits include reduced operating and maintenance costs, smaller floor space, almost complete capture of the beam and operation of the ion source near ground potential for ease of access. For large, complex ion sources such as polarized proton, EBIS or ECR sources, 100% capture and source accessibility are of particular importance.

The RFQ has only one operating parameter: the vane voltage which is established by the r.f. drive power. For heavy ion injectors, the RFQ operating point is adjusted for the particular q/A of each accelerated ion. The focusing and accelerating parameters all scale together, providing rapid tune-up after changes of ion species.

RFQ accelerator technology is still progressing. Solid state r.f. amplifier technology is advancing rapidly and one 425 MHz RFQ has already been powered by a 160 kW solid state amplifier[15]. The stabilization of the field distribution in the RFQ has previously been accomplished by strapping opposing vanes together, and new, less intrusive methods are now being developed[16].

Alternate RFQ structures are being developed, such as the 4-rod type for lower frequencies, as well as multiple beam RFQ's. The development of the

RFQ will stimulate new applications analogous to the development of the compact electron linac structure which made possible many new applications such as hospital-based irradiation devices.

Many thanks to Krsto Prelec for help in preparing this report.

References

1. I. M. Kapchinskii and V. A. Teplyakov, Prib. Tekh. Eksp., No. 2, 19 (1970).
2. K. R. Crandall, R. H. Stokes, T. P. Wangler, 1979 Linear Accelerator Conference, Montauk (1979), p. 205.
3. J. W. Staples, 1986 Linear Accelerator Conference, Stanford (1986), p. 227.
4. S. Yamada, 1981 Linear Accelerator Conference, Santa Fe, (1981), p.316.
5. Conceptual Design of RHIC, BNL Report BNL 51932, May 1986.
6. H. Foelsche, D. S. Barton, P. Thieberger, 13th International Particle Accelerator Conference, Novosibirsk (1986), p. 250. p.229.
7. K. Prelec, private communication (1986).
8. J. W. Staples, 1983 Particle Accelerator Conference, Santa Fe (1983), p. 3533.
9. N. Angert et al, 1984 Linear Accelerator Conference, Seeheim, (1984), p. 374.
10. J. L. Laclare, A. Ropert, LNS Getis 063, LNS, CEN, Saclay
11. N. Ueda et al, 1984 Linear Accelerator Conference, Seeheim, (1984), p. 71.
12. A. Schempp et al, 1988 Linear Accelerator Conference, Williamsburg, (1988).
13. V. S. Artemov et al, 1987 Particle Accelerator Conference, Washington (1987), p. 388.
14. S. Yamada et al, Third Japan-China Joint Symposium on Accelerators for Nuclear Science and Their Applications, Nov 1987, Saitama.
15. D. Schrage et al, 1988 Linear Accelerator Conference, Williamsburg, (1988).
16. A. Schempp, 1986 Linear Accelerator Conference, Seeheim, (1984), p. 251.

DISCUSSION

Keller: I want to point out that we should come together and produce a standard definition of emittance so that we can easily compare different sources. I think it is in the best interest of the entire community.

Staples: Do you want to elaborate a little bit on that?

Keller: I think some of the values you displayed were quite old. Specifically, the quoted values for the GSI xenon source are ten years old, and we have done much better in the meantime with other sources, for example a new duoplasmatron. So I think one should sit down and compare EBIS and ECR sources in terms of current and emittance.

Staples: I agree. As ion sources improve and their emittance decreases, the RFQ design becomes easier, as the transverse acceptance for low q/A ions is relatively small.

Becker: May I ask what is a reasonable platform voltage for an EBIS in such a proposal?

Staples: In order to achieve a reasonable acceptance, the injection voltage would be on the order of 50 kV for a RHIC-type injector based on an EBIS-RFQ combination. However, the longitudinal output emittance, and generally the length of the RFQ, increase with increasing injection energy. Therefore, the final choice of injection energy is a compromise between the inconvenience of placing the EBIS at a high potential, the output longitudinal emittance, and the transverse acceptance of the RFQ.

Ben-Zvi: What are the limitations of breaking an RFQ up into multiple sections?

Staples: Many people have looked into breaking RFQs up into more than one section. For example, as the sensitivity of the RFQ to mechanical errors goes as the square of the length of the RFQ, two short RFQs could sustain looser tolerances than one long one. However, since the transverse focussing in an RFQ is very strong, the betatron wavelength is short, and any break in the quadrupole focussing field would cause a transverse focussing mismatch and subsequent emittance dilution. Any lens placed where the beam bunch is not short must be time-varying as well to match the betatron parameters of the beam as they vary over the length of the bunch.

Herrlander: If the linac chain has frequency jumps, then the beam must be refocussed and rebunched.

Staples: Yes. The transition between an RFQ and a subsequent DTL usually contains transverse and longitudinal matching elements. The bunch is short enough so that the transverse focussing need not be time-varying. A rebunching cavity compresses the bunch longitudinally. If the RFQ-DTL spacing is short enough, such as when the RFQ is placed directly at the subsequent DTL, this matching section could be eliminated.

Benjamin: Would you comment on the BEAR RFQ and any others?
Staples: The BEAR (Beam on a Rocket) RFQ has had an interesting history. LANL produced a six-inch sparker followed by an unmodulated cold model. Then an unmodulated full length engineering model was produced to test the electroforming procedure which joins the four vane-cavity sections together. Two fully operational RFQs were produced by Grumman, with LANL providing the support components. One unit is flight qualified and will be launched on a rocket. The other unit, the one that has been on public display, will be kept by Grumman for possible further beam tests. There are, perhaps, 20 RFQs or so that are operating today with many more proposed or under construction. They span the gamut from small proton injectors to giant low-frequency front ends for heavy ion fusion drivers.
Benjamin: Could you comment on the machining operation?
Staples: The machining techniques were developed at LANL and at other national laboratories, frequently on smaller numerically controlled mills that required that the vane be moved during the machining. Now the machining technology is being moved out to industrial contractors with particular expertise in precision machining of large, complex shapes.
Ben-Zvi: What is the mechanism limiting the transverse acceptance of the proposed RHIC RFQ?
Staples: For heavy ion machines where space charge is not significant, the limitation is the maximum voltage that can be sustained between adjacent vane tips which generates the quadrupole focussing field. At a low q/A value, strong fields are necessary to develop sufficient transverse focussing. As the input ion velocity is generally low, the cells are short, requiring the vane tips to be fairly close to the axis so that the vane tip modulation depth, which produces the longitudinal field, not be unreasonably large. The small adjacent vane tip spacing and distance from the axis, and the limitation of the voltage by sparking which reduces the focussing strength, combine to limit the transverse acceptance of the RFQ for heavy ions with low q/A.
Ben-Zvi: Do you observe longitudinal-transverse coupling?

Staples: Probably yes, but the design process does not address that question directly. At the beginning of the RFQ when the beam is still unbunched and also at the high energy end where space charge forces are of less , significance, the longitudinal-transverse coupling is probably not an issue. In the intermediate sections of the RFQ the coupling could be significant. In fact, we model the evolution of the beam numerically with the PARMTEQ program and find that the transverse emittance is usually well preserved in a good design, even with significant space charge. The nature of the design can have a strong effect on the resulting longitudinal emittance. The PARMTEQ code, written at Los Alamos and now distributed world-wide, serves as a common base of comparison of RFQs designed at various laboratories. The program numerically integrates particles through the RFQ field configuration and uses a somewhat crude but fast space charge algorithm. The code has been verified many times with data taken from actual RFQs, with surprisingly good agreement.

Becker: Is it possible to increase the duty cycle of your RFQ?

Staples: RFQs have been designed with duty factors of up to 100%. However, two major mechanisms limit the design of heavy ion RFQs. The first one is the sparking between the vane tips. Heavy ion RFQs are generally designed to operate near the sparking limit to provide the largest possible transverse acceptance and to shorten the machine. As the pulse length is increased, the sparking probability increases until it saturates at pulse lengths of tens of milliseconds. For long pulse operation, the surface field on the vane tips must be reduced, which reduces the focussing and therefore the transverse acceptance. The second mechanism is simply the effect of heating due to the rf current on the inner surfaces of the cavity, the drive loop and vane coupling rings. Much of the heat is produced in regions where cooling passages are difficult to provide. Thermally induced mechanical changes may alter the electric field configuration in the structure itself, disrupting the acceleration process.

Becker: What about a 20-30 microsecond pulse every 100 milliseconds? This would match an EBIS well.

Staples: RFQs operating as injectors to synchrotrons today typically have a duty factor of 10^{-3}, which is consistent with your requirement. If the duty factor goes above about 1%, a qualitative change in the cooling system is probably necessary. A 100% duty factor machine is different again, like FMIT or perhaps as a superconducting machine.

Ludlam: What is your RHIC RFQ cost?

Staples: That is a loaded question. The actual RFQ cost is in the noise compared to the overall project cost. The RFQ has proven itself as a superior injector to many synchrotrons, and several heavy ion RFQs are now in operation. RFQs have proven themselves as superior to Cockcroft-Walton injectors in most applications in terms of reliability, access to the ion source, and economy of operation.

Faure: I would like to comment that at Saclay we have an RFQ operating with 5 pulses, each 100 microseconds long every second.

Noe: With the long numerically controlled mill, what type of tolerances do you achieve over what length?

Staples: The measured tolerance includes the machining and verification tolerances and the warpage of the vane. At LBL, we normally achieve tolerances of approximately 0.5 mil (0.0005 inches) total accumulated error for a mild steel vane. We check the mill occasionally with a laser alignment device and find that the mill itself is good to about 0.1 mil. An RFQ that I have not mentioned is an aluminum one we just manufactured. This one came out with total accumulated errors of about 0.2 to 0.3 mils, as the aluminum is dimensionally more stable and does not wear the cutter as fast.

Noe: How much looser are the tolerance requirements for the rod designs?

Staples: The quality of the field on axis depends on two things: the location of the electrodes producing the field and on the relative excitation of the four quadrants in the 4-vane type of structure which determines the potential on the vane tips. In 4-vane RFQs, the four quadrants are weakly coupled. When driven at resonance, the dipole and quadrupole potential on the vane tips is a very sensitive function of the deviation from resonance of the four quadrants. This results in very tight assembly tolerances. In 4-rod RFQs, the correct quadrupole potential on the rods is inherent in the design, and the quadrupole field quality on axis is dependent only on the position errors of the rods themselves. This tolerance is generally somewhat looser than that for 4-vane RFQs where resonator symmetry must be maintained. At present, many 4-vane RFQs strap the vanes together with coupling rings or use some other rf stabilizing technique to suppress the undesired dipole component, loosening the assembly tolerances.

Staples: I think that the tolerances of the rods are probably about the same. It turns out that for the HERA machine, the fore rod vs. the four vein structure, the four rod structure actually did not have quite as good an acceptance as the four vein structure. Tolerances are looser. I do not know what the numbers are, Carl, maybe you have that information. The performances were quite similar but the four vein had a little edge on the performance. I do not know what those tolerances were. I suspect, in fact, one of the reasons that the four vein RFQs are so tight, is it really kind of goes back to the old days before we were using vein coupling rings and we had to be very precise on the location of the vein tips to keep the resonant frequency in the four quadrants the same. If in fact the only tolerances the field configuration, that is to say, take a four rod machine and you, the voltages do not vary if you were to change the locations of the four rods, then the next thing that happens is that you start distorting the fields and cells. The tolerances for that is far looser than it is for maintaining of electrical symmetry in the machine.

THE FEASIBILITY OF AN EBIS FOR THE CERN LEAD PROJECT

Reinard Becker
Institut für Angewandte Physik der Johann-Wolfgang-Goethe-Universität
Robert-Mayer-Straße 2-4, D 6000-Frankfurt am Main

ABSTRACT

The Ionization of lead to charge states between 40 and 50 is possible with an EBIS (Electron Beam Ion Source), if an external ion source[1] will be used to inject Pb^+ ions. The requirements for current density (< 30 A/cm^2) and beam power (< 3.5 kW) are modest, especially due to long repetition times. With respect to ion yield, the EBIS will be competititve to an ECR source, if Pb^{54+} will be extracted, thus avoiding the stripper loss before injection into the BSP[2] In that case, the LINAC can be shorter and using Pb^{54+} instead of Pb^{53+} might reduce the charge exchange losses in the LINAC as well as in the PSB.

IONIZATION CROSS SECTIONS

Unfortunately, no data exist on the ionization of lead, especially to high charge states. Therefore Donets measurements of "effective" cross sections[3] for Xenon will be used for scaling. These "effective" cross sections include multi-step ionization, which becomes predominent in muti electron atoms, where many electrons have similar ionization energy. Since the time evolution of stepwise ionization is calculated with formulae of E. Baron[4], using single step cross sections, Donets "effective" cross sections are used to determine a fudge factor C in a Bethe[5] type cross section formula:

$$\sigma_{k,k+1} = C \frac{\ln(E_e/E_k)}{E_e * E_k} \qquad (1)$$

where E_e and E_k are the electron and the ionization energy, respectively and C is fittet from a comparison with Donets scaled [6] cross sections

$$^{10}\log \sigma_{k,k+1} = -16.227 - 3.8 \cdot 10^{-5} E_e - (0.0902 - 1.2 \cdot 10^{-6} E_e) k \qquad (2)$$

to

$$C = 7.7 \cdot 10^{-12} \text{ cm}^2 \text{ eV}^2 \qquad (3)$$

© 1989 American Institute of Physics

It may be seen from Fig. 1 at 8.5 keV and Fig. 2 at 18 keV, that this value for C gives reasonable agreement between measured and calculated ionzation times for the charge range 10<k<44. For higher charge states, this value of C is too high for Xenon, because the L-shell with a clear separation of ionzation energies gets involved, whereas for lead, the electrons in this charge state region are still in a multi electron environement. The time evolution of Xenon charge states is shown in Fig. 3 at 8.5 keV and in Fig. 4 at 18 keV electron energy, using eq. (1) and (3), Carlsons[7] calculated ionization energies and the explicit formulae of E. Baron[4]. With this confirmation of the fudge factor C, the ionzation cross sectrions of lead can be estimated certainly within a factor of 2, which has no influence on the ion yield, but determines the required current density of the electron beam.

ION YIELD

The number of electrons within the trap region of an EBIS limits the maximum number of ions[8], which can be extracted in charge state Z:

$$N^{Z+} \leq 10^{13} \, P_{[\mu A/V^{3/2}]} \, Ee_{[eV]} \, L_{[m]} / Z \quad (4)$$

The <u>perveance</u> <u>P</u> of an electron beam easily reaches $\mu A/V^{3/2}$; it can be made higher for lower voltage and/or low current density. The theoretical limit is 25.4 $\mu A/V^{3/2}$ for Brillouin flow focusing and 32.2 $\mu A/V^{3/2}$ in the case of immersed flow[9], but practical limits are in the range of 5 to 10 $\mu A/V^{3/2}$ even with deceleration of the electron beam: As a reasonable value, P = 2 $\mu A/V^{3/2}$ will be used to estimate the ion yield according to eq. 4.

The <u>electron</u> <u>energy</u> <u>Ee</u> usually is chosen to give the shortest ionization time (~ 3 times the ionization energy of the wanted charge state), but at the very low repetition rate of the SPS, it seems to be more reasonable, to renounce on speed of ionization and to select the electron energy in such a way, that the wanted charge state cannot be depopulated by ionization. This will then result in all ions leaving the EBIS in the same (wanted) charge state. This procedure is demonstrated through figures 5-8: Fig.5 gives the evolution of charge states at optimum energy for Pb^{40+}, namely 5.3 keV. The abundace of Pb^{40+} reaches only 17.8 %. All ion charge states, which preceed depopulation of a new shell or subshell of electrons are enhanced, clearly seen for Pb^{14+}, Pb^{36+}, Pb^{46+}, and Pb^{54+}. Fig. 6 shows stepwise ionization at 5 keV: Since Pb^{54+} can't be ionized further, it reaches full abundace in about 1 second at 30 A/cm^2. Fig. 7 and 8 give the according results of 20 A/cm^2 at 2.6 keV for Pb^{46+} and of 3 A/cm^2 at 1.75 keV for Pb^{36+} being sufficient for an ionization time of 1 second.

The length L of an EBIS must be considered on a logarithmic scale: 10 cm are surely too short, 10 m are too long. The length must be limited with respect to mechanical tolerance requirements, coming from flux staightness and misalignment considerations: These requirements scale with the beam radius, therefore again, by the modest current densities required, tolerances are in the mm range, for surrounding the electron beam with an electrode as well as for the flux straightness. This will greatly facilitate to set up the electrode structure and to manufacture a solenoid with useable trap lenth of at least 2 m.

Table I EBIS source parameters to produce favourite lead charge states

		Pb^{36+}	Pb^{46+}	Pb^{54+}
Perveance	[$\mu A/V^{3/2}$]	2	2	2
Length	[m]	2	2	2
Electron energy	[keV]	1.75	2.6	5
Electron current	[mA]	146	265	707
Electron beam power *)	[kW]	0.256	0.689	3.54
Electron current density	[A/cm^2]	3	20	30
Electron beam diameter	[mm]	2.49	1.30	1.73
Maximum number of ions/pulse	[10^9]	1.94	2.3	3.7

*) Some deceleration is associated with collector extraction, therefore the dissipated beam power will be less.

Since Pb^{36+} and Pb^{46+} must be stripped to Pb^{53+} before the injection into the PSB, only the production of Pb^{54+} with an EBIS may be considered as an alternative to an ECR-source. The yield of $3.7 \cdot 10^9$ ions per extraction pulse is a theoretical upper value, assuming complete compensation of the electronic space charge by the ions and no losses during extraction. Most existing EBIS have only reached 10-30 % of this theoretical limit, but again, the modest current density requirements here make it most likely to reach a high yield.

An EBIS for the production of Pb^{54+} needs an extrenal ion source, which must supply about 100 μA of Pb^+ with low energy spread. These ions will be injected into the trap region through the collector, decelerated and trapped inside the electron beam by appropriate potentials. This method has proven to be very reliable and versatile and a lot of experience is available at SACLAY[1].

CONCLUSION

An EBIS with modest current density (30 A/cm^2) and electron beam power (<3.5kW) will be able to deliver 10^9 ions of Pb^{54+} in about 1 second. By its internal storage and accumulation of high charge states it matches well the low duty cycle injection into a synchrotron with long repetition time.

REFERENCES

1. Faure J. et al. NIM
2. CERN PS/DI/DS/Min 88-6
3. Donets E.D. et al. JINR R7-12905, Dubna 1979
4. Baron E. IPN-73-05, Orsay 1973
5. Bethe H. Ann. d. Physik 5 (1930) 325
6. Becker R. Proc. ECR-Workshopm, Darmstadt 1977
7. Carlson T.A. et al. ORNL-4562, Oak Ridge 1970
8. Becker R. et al. Proc. All Union Conf. Part. Acc., Dubna 1980
9. Brewer G. in A. Septier (editor) Focusing of charged particles II Ac. Press, NY,London (1967)

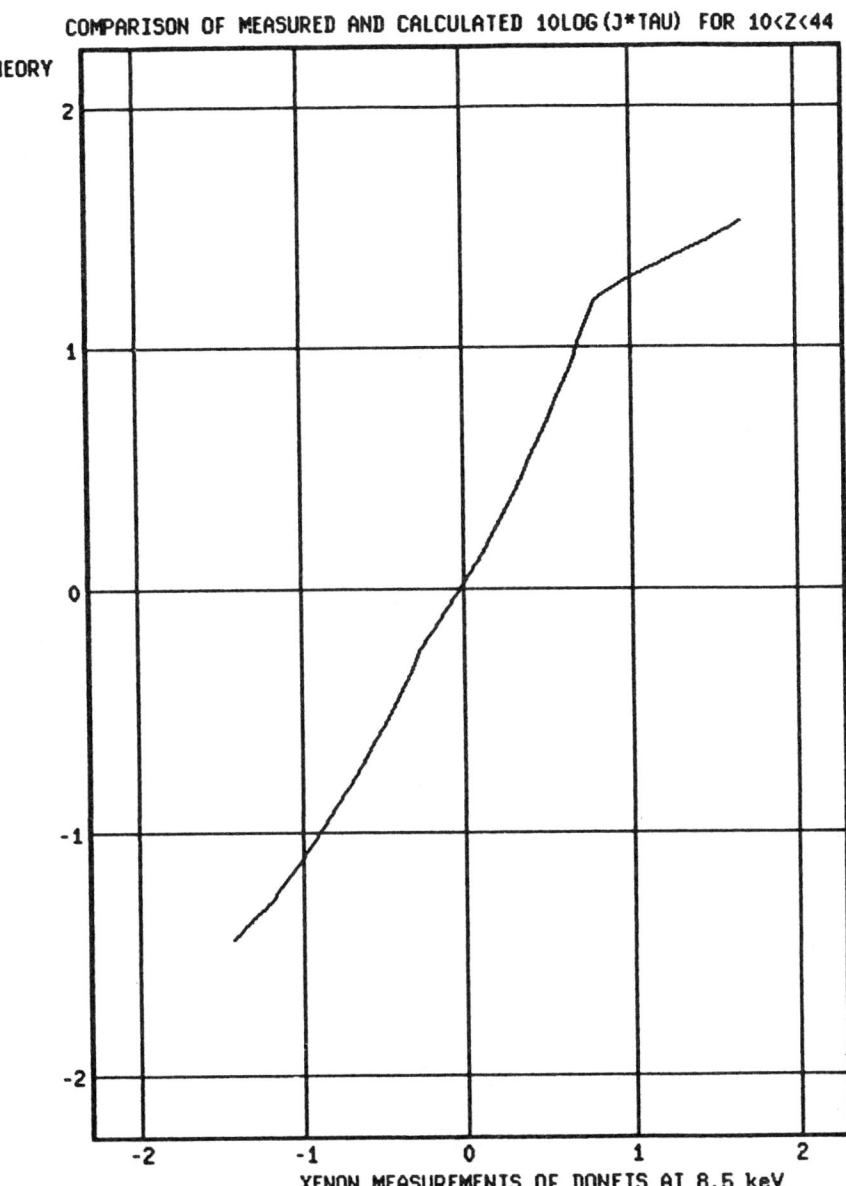

Fig. 1 Comparison between measured and calculated ionization times for Xenon charge states between 10 and 44 at 8.5 keV electron Energy

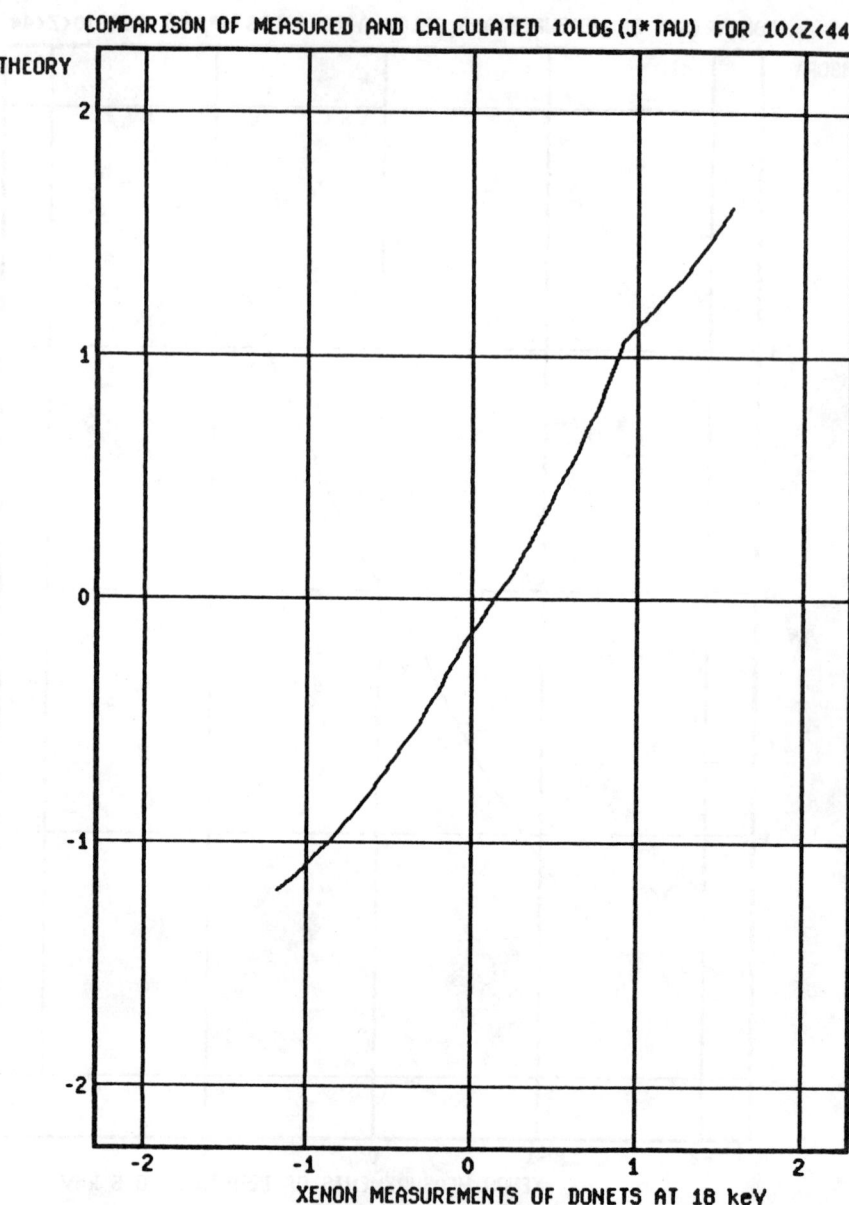

Fig. 2 Comparison between measured and calculated ionization times for Xenon charge states between 10 and 44 at 18 keV electron energy

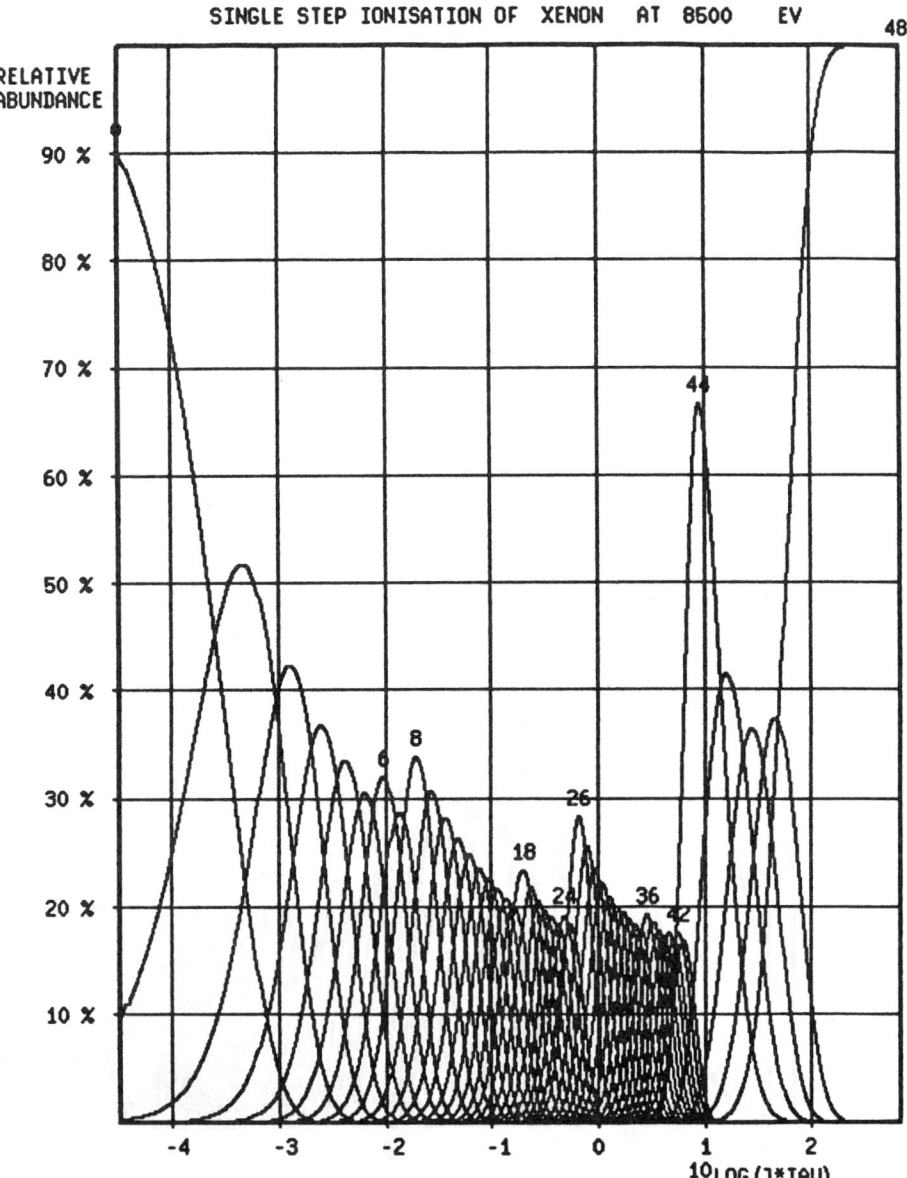

Fig. 3 Evolution of Xenon charge states under electron bombardment with 8.5 keV

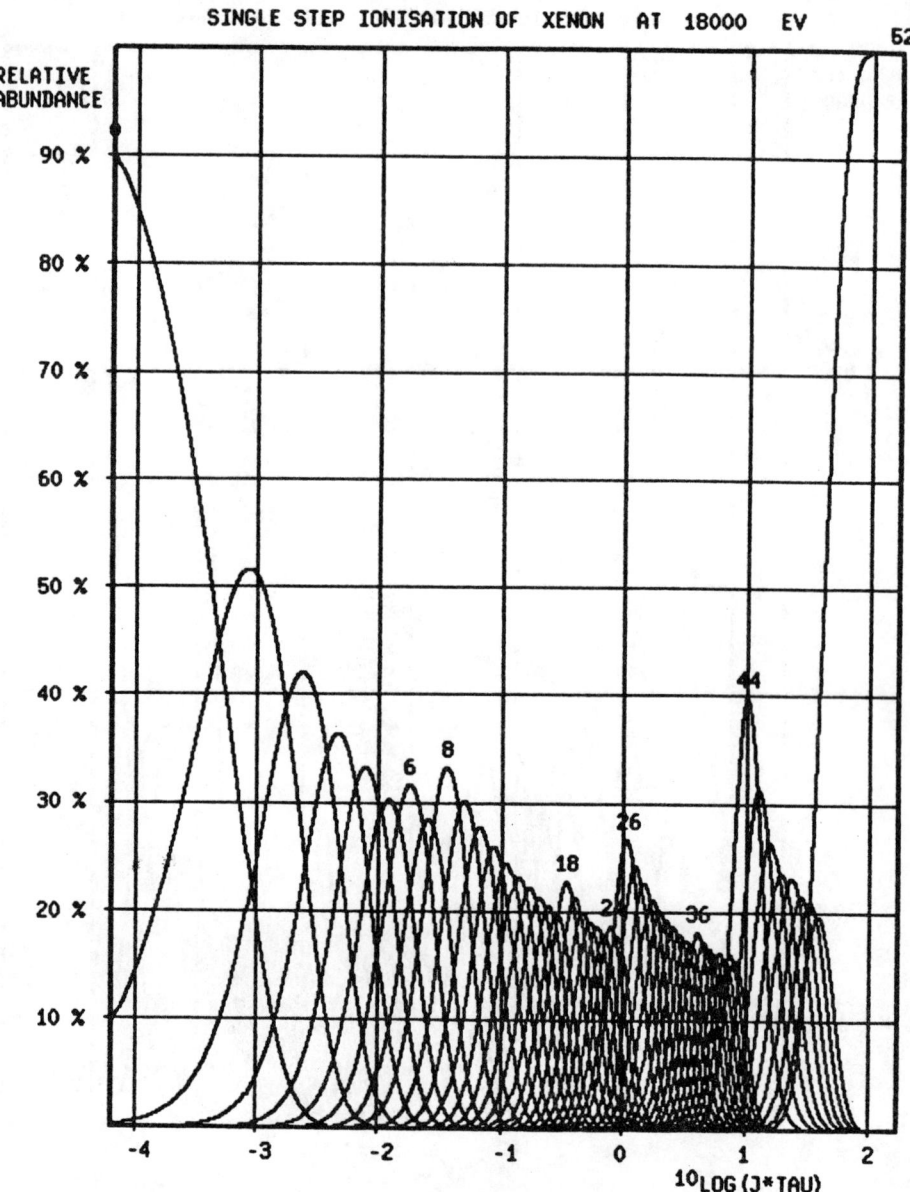

Fig. 4 Evolution of Xenon charge states under electron bombardment with 18 keV

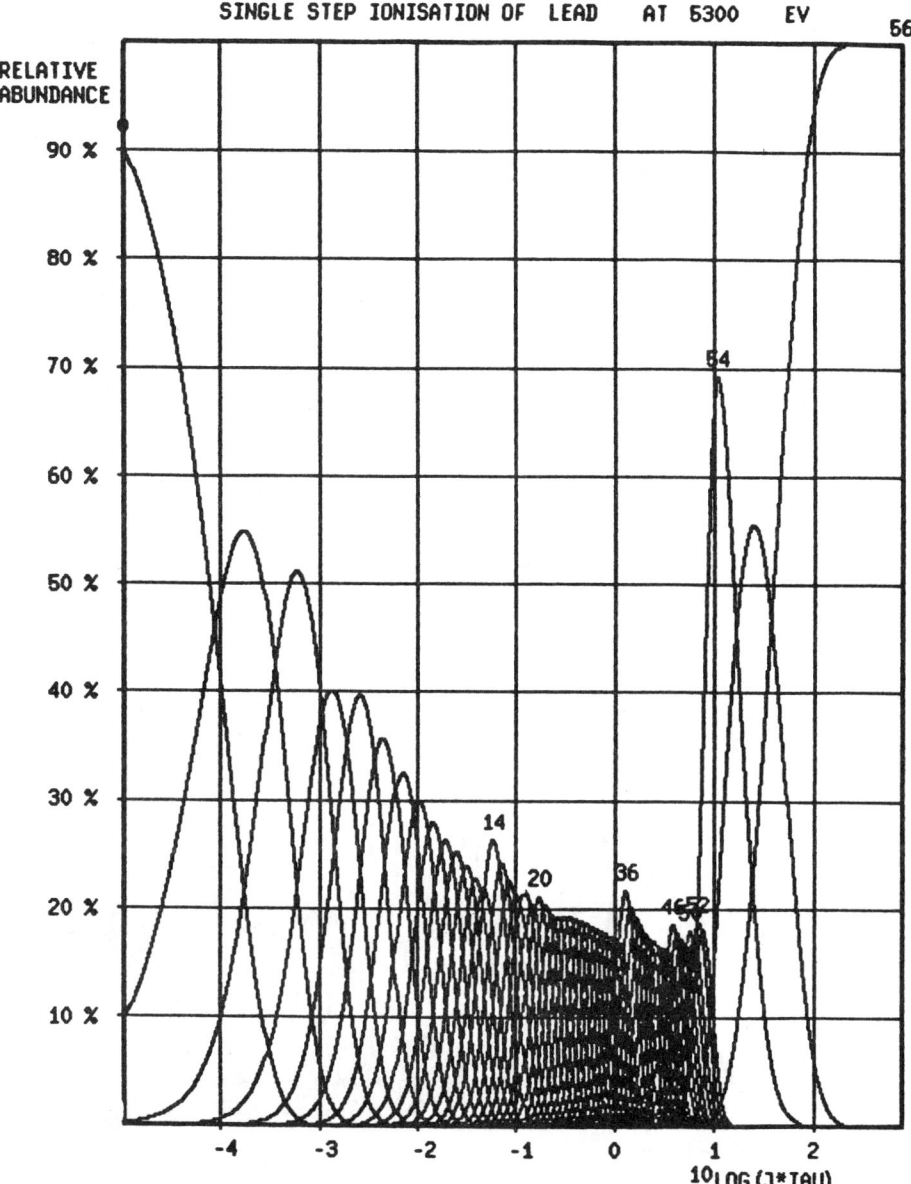

Fig. 5 Evolution of Lead charge states under electron bombardment with 5.3 keV

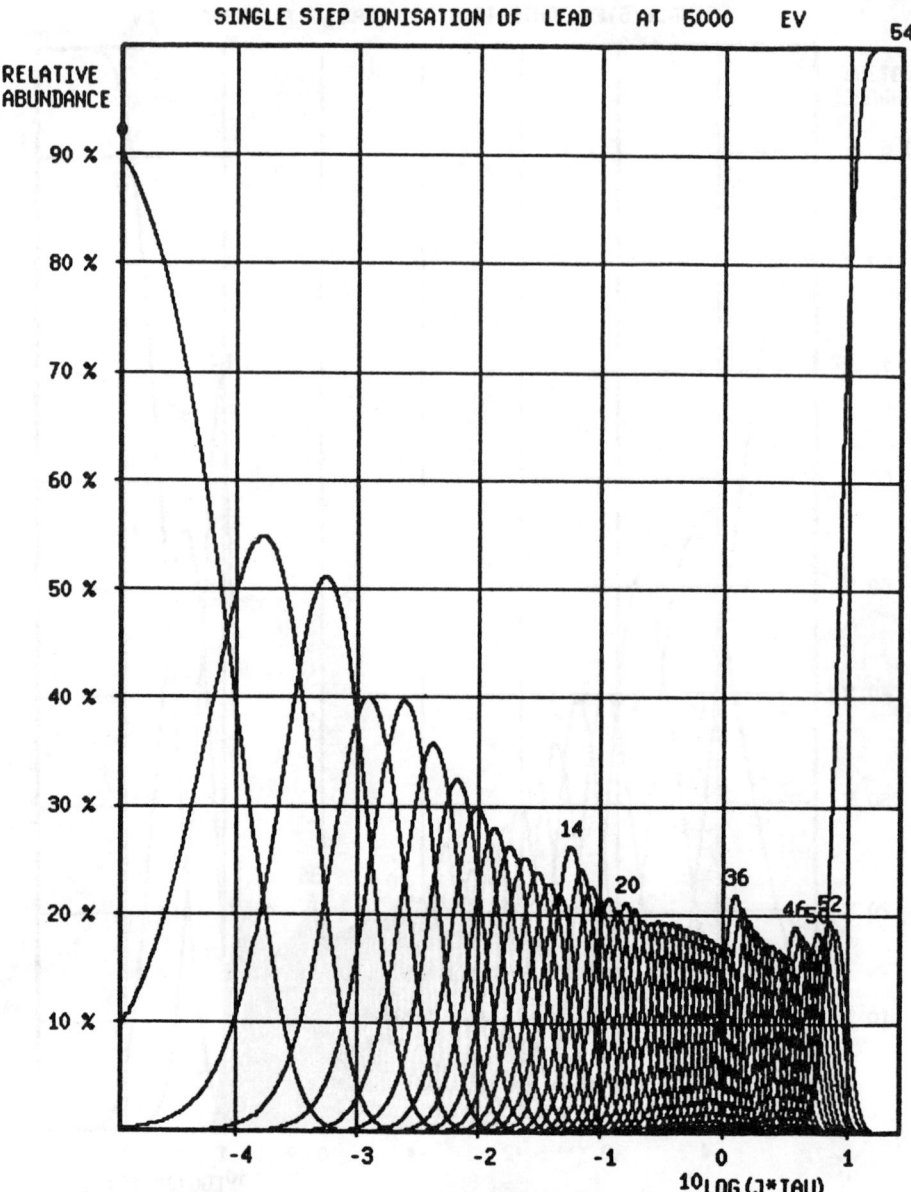

Fig. 6 Evolution of Lead charge states under electron bombardment with 5 keV

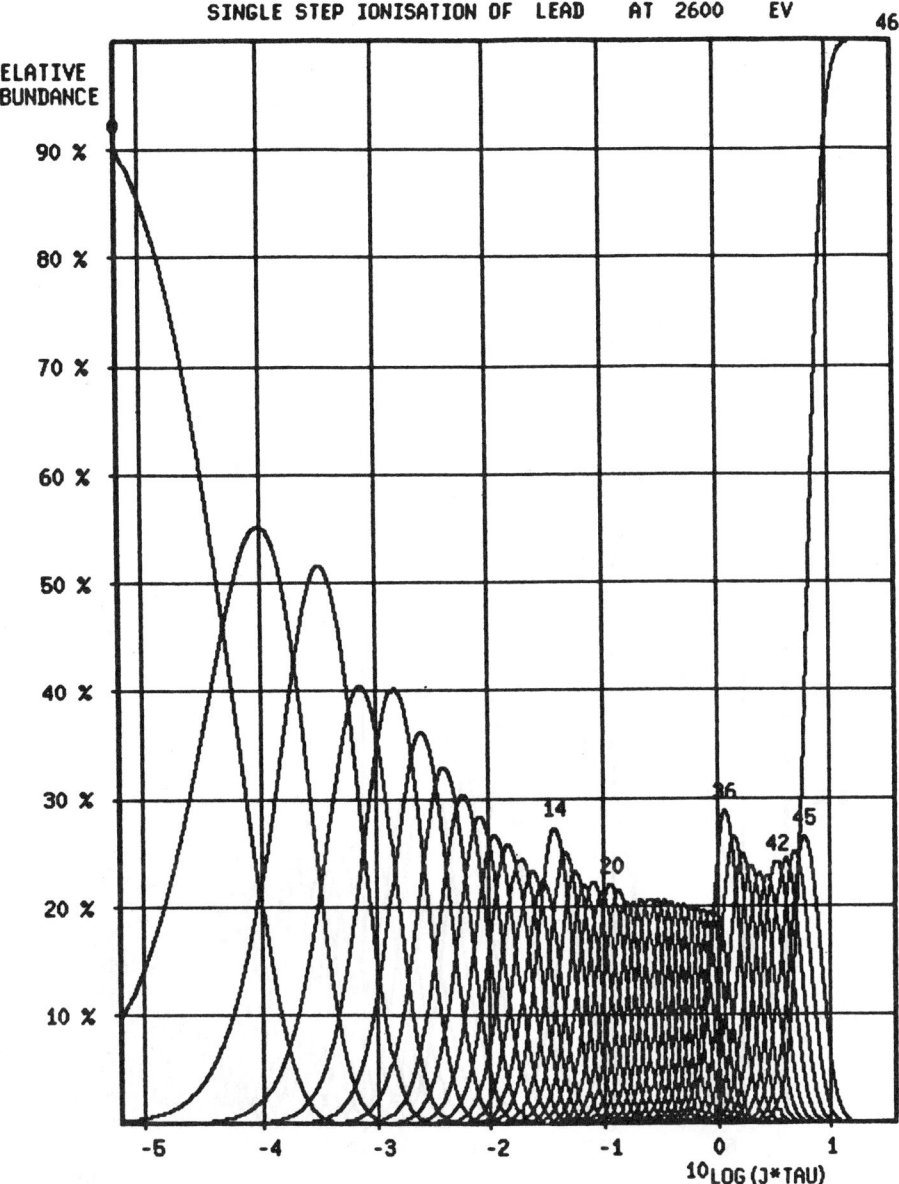

Fig. 7 Evolution of Lead charge states under electron bombardment with 2.6 keV

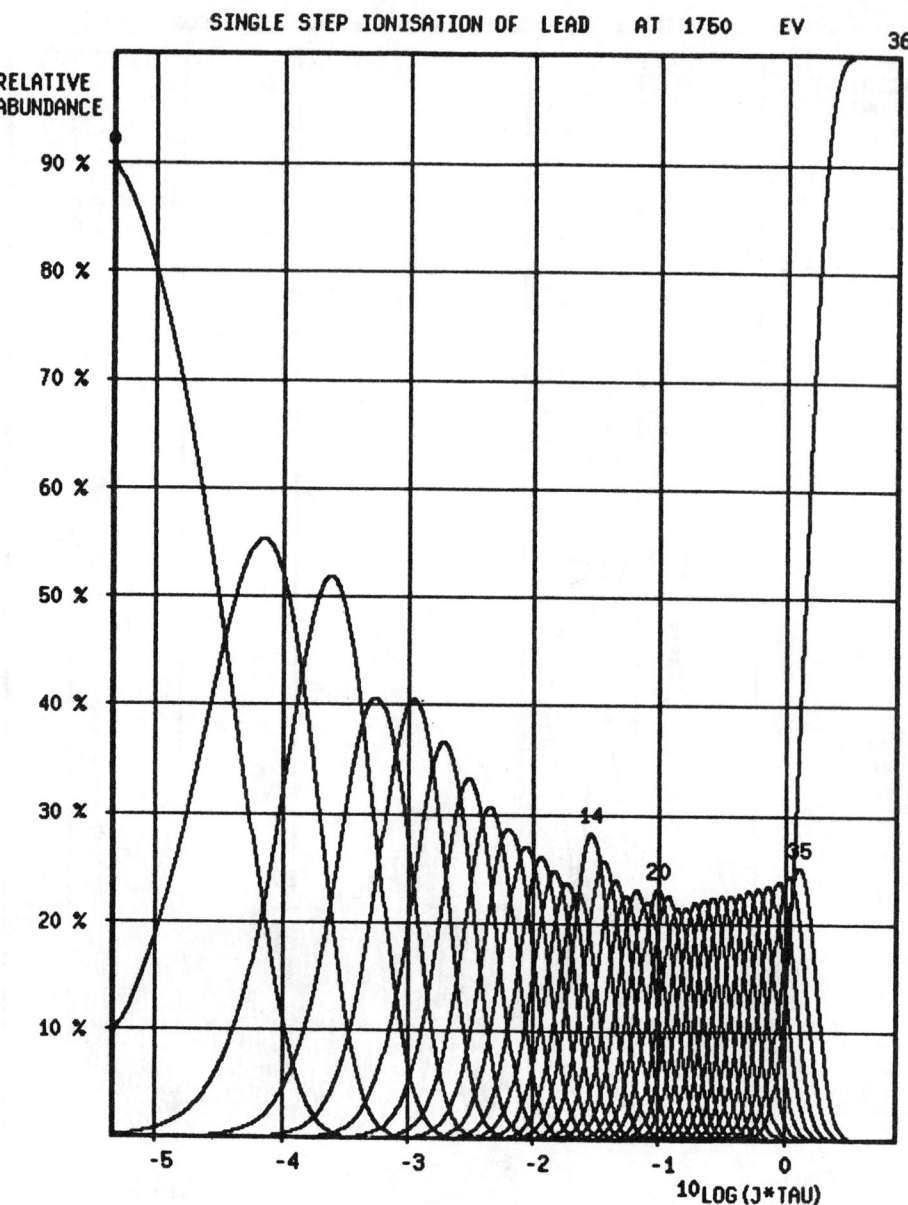

Fig. 8 Evolution of charge states under electron bombardment with 1.75 keV

PANEL SESSION: FUTURE PROSPECTS AND LIMITS

J. Faure (Moderator), R. Becker, C. Herrlander,
M. Kleinod, M. Levine

Faure: In order to start this discussion, I would like to say that last night after the evening session I had a dream. This dream consisted of saying that we have cooling taking place in DIONE, and we have 10^{10} charges. I would like to show you what it takes to go up to 10^{12}. I would like to go up to Ar^{18+}, Kr^{34+}, Xe^{44+}, and U^{64+}. That means the 10 keV electron beam, with 2000 A/cm^2 and a confinement time of one second. That means maybe a little bit of Kr^{36+} for the atomic physicist and Xe^{54+} for the same people. I remind you that Donets obtained these particles and we have at Saclay 10^9 - 10^{10} charges and 1000 A/cm^2. Now, I have looked at what happens if we want 10^{11} - 10^{12} charges. We have 1 m, 2 A or 20 A, with 55% of neutralization. This again should be with a cathode of 8 mm for the first case and 25 mm for the second one. We are going from 825 G up to 6 Tesla, and the compression is given only by the ratio of the magnetic field. The emittance from the magnetic field should be $\pi \times 10^{-8}$ mrad, normalized. It is ten times larger for the 10^{12} charges. And from the potential well, the oscillating ions have the same values. First you have a dream, then you have to make a first step: 2 A, a second step, 20 A, but this is just to show what takes place to go up to 10^{12} charges.

Becker: I have another solution which has already been suggested many years ago. You have to imagine that you put in a beam of one third millimeter into a bore like this. So why not put in a hundred beams or a thousand beams, because this surely is a way you can work on in the future, so that someone should start with perhaps with two beams. I would say we should not really push the limits of one single beam, but we can be rather modest on single beams and multiply this way.

Levine: I prepared some viewgraphs to answer the question. I propose a figure of merit for EBIS which is essentially what we have been talking about. Herrmann implies that "J" is proportional to the magnetic field, the current emissivity from the cathode, and inversely proportional to the effective cathode temperature. My answer would be to go after "J" of the cathode. If you want to dream, let us dream 100 A/cm^2 and not two. If we want to be realistic, we can talk about 6 A/cm^2 or 20 A/cm^2. I have been saying that n_e depends on the instabilities, and I do not know what is correct or not, but the problem is that if you build a machine that does not work, you suffer. And when you get a little older you do not like to build machines that take too long to perform. If you have instabilities they are awfully hard to cure, and we do not have the money to explore the space too broadly. So if you have to sit down and spend a million dollars to build a machine, you really do want to be on the safe side. There are three classes as I see it, and this is a very arbitrary breakup. There are the backward and forward waves that go like length cubed and the current, and the cure would be to make them short. With this "L cubed", going up 20% doubles the gain of whatever instability you are talking about. The other thing is to use absorbers and a lot can be done in this area. I have no expertise in it and we just have to tap into another community to find out what to do. The second thing is the Penning electrons; they stimulate a number of oscillations and they also, I think, can be controlled. Dr. Becker has suggested putting the collector at high voltage. One can create banana orbits in the Penning electrons which helps them to diffuse across fields rather rapidly; there are suggestions that a semi-conductor interceptor can be used to sweep them out of the system. And I do not know the answers again to this. But this is an EBIS problem. And the third thing is the controversial Litwin, Vella, Sessler modified two stream instability that has also been looked at. I am not a theoretician; I look at what they have put out and I have heard it is a refereed journal so I just take it like that until someone tells me differently. It corresponds to what I have heard in the plasma community that two streams tend to be unstable. The safest thing is to make the trap about the length of the plasma oscillation, then I know there can not be any instabilities. And when you are investing all that effort, and it can be 5 or 10 years if you make a mistake, you want to be safe and you want to be sure. So I would say that one should look at this relationship which can be taken from their paper, or just taken from the concept that you do not want more than one plasma wave in the length of your trap, which is essentially what the thing says. And that means that the plasma frequence is proportional to the square root of "n", so "n" goes as $1/L^2$ so that the total number you can get into the system is just the density times the length times the current, and that is inversely proportional to the length. If you put all of this together, and I do not know if anyone believes it, then what you come out with is this problem: What we want to do is raise this up. Maybe we can do something there to cool the cathode, to cool the electrons out of the cathode. That is not an impossible concept. Maybe we can get a factor of three there. The length is pretty well determined in terms of the voltage and I will not discuss that. We can do things about that, maybe chop up the beam into several lengths. And then the other thing, the one that I think is most interesting is the emissivity of the cathode. I think that is where I would like to work to improve the system.

<u>Hershcovitch</u>: My main question is regarding scaling lows. I just noticed that when you put in the numbers for DIONE, if you just calculate the very basic scaling lows regarding L, N, and τ, indeed you are getting what you expect from 600 mA total current. However, when we go to the figure of Donets, that assumes stepwise ionization. And yesterday it was brought up by Val Kostroun that when we go to the very heavy ions, we seem to see better

results. A lot of it may be attributed to the fact that one can get multiple ionizations. Will that make things more optimistic?
Becker: If you look at the numbers I have given on the proposal on the poster, it is just a factor of 30 enhancement as compared to the simple Lotz formula, or whatever you call it. And it comes from multiparallel steps of 2, 3, 4, 5 steps, but it also comes from that ionization is easier if you have autoionization before. So this makes cross sections larger by a factor of 2, 3, 4, 5 as Alfred Muller has shown by measurements in Giessen. So we should take, since we do not know better, Donets' cross section data on xenon. And they show that, up to the four last shells, you have such a very high multi-electron environment that you have much faster ionization going on. The factor to the standard Lotz formula is not 4.8×10^{-14} but is the region of 2×10^{-12}.
Hershcovitch: Ross Marrs told me once that you cannot maximize every single parameter; usually you can do good with two out of three.
Levine: But the ionization cross section is one parameter that we have no control over. Only God can maximize that one!
Hershcovitch: But if this is indeed the case, one can relax "J", then one can work better with the other parameters.
Knapp: We do observe, based on our predictions of the time that it takes for each neonlike goal, it certainly is different by no more than a factor of 2 from what you would expect from stepwise ionization. That is somewhat misleading because all the time is in the last few charge states.
Kleinod: Shouldn't we think that God helps the others also, not only us, but also the ECR people? To me it was very interesting to hear the figures that Tim Antaya gave, especially from the Julich source. Unfortunately I have had got any contact with them. They claim to get Ar^{18+}, 0.5µA. If you evaluate this to how many particles per second, you end up with about 2×10^{11} particles per sec. So I think we should not think how we get a little bit faster to Xe^{44+}, or whatever. We have to get the yield. That is one of our main problems. The way they did it, as far as I understood, was heavy pumping at the extraction area so they could lower the losses from charge exchange. We always thought this might be the final limitation to them or the last thing where we differ. So we really have to keep this in mind: We first have to show that even for Ar^{18+} our sources also go up to the limit that we are talking about. Then perhaps we go further on and try to think about U^{60+}.
Hershcovitch: Again, if one wanted an injector for a synchrotron, we have only half a millisecond acceptance time. That is why thinks look very different.
Becker: Nevertheless, he has put a finger on the right point. I suppose we should show, and you probably can do this first on your source, that (1) cooling, as Mort has shown for long trapping times on an EBIT, works for an EBIS as well. So this is one basic step. The second basic step then is to show (2) we are able to extract charge states as high as fifty or so. It is reasonable to

do that, because Donets probably has not often done Xe^{54+}. And then there is (3) a real challenge for the EBIS camp is to show that we can work with reasonable current and high yields. Should we do the multiple beam business, the single beam, or look in the business with the hollow beams of Dick True? We have to make progress in this direction.

Tawara: From the atomic physics side we can stress the differences of the EBIS over the ECR source. The EBIS source is relatively free from the metastable states. This is very important for atomic physics research. There are many groups who are using the ECR source to get ionization cross section of highly charged ions. The problem is most of this data is contaminated. Somehow I think we must be careful about when you apply this source.

Faure: As soon as the source was installed in SATURNE it was very difficult to have atomic physicists and nuclear physicists using the same source at the same time. I would say that our program in SATURNE is to realize the dream I was speaking about. I hope we will cool our ions. If we arrive in one or two years, maybe we can think of the next step, but not before. I cannot say right now how to design a new source saying we will have 10^{11} charges or 10^{12} charges. We first have to understand everything inside the source. Maybe we should shorten it up to 30 cm and increase the intensity to keep the number of charges. Anyway, it is very difficult to say.

Levine: I would like to say that anyoen who builds an EBIS source without a radial port is tying at least one hand behind his back, maybe both hands. It is extremely important. The atomic physics is both important and a very good diagnostic tool for understanding the device, and we do need data about what is going on in the machine. People really have not shown that they can get that kind of data from an EBIS without a radial port.

Becker: I suppose there is always a group of atomic physicists waiting for an ion source which has failed to feed an accelerator! At least this has been a hope for EBIS builders so far. You try to improve it and you find someone who is excited about the ions.

Levine: Jeremy Good said that the superconducting magnetic with radial ports is not that much more expensive.

Stockli: On the other hand there was a good reason that Bob Schmieder backed out on that. For diagnostics it would be very useful. If you look at the problems with which Bob struggled, and go and order a magnet from CCL with a radial port, you can expect a delay which is seriously longer than the normal delay in which they deliver. It has not been done before, but it can be done; that is for sure. One of the problems of putting a radial port in there is that we still believe that the axis should be reasonably straight. Then the problem you end up with is that you have to have one common bore which is straight, you have to put holes through there, seal it again because you have LHe on both sides, and you need vacuum access. You can make two solenoids, which I think is much easier to do, but then you are not sure about the straightness of the whole system anymore.

Becker: One could easily do a series of solenoids, as the LBL EBIS was doing. If the distance between the solenoids is small compared to the diameter, you can just calculate the homogeneity, which you can achieve. And then in the center, you probably have sufficient access. But it complicates everything. I worried in former times about ion heating. Now Mort and his group has shown that ion cooling works, so we should go home and try it and see how it improves things. If ion cooling works in our devices then we can make the next step. And of course as Kleinod pointed out, the important step is still to see how we come close to a theoretical limit of an EBIS.

Kleinod: I learned today that they asked us for something like 3×10^{11} particles for RHIC. But that is what people demand for CERN. So this would just be a way of learning it. I think this has to be proven, even though they will not take our EBIS. We just have to show that we get the yields of something comparable. We feel we can handle it and I think that everyone that has a source like this should try to see if it is not possible to go in this direction. I think those figures are not too high for energy or current density.

Becker: For people who did not recognize this viewgraph, it comes from a proposal of the feasibility of an EBIS for the CERN project. Taking these enhanced effective cross sections of Donets, and I must say since we have no other data, it turns out that making that Pb^{54+}, is very easy with 30 A/cm^2. So you would dare to do a 2 m long solenoid because the constraints on the dimensions are simple. Of course in such a case one would immediately learn what is going on with length and stability.

Faure: Does someone have a metal source to lend me? I can do it on DIONE because it is very easy to decrease the density. I have a question about the radial port. I agree it would be very useful to look at the x-rays and I know some people were looking through the cathode. I am not sure if you have the problem of the bremsstrahlung from the collector, or what the reason was. Did that work on the Frankfurt EBIS?

Becker: No. We are not experienced x-ray people. But we can look through it all and it should be possible to diagnose the beam as well. We have a 1 mm free path around the beam and inside of this you should see, from longitudinal direction, the electron beam. There is no question.

Beebe: The other thing I was wondering is, if that is inconvenient for some people, then perhaps extracting the electron beam in a different manner so that one could look down from the collector with an x-ray detector, would be better. There are some problems, but maybe when you are extracting you do not have to look at the ions at that same instant.

Herrlander: We tried twice in Stockholm to make quick and dirty experiments look this way without any particular success. We just looked in the longitudinal direction.

Beebe: Has anyone ever considered putting a solid state detector inside the vacuum chamber, inside the solenoid, to look for x-rays?

Herrlander: It is impossible.

Levine: Could I suggest another subject? We are looking for applications of EBIS. People are talking about injecting into storage rings. Maybe we ought to be in competition with storage rings. We actually could, in an EBIS trap, get counting rates that are in some ways more favorable for measurements than in storage rings.

Herrlander: I think we are aiming at different types of physics. That is probably the main difference. Once you build a storage ring then you have committed yourself to a kind of physics that is slightly different from the trap physics. We have been discussing adding a trap to our program at Stockholm. We are just too tied up in building the ring that we do not have the chance at the moment to go into details.

Becker: I would say in summing up it is quite true we have made progress. In that sense we are hopeful of getting high charge states from an EBIS source. The main problem of EBIS' in the past, and probably the future, is still unresolved; and that is the question of yield. Even if I scoped a fancy picture of having a machine gun like array of 1000 electron beams inside the cold bore, it must be done at least with one displaced beam first. Yes, we can do it with two, and so on. There should be no real problem doing this in immersed flow, so any development of high current intensity is welcome. This, I suppose, is just the thing Brookhaven is doing. So if one can really start out with a 1000 A/cm^2 inside of a homogeneous field, then it is quite easy to have parallel beams. Basically we have the question: What is the limit to one beam? What is the limit to the length that we can focus it and use it as a source, as a trap for ions, and to extract ions? Then, of course, the question is: Do we have better schemes, for instance hollow beams, which are more favorable with one beam instead of many beams. The other questions of higher ions and ionization cross sections and shorter cycling, are more or less battlefields on the side. One should work on them, but they are not really crucial for accelerator applications, otherwise an EBIS will only become a future for atomic physics people.

OTHER DEVICES

CRYRING — A HEAVY-ION STORAGE AND SYNCHROTRON RING

C.J. Herrlander
Manne Siegbahn Institute of Physics, S-104 05 Stockholm, Sweden

INTRODUCTION

CRYRING is a project aiming at a facility for research in atomic, molecular and nuclear physics using a cryogenic electron-beam ion source CRYSIS together with an RFQ linear accelerator as injector into a synchrotron-storage ring for very highly charged, heavy ions [1]. (CRYSIS stands for CRYogenic Stockholm Ion Source and CRYRING for CRYsis-synchrotron-storage-RING). The lay-out of a first phase of the planned facility is presented in Fig. 1. At the ring research programs on photon-ion, electron-ion and atom/ion-ion interactions will be possible over a wide range of relative energies and particle species. The quality of the beam of circulating ions will be improved by electron cooling.

In a second phase CRYRING will be equipped with a high-voltage (around 300 kV) ion accelerator as an alternative injector but also as the second source needed for atom/ion-ion collision experiments. An extracted beam is also planned for this phase.

CRYSIS

CRYSIS is an ion-source of a cryogenic EBIS type which has been built in close cooperation with the Institute de Physique Nucleaire in Orsay [2,3]. It is presently used as a stand alone machine for low energy atomic collision physics research. To improve the performance of CRYSIS an ion injector, INIS, which in principle is a small isotope separator with an r=50 cm analysing magnet will be connected. In Fig. 1 is also demonstrated the areas dedicated to research programs using the ion-beam from CRYSIS as well as the beam from INIS (laser-ion interaction studies). The CRYSIS-INIS system is subject to a separate contribution to this symposium (4).

RFQ

A radio-frequency quadrupole linear accelerator is under construction and will be used as an injector to the ring, accelerating the heavy ions from CRYSIS (and possibly in a later phase also ions from the high-voltage accelerator) to an energy of 300 keV/u. The RFQ will be of the four-rod type developed at the Institut für Angewandte Physik in Frankfurt/M. It is reported to this symposium in a separate contribution [5].

THE RING

The main parameters of the ring are summarized in Table I. The Lattice [6] will have six superperiods as demonstrated in Fig. 1.

Table I. Main parameters of CRYRING

Circumference	51.63	m
Q_x	2.30	
Q_y	2.27	
$\Delta Q_x/(dp/p)$	-1.3	
$\Delta Q_y/(dp/p)$	-3.2	
ϵ_x at inj.	200π	mm·mrad
ϵ_y at inj.	100π	mm·mrad
$\Delta p/p$ at inj.	$5 \cdot 10^{-3}$	
$\beta_{x,max}$	6.3	m
$\beta_{y,max}$	6.5	m
$D_{x,max}$	2.1	m
E_{inj}	0.3	MeV/u
for q/A < 0.25	0.3 q/a	MeV/u
E_{max}	$96(q/A)^2$	MeV/u

No focusing elements will be placed in six of the twelve straight sections. Out of these, three are used for injection, RF cavity and electron cooling. The remaining three will be used for beam diagnostics and experimental set-ups, and (in phase 2) for extraction.

The advantages with this lattice are:
- It is a simple lattice with only two families of quadrupoles.
- Six superperiods is beneficial to the stability of the machine since the number of systematic resonances are few.
- The smooth β-functions result in a relatively low intrabeam scattering.
- It is a very flexible lattice. Other solutions with other working points and/or three superperiods can be found.

The main disadvantage is the large despersion in the straight sections. Despite this the efficiency of the multiturn injection seems to be satisfactory. The design of a ten turn horizontal injection system has been completed including computer simulations and the mechanical design of the injection section [7]. Using a Monte Carlo method the efficiency is found to be about 85%. The closed orbit will be shifted locally at the injection section with electrostatic kickers furnished with a HV supply producing a voltage pulse with a linearly decreasing tail [8].

Two other sorking points with smaller dispersion have been studied to some extent. One is $Q_x = 3.3$, $Q_y = 1.8$, which still utilizes six superperiods. The other is a three superperiod solution, which requires three quadrupole families. In this case $Q_x = Q_y = 2.75$.

MAGNETS AND POWER SUPPLIES

The main magnet data are collected in Table II.

The dipoles are built up from 1.5 mm laminations and are giving a 30 degree bend each. The useful area of the field is increased with minor Rose-shims and the entrance and exit edges have Rogowski contours. In order to get more space the coils are removed from the central plane and are mounted at a distance of 200 mm from each other. The laminations for the quadrupoles are being punched in one piece in order to improve the symmetry of the field. The pole ends are cut at an angle of about 45 degrees.

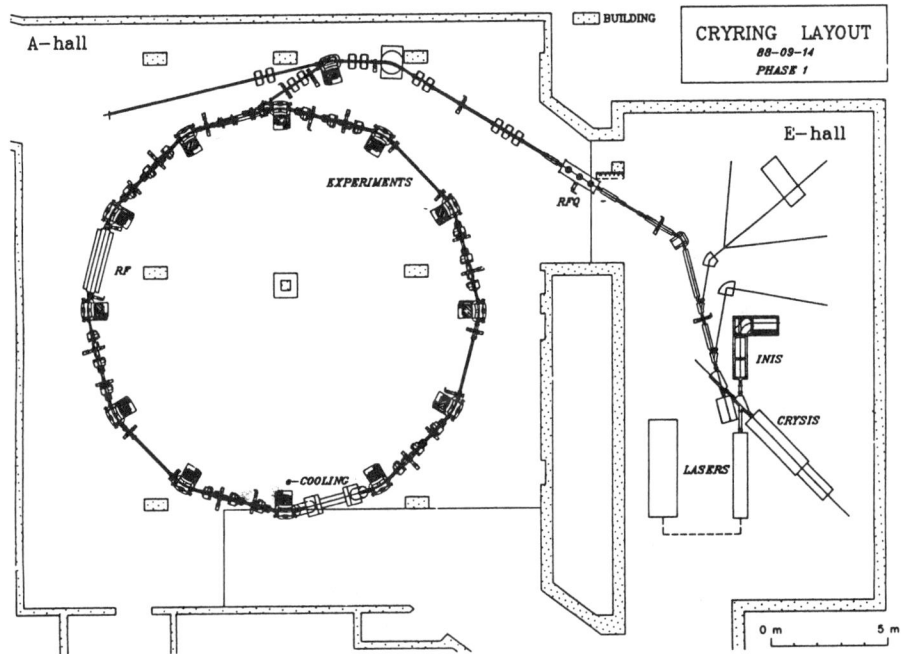

Fig. 1. Layout of CRYRING

The influence of the vacuum chamber walls due to eddy currents at the fast ramping has been carefully calculated. The heating of the chambers calls for cooling [9].

Correcting dipoles have been calculated and a prototype will be tested before the mechanical design is settled. Extra laminations of the quadrupoles are purchased to be used for skew quadrupoles.

Table II. Magnets

Dipoles		
number	12	
bending angle	30°	
bending radius	1.2	m
gap height	80	mm
max field	1.2	T
beam aperture	55 × 100	mm²
Quadrupoles		
number	18 (12QF, 6QD)	
aperture	125	mm
eff. length	0.3	m
max field grad	5.0	T/m
beam aperture	100	mm
Sextupoles		
number	12	
aperture	125	mm
eff. length	0.2	m
max field par	12	T/m²
beam aperture	100	mm

The design of the dipole and the quadrupoles power supplies allows the possibility of running the ring in a fast ramped mode of up to about 7 T/s. This will have positive implications for many scientific programs discussed, offering high duty-factors particularly when running the ring with ions that have a high loss rate in the storage mode.

UHV-SYSTEM

In storage mode CRYRING is supposed to operate at a pressure of 10^{-12} torr in the ring [10]. Originally this goal was planned to be achieved with a number of ion pumps and titanium sublimation pumps. However, the possibility of replacing the TSP:s with NEG pumps has been studied and found realistic, at least for pressures down to 5 ptorr. Special care has to be taken at several points in the machine, at injection and cooling, e.g., where extra pumping capacity will be needed. Also between the RFQ and the ring, differential pumping has to be applied to reduce the pressure by several orders of magnitude. The UHV technique has been studied in different model experiments comprising a section of the ring.

ACCELERATION SYSTEM

CRYRING is expected to work in a very wide energy range, from 0.3 to around 24 MeV/u, corresponding to a frequency span of about 8. In the future, moreover, ions will also be injected directly from the 300 kV electrostatic accelerator, giving ions with a lower energy per nucleon. For a first harmonic bunching and accelerating cavity, this means an even higher frequency span. Since the circumference of the ring is relatively small, the possibility of using a non-resonant driven drift-tube has been studied. A model cavity has been built and studied in detail up to drift-tube voltages of 600 V [11].

The gain variation can be kept within 6% between 10 kHz and 1.5 MHz and easily within the control range of a voltage levelling circuit. This flat gain characteristic was obtained by adjusting the inductance and capacitance in the plate circuit of both stages of the amplifier.

Control of the system is relatively easy, it requires only two feedback loops — a levelling circuit for the voltage and a beam position pick-up signal to keep the frequency (or phase) such that the beam is centered radially.

The DD-T would allow beam bunch manipulations, it means in the synchrotron operation, non-sinusoidal wave forms could be used to "flatten" the buch and improve the space charge limit, or shorten the bunch for fast extraction. In higher mode operation, sub-harmonic modulation could be used to preferentially accelerate one buch for extraction.

The frequency range is wide enough to cover both synchrotron and low energy storage modes which requires deceleration to low energies without mode changes.

Table III. Driven drift-tube parameters

Frequency range	f_0	10 kHz to 1.5 MHz
Length	ℓ	2.67 m
Accelerating voltage	$V_{acc.}$	0 to 1200 V
Drift-tube voltage	$\hat{V}_D$	0 to 3700 V
Power	P_{rf}	~ 10 kW

ELECTRON COOLER

The electron cooler for CRYRING [12] will cool highly charged heavy ions with energies from 3.6 MeV/u, requiring 2 keV electrons, up to the highest ion energies in CRYRING, corresponding to 13.5 keV electrons. It shall also be able to work as an electron target for atomic-physics experiments at electron energies up to 20 keV. The main design parameters, still preliminary, of the cooler are given in table IV. Other particular features of this cooler are the high vacuum required to avoid charge exchange of the ions and the fast ramping of the cooler field together with the rest of the magnets in the ring.

Table IV. Main design parameters of the cooler

Electron energy	2–20 keV
Perveance	0.1–5 $\mu A/V^{3/3}$
Electron current	1–3000 mA
Beam diameter	40 mm
Magnetic field	0.2 T
Cooling length	1.1 m

DIAGNOSTICS

CRYRING will be equipped with different diagnostic elements to determine the beam position and profile [13]. In the beam lines outside the ring destructive beam-profile monitors will be used. Each one of these consists of an Al_2O_3 disk onto which 32 gold strips are evaporated. Each measuring points will have two such detectors for measuring the beam size in x and y direction, respectively.

In the ring electrostatic beam pick-ups will be used as the main tool for diagnostics. Eight pairs of such detectors are planned around the ring mainly to be placed in the straight sections containing the focusing magnetic elements. The electronic systems like preamplifiers, amplifiers, multiplexor, ADC and a VME computer system are being tested.

As a complement to the electrostatic detectors a Schottky-noise pick-up is planned to be included in one of the free straight sections. The signals will be analyzed in a spectrum analyzer. Also, a fast fourier transform (FFT) program has been acquired for the separate diagnostic computer system, which will be controlled from the main control system computer.

CONTROL SYSTEM

The control system for CRYRING [14] is in all essential parts a copy of the CERN-LEAR control system. It is built in three hierarchic levels mainly consisting of:

a. A central computer (PDP 11-73) with operating console
b. A CAMAC-based data-distribution system
c. About one hundred micro-processors for direct control of the different systems.

Level a. and part of level b. are already in operation using a copy-transcription of the LEAR software program. The program for the static parts of CRYRING is operational, whereas the parts needed for the dynamical (fast ramping) controls still are under development. It is planned to have CRYSIS and its injector INIS as the first parts connected to the control system in 1988.

FUTURE PLANS

The plans for the next two years comprises the building of different components and subsystems mounting of the equipment, and — hopefully — starting-up phase 1 of the system in 1990.

After 1990 mainly two development programs are foreseen for a second phase: An extraction system using a third order resonance regime and the introduction of a 300 kV electrostatic ion-accelerator. Extraction of the beam from CRYRING aims mainly at different experiments using very heavy ions impinging on solid targets. The electrostatic accelerator will serve two purposes. It will be a second injector used for light ions (q/A larger than 0.25), and for low charge-state heavy atomic and molecular ions (q/A less than 0.25). It will also be a source for ion-beams used in collision experiments with the CRYRING beam in crossed and merged beams.

REMARK

CRYRING is being built by a group of about 40 physicists, engineers and technicians (including the group at Institut für Angewandte Physik in Frankfurt am Main). Credits should be given to all the team-members for their devoted efforts, which are highly appreciated.

REFERENCES

1. C.J. Herrlander, A. Bárány, K.-G. Rensfelt, and J. Starker, Physica Scripta
   C.J. Herrlander, K.-G. Rensfelt, and J. Starker, Proceedings EPAC, Rome, June 7–11, 1988.
2. J. Arianer, A. Cabrespine, and C. Goldstein, Nucl. Instr. and Meth. **193** (1982) 401.
3. S. Borg, H. Danared, and L. Liljeby, Proc. Third Int. EBIS Workshop, May 20–24, 1985 (Ed. V.O. Kostroun and R.W. Schmieder) Cornell Univ., Ithaca, N.Y., p. 47.
4. L. Liljeby, this symposium.
5. A. Schempp, H. Deitinghoff, H. Klein, A. Källberg, A. Soltan, and C.J. Herrlander, this symposium and proceedings EPAC, Rome, June 7–11, 1988.
6. J. Jeansson, and A. Simonsson, Proceedings EPAC, Rome, June 7–11, 1988.
7. A. Simonsson, Proceedings EPAC, Rome, June 7–11, 1988.
8. M. Kvarngren, and M.A.K. Eriksson, Proceedings EPAC, Rome, June 7–11, 1988.
9. A. Nilsson, and P. Carlé, Proceedings EPAC, Rome, June 7–11, 1988.
10. Th. Lindblad, L. Bagge, J. Bjon, and S. Levén, Vacuum **37** (1987) 293.
    L. Bagge, and J. Bjon, priv. comm.
11. K. Abrahamsson, G. Andler, and C.B. Bigham, Nucl. Instr. and Meth., to be published. Proceedings EPAC, Rome, June 7–11, 1988. KP 156.
12. H. Danared, Proceedings EPAC, Rome, June 7–11, 1988.

13. Th. Lindblad et al., priv. comm.
14. J. Starker et al., priv. comm.

DISCUSSION

Antaya: What do you think will be the first experiment with CRYRING?

Herrlander: Ions colliding with photons from a laser. I think that is one of the easiest you can do. Otherwise, the first experiment will just be to inject ions. It might be difficult enough.

Becker: Is this number right, that you want to have up to 5 µperv at the electron-cooler?

Herrlander: OK. I got this overhead before leaving. And this morning when I looked for it, I started to think the same thing. I can help you get a copy of the Rome paper, but the table is the same.

Liljeby: I have a copy.

Herrlander: Then you must know.

Liljeby: In the gun there is post-acceleration, so it is not the same.

Becker: In order to reach 5 µperv you must really decelerate. You cannot, I suppose, do a resonance focus gun with a high perveance.

RFQ-INJECTOR FOR CRYRING

A. Schempp, H. Deitinghoff, H. Klein
Institut für Angewandte Physik, J.W. Goethe Universität
D-6000 Frankfurt 11, Postfach 111932, West Germany

A.Källberg, A.Soltan, C.J.Herrlander
Manne Siegbahn Institute of Physics
S-104 05 Stockholm, Sweden

ABSTRACT

An RFQ is built as an injector for the CRYRING project in Stockholm. It will accelerate heavy ions from an EBIS ion source with q/A > 0.25 from 10 keV/A to 300 keV/A. The 4 Rod-RFQ will be 1.6 m long with diameter of 35 cm and operate at 108.5 MHz. The transmission will be greater than 90%. The status of the project and first results are presented.

INTRODUCTION

At the Manne Siegbahn Institute of Physics the CRYRING facility is being built for atomic, molecular and nuclear physics [1,2]. The layout of CRYRING is shown in Fig. 1. It consists of a cryogenic electron beam ion source (CRYSIS), a Radio Frequency Quadrupole (RFQ) and a synchrotron/storage ring. CRYSIS will deliver pulses of highly charged ions e.g. Ar^{18+}, Kr^{34+} and Xe^{44+}. The RFQ accelerates the ions from 10 keV/u to 300 keV/u and is designed to accept ions with charge to mass ratio greater than 0.25. The ring will be able to accelerate the particles and store them at energies from 24 MeV/u (for q/A = 0.5) down to at least 100 keV/u.

An RFQ[3] has been chosen as accelerating structure as it is a compact and efficient unit for accelerating low-energy ions and it introduces minimum emittance growth. It allows CRYSIS to be put on a moderate voltage platform (50 kV) instead of the huge and unpractical 1.2 MV platform needed for electrostatic acceleration to 300 keV/u. We have decided to build a 4-Rod RFQ (Fig. 2) the structure developed in Frankfurt[4]. It is well suited for low current, heavy ion acceleration, where a rather low frequency of around 100 MHz has to be chosen.

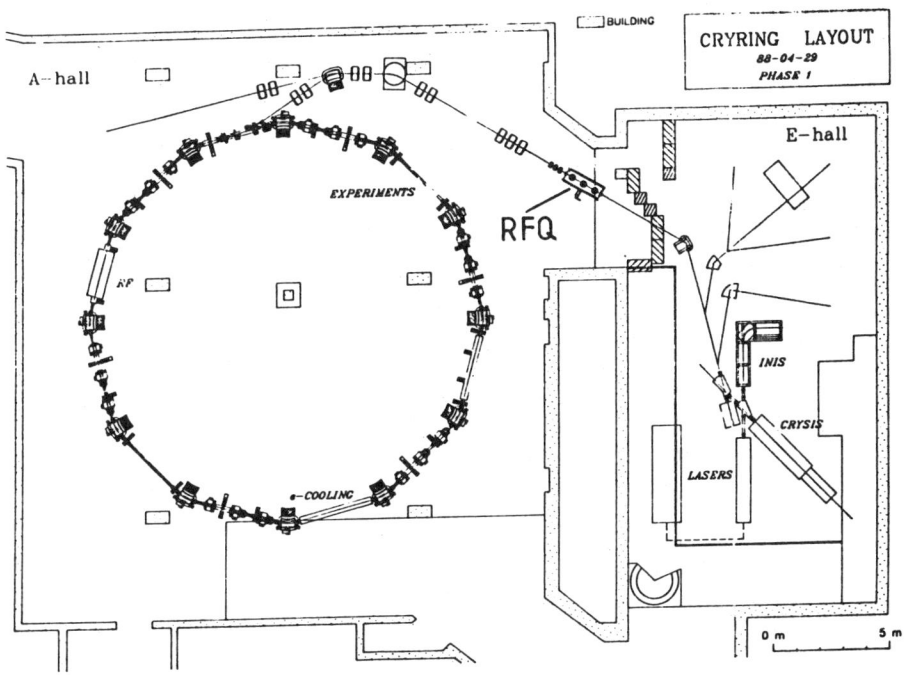

Fig. 1 Layout of the CRYRING facility

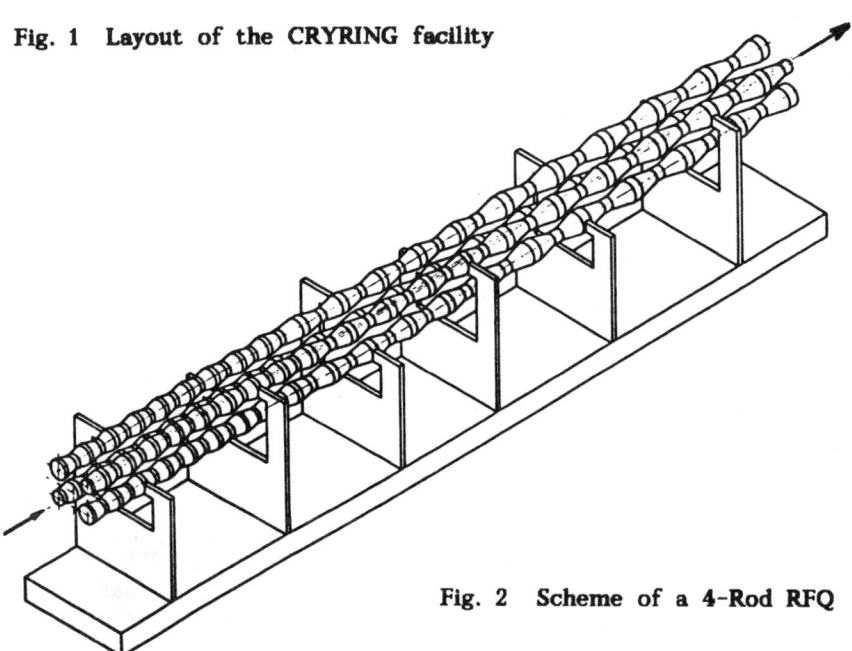

Fig. 2 Scheme of a 4-Rod RFQ

Design features are small size (tank diameter 35 cm, length 1.6 m), good RF efficiency and relatively modest mechanical tolerances. The injection into the ring[5] requires a small energy spread of less than 1 %. To obtain this small energy spread, a separate debuncher will be placed 1 meter after the RFQ. Due to the very high vacuum in the ring, attempts have been made to obtain a good vacuum in the RFQ. This has lead to a resonator design with metal seals and the unit can be moderately "baked" to 150° C.

BEAM DYNAMICS

The normal way of designing the parameters of an RFQ as originally done in Los Alamos[6] with a gentle bunching of the beam would in our case lead to an RFQ, which is more than 2 meters long. This length would cause manufacturing problems and too much rf power would be

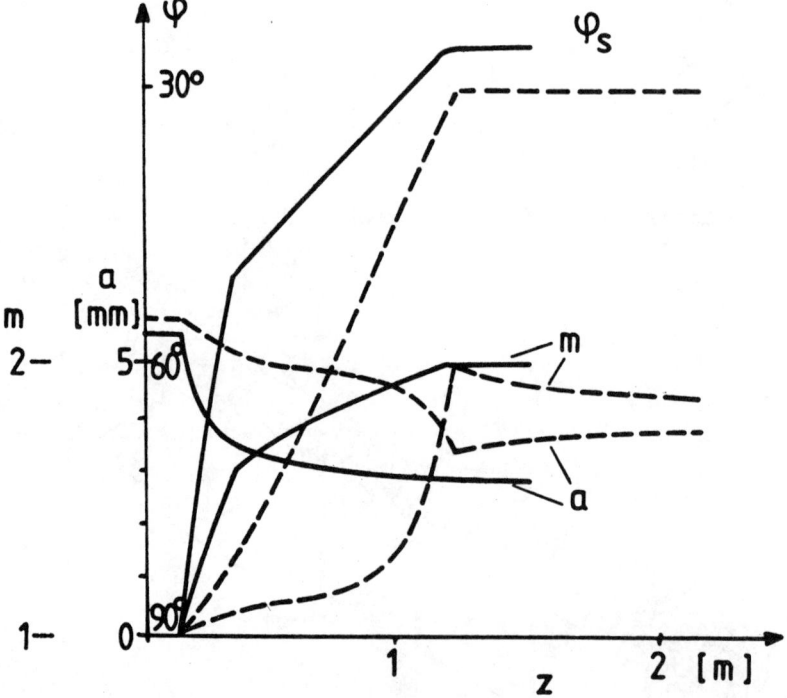

Fig. 3 The variation of the electrode parameters synchronous phase φ_s, modulation m, and aperture a along the z-axis. The dashed lines correspond to the standard method of RFQ design and the solid lines to the new method used in our design.

needed. To make a shorter RFQ we have used a novel design procedure[7]. At first, there is a prebuncher, consisting of four cells with a small modulation followed by a 12 cm long drift section without modulation. The prebuncher enables the following rapid increase in synchronous phase φ_o, and modulation m from $90°$ to $50°$ and 1.0 to 1.6 respectively, without increased particle losses. The increase in φ_o and m takes place in only 25 cells (21 cm). Then follows a slower linear increase to the final values $\varphi_o = 25°$ and m = 2.0 that are reached near the end. Fig. 3 shows a comparison between the two ways of designing an RFQ. The prebuncher also reduces the energy spread of the beam, as most of the particles are near the synchronous phase during the rapid increase of the longitudinal field, thus avoiding most of the energy spread.

It should be mentioned that this special RFQ design for CRYRING is resulting in a shorter RFQ structure even compared to solutions for a higher frequency of 200 MHz [8] and will operate at a rather conservative field strength so there is room for developement for higher gradients and ions with lower specific charge to mass ratios. But also availability of rf equipment and detailed calculations of beam matching will influence the final choice of operating frequency. In Fig. 4 plots of the beam in the transversal and longitudinal phase spaces according to the PARMTEQ calculations are shown. The new design increases the smallest distance between the rods, which normally is found where the final values of φ_o and m are reached. Thus an increased voltage can be applied between the rods without sparking.

Our RFQ is designed for a voltage of 70 kV between the rods, which would be needed for ions with q/A = 0.25. This voltage corresponds to a surface electric field of approximately 1.6 times the Kilpatrick limit for sparking[9], which should be safe for the short RF pulses of less than 1 ms and give a possibility to increase the voltage and accept ions with lower charges. Some parameters of the RFQ are given in table 1.

TABLE I PARAMETERS OF THE CRYRING RFQ

Charge to mass ratio (q/A)	≥ 0.25	Frequency	108.48 MHz
Input energy	10 keV/u	Output energy	300 keV/u
Electrode voltage	70 kV	Length of electrodes	1.54 m
Synchronous phase	$90° - 25°$	Maximum modulation	2
Number of cells	103	Aperture radius	5.5 − 3.0 mm

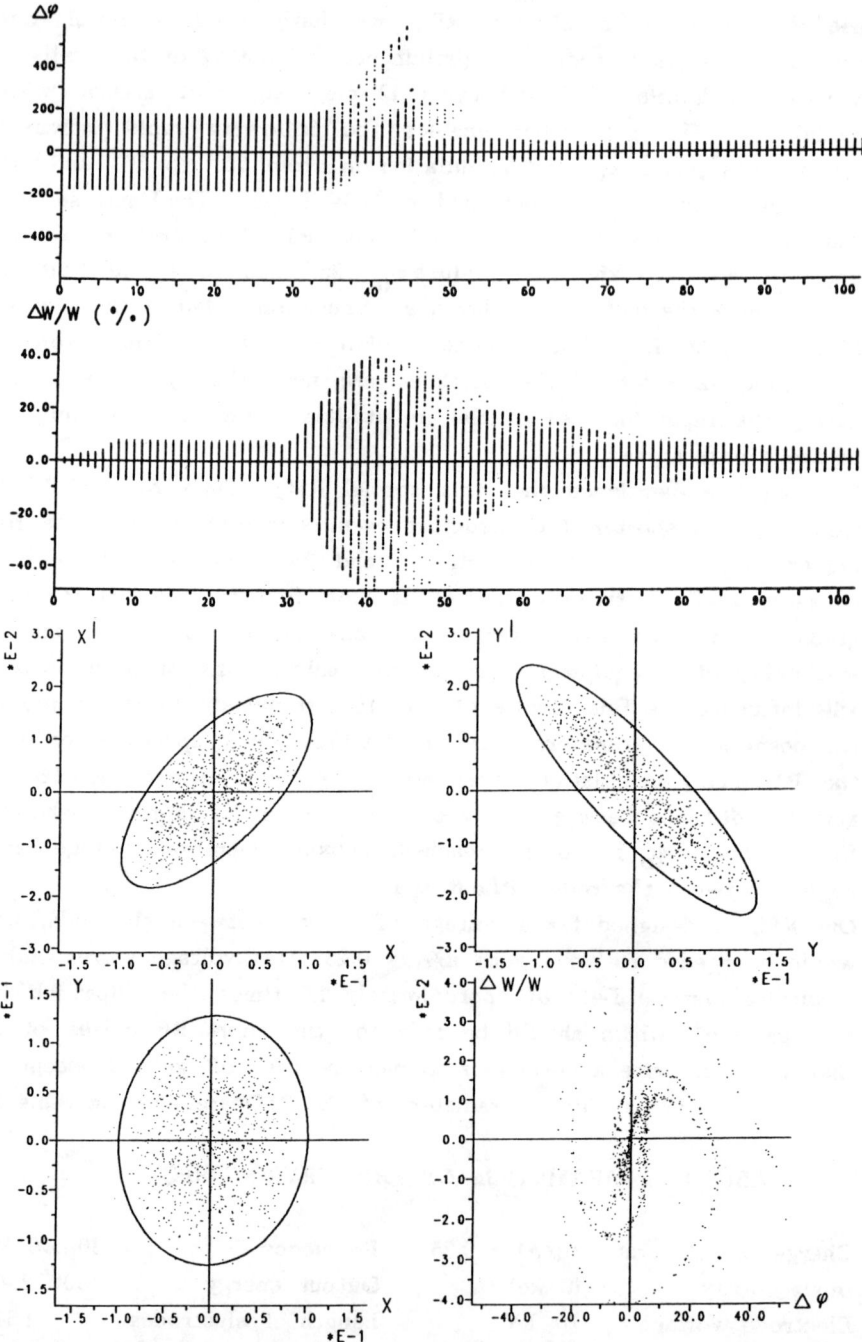

Fig. 4 Calculated phase space plots for the beam after the RFQ

RFQ RESONATOR DESIGN

The 4-Rod RFQ is named after the electrodes, which are rods with conically varying diameter. In the first order approximation the variation of the electrical field on the axis is the same as that of the modulation. While in the design work and in RFQ resonators with vane-shaped electrodes the field can be derived from a two term potential, for the cylindrical electrodes higher order terms have to be considered. Thorough investigations have shown, however, that the higher harmonics are not important, if the aperture is not fully used. Stepwise appropximations of the ideal sinusoidal modulation has been done for the 4-Rod RFQ tested at DESY, has shown little influence on beam quality [4].

The rf resonator driving the electrodes consists of a chain of coupled $\lambda/2$ transmission line resonators. The supporting stems, which correspond to the inductivities, are arranged linearly on a common base plate. In this arrangement currents are confined in the resonant stem-electrode structure, which has two important consequences. Firstly, the tank has practically no influence on the resonant frequency, secondly, all parts of the resonant copper structure can easily be cooled efficiently by water-flowing through bores in the stems as shown in Fig. 5.

The electrodes are made on a lathe and brazed to the stems. Alignment of the electrodes and a rough frequency tuning of the structure is done outside the vacuum tank. The tuning is done with copper inserts in the end cells. The number of stems and the size and shape of the stems determine the frequency and the efficiency of the resonator.

The shunt impedance R_p, which is a measure of the efficiency, is defined as $R_p = U^2 L_c/N$, where U is the electrode voltage and N is the rf power for the RFQ cavity of length L_c. In Fig.6 calculations of R_p as a function of the number of stems and the diameter of the stems are shown. Assuming 60 % of these values as a conservative estimate, 60 kW of RF power is needed for the design voltage of 70 kV. Due to a very low duty cycle of less than 0.1 % the average power level is very low. Figure 7 shows a view of the rod assembly during tuning.

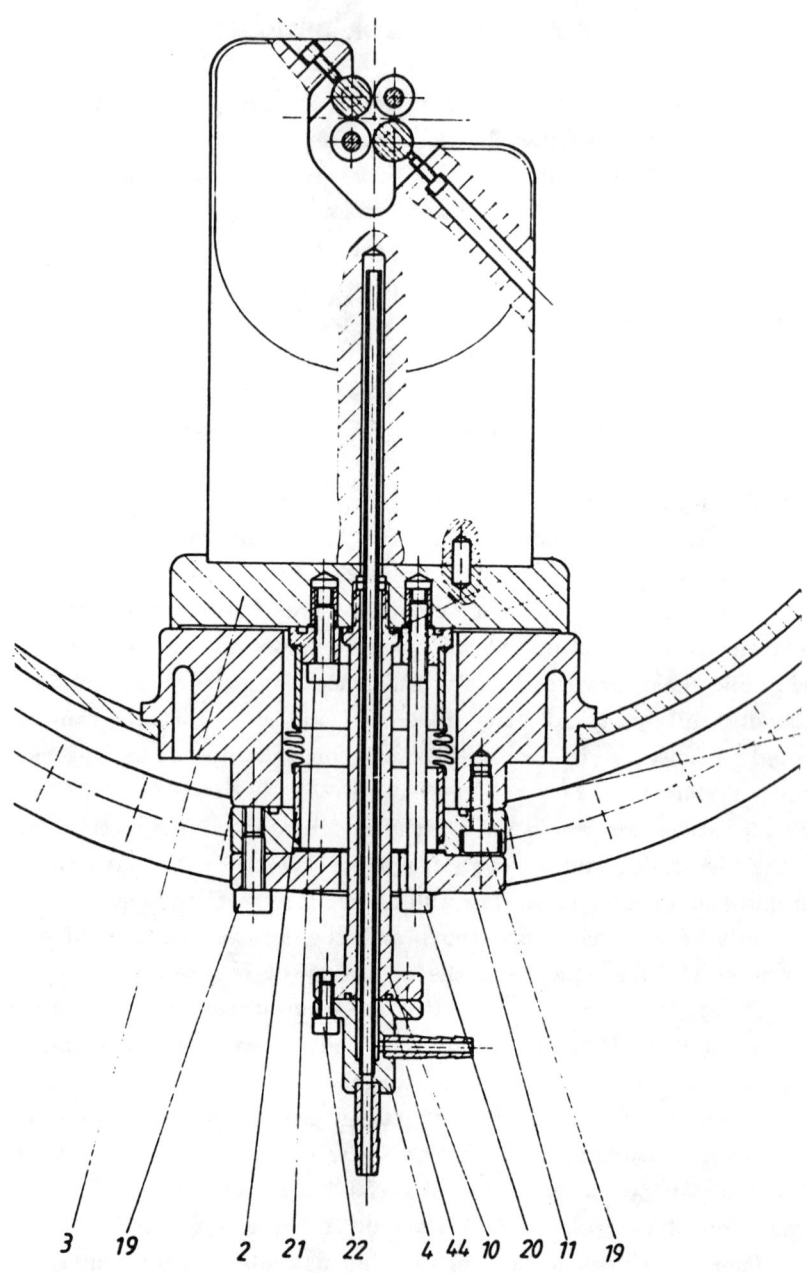

Fig. 6 Cross section of the 4 Rod RFQ support stem

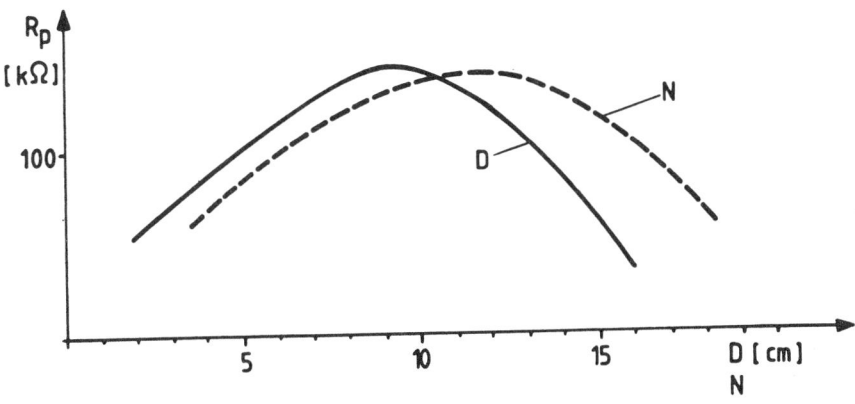

Fig. 5 Calculated R_p values as function of stem diameter D and number of stems N

Fig. 7 View of the 4-Rod RFQ insert

STATUS OF THE PROJECT

The design of the RFQ has been completed and the beam dynamics parameters have been fixed. The tank has been manufactured and will be copper-plated at GSI. This had been delayed due to a reconstruction of the copper plating facility at GSI and is now scheduled for Feb. 89. The RFQ will be assembled and tuned in Frankfurt. We hope to make first rf tests in March and to have the whole unit ready in April 1989.

REFERENCES

1. C.J. Herrlander et al., PAC 85, IEEE, Ns 32, No. 5, (1985) p. 2718
2. C.J. Herrlander, K.-G. Rensfeldt and J. Starker,
   Proc. of EPAC 88, Rome, June 1988
3. I.M. Kapchinskiy and V. Teplyakov, Prib. Tekh. Eksp. 119, No. 2 (1970) 19
4. A. Schempp, M. Ferch, H. Klein, PAC 87, IEEE 87CH2387-9 p. 267
5. A. Simonsson, Proc. of EPAC 88, Rome, June 1988
6. K.R. Crandall, R.H. Stokes, and T.P. Wangler,
   Linac 79, BNL-51143 (1980), p. 205
7. A. Schempp, Design of compact RFQs, EPAC 88 Rome, June 1988
8. J. Staples, these Proceedings
9. W.D. Kilpatrick, Univ. of California, Berkeley UCRL-2321 (1953)

DISCUSSION

Stockli: Can you tune or shim the rods of the RFQ, or are they fixed?
Herrlander: They can be machined in cases like that and then screwed together. You then line them up and solder or braze them to their supports in an upline position.
Stockli: Hence there is no tuning or shimming after brazing?
Herrlander: As I said, it is done outside the tank. Everything is done outside the tank almost, except acceleration.

ECR ION SOURCE STATUS 1988

T.A. ANTAYA
Michigan State University, East Lansing, MI 48824

C.M. LYNEIS
Lawrence Berkeley Laboratory, Berkeley, CA 94720

INTRODUCTION

Electron Cyclotron Resonance ion sources (ECRIS), like Electron Beam Ion Sources (EBIS), produce multiply-charged ions via electron impact ionization. As with EBIS, they are stand- alone ion sources with their own power supplies, vacuum systems and control systems-- they can be as complex as small accelerators. In contrast to EBIS, the electron acceleration process and the ion confinement strongly couples through the magnet field, and this makes ECRIS both simple to build and at the same time difficult to understand quantitatively.

In 1988, ECRIS find important application in nuclear, atomic, and high energy physics. Also there is an emerging new application-- efficient bulk ionization, which will extend the application of ECRIS to ionizers for isotope separation and polarized beams.[1-3] The initial driving force in the development of ECRIS was the need for more highly charged ions. The first ECRIS injection into a cyclotron occurred in Karlsruhe in 1981, and was quickly followed by Cyclone in Belgium, SARA in France, KVI- Groningen in the Netherlands, and Berkeley.[4-8] ECRIS have now largely replaced arc discharge ion sources for this purpose in most cyclotron facilities,[9] where gains are made in both intensity and energy, and maintainence costs are substantially reduced. ECRIS have made the stand-alone cyclotrons so competitive that the coupled operation of the two cyclotrons at MSU has been bi-passed in favor of operation of the K=1200 booster cyclotron stand-alone with ECRIS injection.[10] In 1985 modifications began at CERN, to accomodate beams from an ECRIS-RFQ injector.[11] Oxygen ions were accelerated to 200 GeV/u in the SPS in 1986 and sulfur in 1987.[12] Further increases in mass at CERN probably require reconstruction of Linac 1, and it has been estimated that lead beams would require 30 eµA of 25-30+ ions.[13] In 1989, an ECRIS-based positive ion injector will replace the tandem injector at the ATLAS facility at ANL.[14]

The potential for atomic physics measurements in new regimes using ECRIS ion beams was also recognized early. Experiments were begun in 1979 at LaGRIPPA in Grenoble,[15] and ECRIS have been built in Oak Ridge and Giessen specifically for atomic measurements,[16,17] while other facilities such as the LBL 88 Cyclotron, KVI Gronnigen and Louvain-la-Neuve allow atomic studies on a time shared basis.[18-20]

ECRIS DEVELOPMENT FOR HIGHLY CHARGED IONS

Approximately 30 ECRIS have been built for the production of highly charged ions. By and large these sources are two stage devices. In a two stage ECRIS, low charge ions generated in the first stage diffuse across a pressure gradient to the main stage where they are further ionized. The output of two stage ECRIS is generally peaked at high charge states. Fig. 1 show a nitrogen charge state distribution obtained from the RTECR at MSU.[21] The charge state distribution is peaked at 5+, with approximately equal currents of 6+ and 1+. Fully stripped nitrogen is obscured by the helium 2+ peak. The helium in the spectrum is deliberate, enhancing the currents of highly charged nitrogen ions. This so called "gas mixing effect" is discussed below.

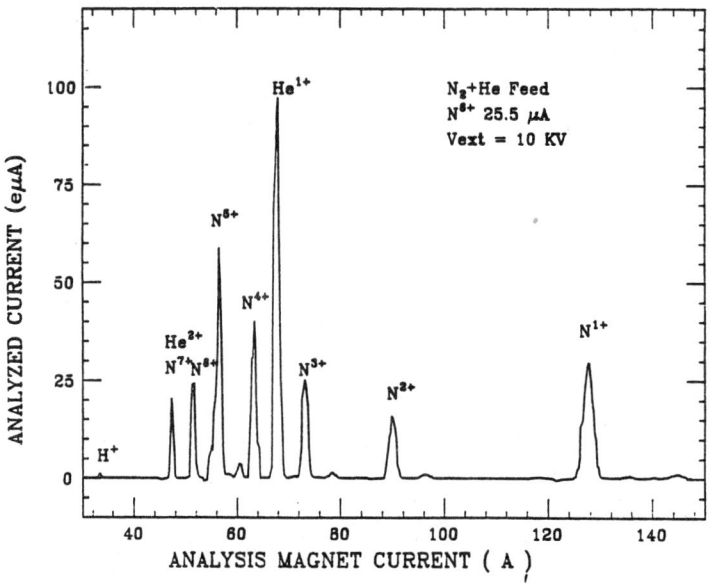

Fig.1. The production of highly charged nitrogen ions in the two stage RTECR at MSU is shown for the feeding of a nitrogen plus helium gas mixture into the first stage.

Most ECRIS accelerator applications use intensities of the order of 10^{12} pps. Therefore most ECRIS development occurs at the eµA level or higher, and little account is taken of more highly charged ions of significantly lower intensity. It is then difficult to establish the maximum charges obtainable from the present two stage ECRIS; in Table I selected performance data are summarized. Table I is intended to be representative rather than exhaustive. The ECRIS development effort in Grenoble under R. Geller still comprises a large fraction of the total ECRIS development effort, and this is reflected in Table I.

Table I. Some Benchmarks for Two Stage ECRIS Performance

SPECIES	CHARGE	INTENSITY (pps)	SOURCE
^{20}Ne	10+	$7.5\ 10^{11}$	Juelich 14 GHz (22)
^{40}Ar	16+	$2.0\ 10^{11}$	Juelich 14 GHz (22)
^{40}Ar	18+	$3.5\ 10^{8}$	Grenoble 18GHz (23)
^{86}Kr	20+	$3.0\ 10^{11}$	MSU 6.4 GHz (21)
^{129}Xe	29+	$2.2\ 10^{11}$	Grenoble 10GHz (24)
^{181}Ta	29+	$6.5\ 10^{11}$	Grenoble 18GHz (25)
^{209}Bi	34+	$1.0\ 10^{10}$	LBL 6.4 GHz (26)
^{238}U	36+	$1.7\ 10^{11}$	Grenoble 18GHz (25)

BEAMS FROM SOLIDS

The development of techniques which allow the production of high charge state ions from solid materials has been of great importance. Since a very high percentage of the elements more massive than argon are solids at room temperature, the ability to use solid materials as source feeds is vital for ion sources used with heavy-ion accelerators. The two main methods are direct insertion of solids into the plasma and use of ovens to vaporize solids.

Direct insertion has been studied in detail in Grenoble with the CAPRICE source for a wide variety of materials ranging from aluminum to gold.[27] In CAPRICE, a solid rod is positioned close to the ECR surface where it is vaporized by hot electrons in the plasma. The plasma is maintained by adding a support gas such as oxygen or nitrogen. To maintain a stable plasma the rod's position is automatically controlled with a feedback loop. In a one week run using tantalum in CAPRICE, the average consumption was approximately 1.0 mg/hr.[28] Direct insertion has been used in the ORNL-ECR to produce iron, nickel, and chromium beams,[29] in the LBL-ECR to produce niobium beams for atomic physics,[30] and in the MSU RTECR to produce flourine, magnesium, silicone, vanadium, iron and tantalum beams.[31] Even without feedback, the stable beams can be produced for periods of several hours.[32]

A variety of metallic ion beams have been produced from the LBL ECR using a resistance heated ovens.[33] A low temperature oven which operates up to 700 deg. C is used for a variety of elements such as Li, Mg, K, Ca, and Bi. A high temperature oven which can operate up to 2000 deg C is used for higher temperature elements such as Sc, Fe, Ni, Cu, Ag, La, and Tb. Because the oven temperatures are regulated externally, stable oven operation and efficient material usage are possible. The ovens are inserted radially into the second stage so that vaporized metal atoms stream through the ECR plasma and are ionized by electron impact. Typically, the plasma is maintained by running either oxygen or nitrogen as a support gas in the first stage. This is similar to the use of a mixing gas when operating the source with gases heavier than oxygen. The amount of metal in the plasma is adjusted by varying the oven temperature.

As can be seen in Fig. 2, the performance of the LBL ECRIS for oven feed is really equivalent to gaseous feed operation.

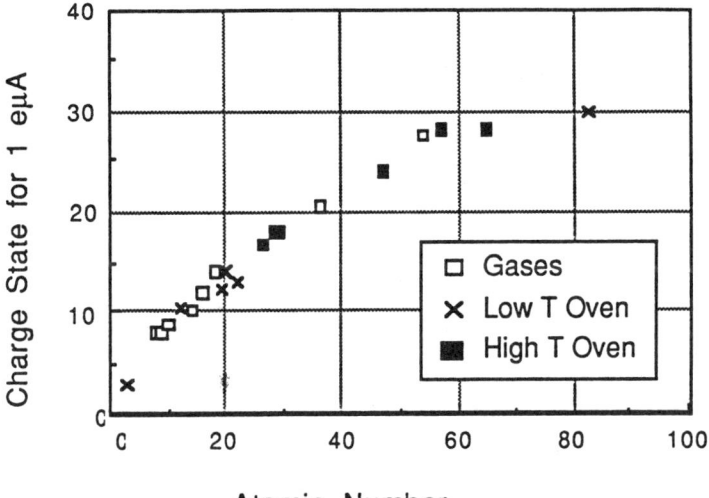

Fig. 2. Charge states for the 1 eµA level for elements run from gas feed and from the low and high temperature ovens at LBL.

THE GAS MIXING EFFECT

ECR builders meeting at the Kyoto ISIAT Symposium in 1983 and the Oxford Conference on Highly Charged Atoms in 1984 reported observing that the addition of a lighter gas to the plasma significantly increased the currents of highly charged ions of a heavier species. One rather surprizing finding at KVI-Groningen was discussed- that the installation of a dirty extraction end plate to the source there significantly increased the current of Ne^{9+} ions.[34] While the reasons for studying dirty electrodes remains obscure, the effect could also be obtained by directly mixing oxygen with neon, suggesting that the normal outgassing of the electrode, when exposed to plasma, was the cause. Gradually the systematics of the effect have become established, and all sources use gas mixing to some degree to inprove currents of highly charged ions. Fig. 3 shows the direct dependence of gas mixing effectiveness on mass for He, N_2 and O_2 mixing with argon in the MSU RT-ECR, for the production of argon ions. All three mixing gases improve the charge state distribution over pure argon feed, with oxygen slightly better than nitrogen, and both significantly better than helium.

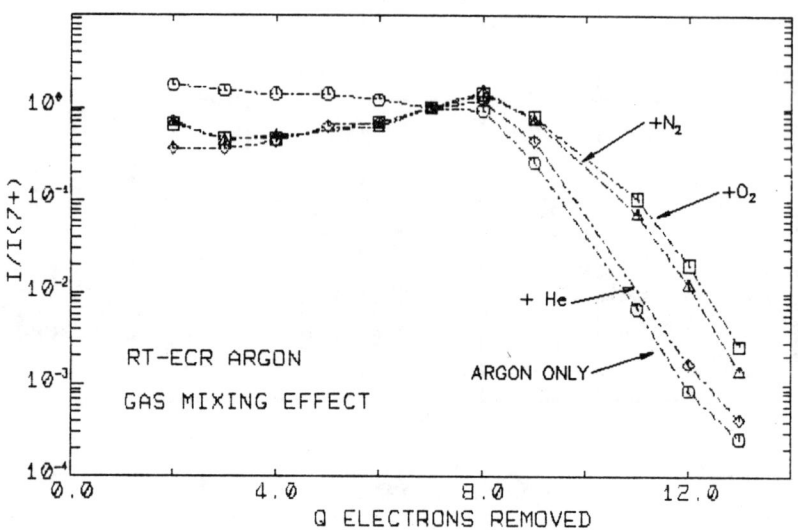

Fig. 3. The effect of the addition of light gases to an argon plasma in the RTECR at MSU is shown.

Several models have been proposed for the gas mixing effect,[35,23] but the most likely explanation was recently reported By Antaya at the Grenoble ECR Workshop in 1988, on the basis of argon energy spread measurements made on the RTECR at MSU.[36] It was found that the argon ion thermal energy in the RTECR increases with the degree of ionization, with the linear rate of 6.5*Q eV for Q>5+. Further, the thermal energy of argon 9+ ions was observed to decrease by 1/3 when oxygen was added to the argon plasma. Such cooling, the result of ion-ion collisions in the plasma, would be expected to significantly alter ion confinement times (longer for argon, shorter for oxygen), since the confinement times are set by ion-ion collision rates.

WALL COATING EFFECTS

At the ECR Conference at MSU in 1987, the LBL and Grenoble groups made first reports on the beneficial effect of wall coatings on the production of highly charged ions in ECRIS. The origin of this effect may be related to the emission of electrons from the walls of the plasma chamber when bombarded by energetic electrons from the plasma.

The day to day performance of the LBL ECR ion source is strongly influenced by the conditions of its walls. In particular, the addition of SiO_2 coating to the walls significantly enhances its performance for high charge state ions such as N^{6+}, O^{7+}, and Ar^{14+}.[26] Fig. 4 clearly demonstrates the basic effect of coating the walls with SiO_2. Most surprising was the discovery that under certain conditions the LBL ECR could produce more intense high charge state currents operating without the first stage than has ever been done operating it as two stage source. This single stage operating mode offers the possibility of building high performance single stage ECR ion sources without the added cost and complexity associated with a first stage.

The enhanced performance in single stage operation after coating the walls with SiO_2 appears to result from an increase in the production of cold electrons at the wall of the plasma chamber. As pointed out previously, the coefficient for secondary emission for electron impact on SiO_2 is between 2 and 4-- considerably above 1.3 for bare copper.[26] Geller et al found that coating the walls of Ferromafios with good electron donor materials with low work functions resulted in improved performance.[23] In equilibrium the production of cold electrons in the plasma must equal the losses to maintain plasma neutrality. In the second stage of an ECR source, cold electrons are injected from the first stage, produced by electron impact ionization of neutral atoms and ions, and produced by secondary emission at the walls. The losses of cold electrons come from loss of confinement, recombination, and ECR heating. The optimum plasma density in an ECR source seems to be set by stability requirements related to the critical density.[23] If the walls serve as a sufficient source of cold electrons to achieve the optimum

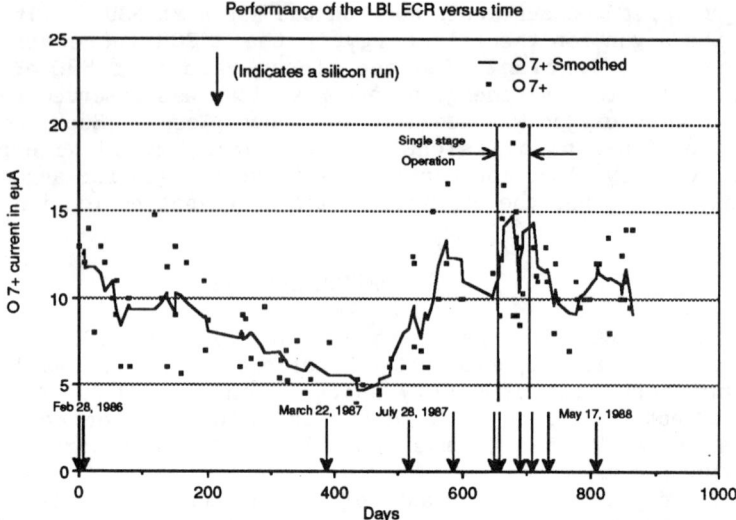

Fig. 4. The O^{7+} performance of the LBL ECR is plotted versus time. The slow decline in average performance from February 1986 to March 1987 coincides with a period in which silicone was not run. In December 1987 and January 1988 heavy usage of silicone made it possible to produce up to 20 eµA of 7+ with single stage operation.

plasma density, then it is no longer necessary to supply cold electrons form the first stage.

ECRIS FREQUENCY SCALING

The existence of a frequency dependence on the maximum plasma density in ECRIS, arising from the plasma cutoff limit, and the value to ion production of raising the plasma density, have been confirmed in a set of studies at 10, 16.6 and 18 GHz in Grenoble.[37-39] In these studies, it was found that the total extracted current increased by the ratio of the square of the resonance frequency, and that the average charge state extracted increased with frequency. The scaling of the the total extracted current with square frequency indirectly confirms the plasma density scaling with frequency. The increase in the average charge of the total extracted current with frequency indirectly confirms that the ionization rate scales with increasing density. In fact the best ECRIS performance to date for several species have been obtained at 18 GHz (see Table I).

Clearly, further work in this area is warranted. A further frequency step increase to 30 GHz would result in an equivalent increase in the cuttoff density over that at 18 GHz. Making this large qualitative step would be important for additional studies of ion production scaling with density in ECRIS.

NEW ECRIS DESIGNS UNDER CONSTRUCTION

Present trends in ECRIS design can be illustrated by looking at three new source designs: the SCECR at MSU, the AECR at LBL, and ECR4 at GANIL. All three have been design to operate at 14 GHz or higher, all have simplified vacuum vessels and two mirror (dominantly main stage) magnetic fields. All three are presently under construction, with first operation expected during 1989. Nevertheless, these three source have substantial design differences, and it is expected that the operation of these new sources will affect the designs of future ECRIS for highly charged ion production.

MSU has undertaken to build an ECRIS with a resonance frequency range of 5-35 GHz, for further study of frequency scaling.[40] The corresponding resonance field range is 0.18-1.25 Tesla. This source, the SCECR, is shown in Fig. 5. A full superconducting coil set is used to produce the required radial and axial field profiles. It will then be possible in a single geometry to study scaling at and beyond existing levels with a magnetic field that can be fully optimized at each frequency. The upper limit for first harmonic operation is set to reach existing gyrotron tubes at 28-35 GHz. The SCECR will become the primary ion source for the K800 cyclotron at

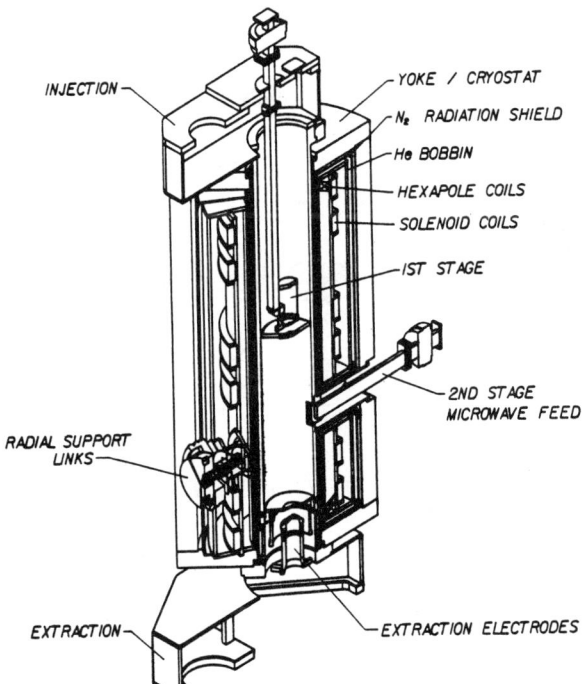

Fig. 5. The present design of the SCECR ECRIS now under construction at MSU.

MSU.[10] The main design parameters of the SCECR are taken from the Fall 1988 operating configuration of the RTECR. Initial operation of the SCECR at 6.4 GHz is expected in Spring 1989, with increases in frequency made subsequently as transmitters become available.

The project to build the AECR at LBL and couple it the 88 Inch Cyclotron began in 1988 and should be completed in two years.[41] The design of the AECR is illustrated in Fig. 6. The axial magnetic field is produced by three groups of copper coils. These groups are sub-divided into three independently adjustible elements, for fine adjustment of the magnetic field. The iron plates between coils 2 and 3 serve to increase the mirror ratio. In Fig. 6, the axial magnetic field is plotted assuming 250 Amps in all coils. This field profile is similar to that used currently in the LBL ECR, which was optimized experimentally. The first stage will operate on the "uphill" gradient of the axial magnetic field, as is the case in the LBL ECR. During the first year of operation, both stages will be driven by a single 14.5 GHz 2.5 kW klystron. The choice of 14.5 GHz was made because commercial klystron amplifier systems are available at this frequency. Even thought it may not be optimal to divide the power bewteen the first and second stage, by coating the wall with SiO_2, it is expected that excellent results for gases such as N_2, O_2 and Ar will be obtained in single stage operation.

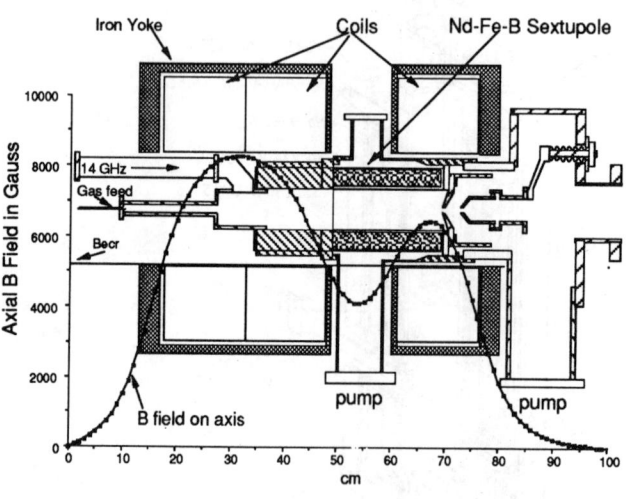

Fig. 6. Elevation view of the AECR source now under construction at LBL. THe axial magnetic field corresponding to 250 Amp current in the coils is superimposed on the drawing.

ECR4 at Ganil, is a 14.5 GHz ECRIS designed to operate with low power consumption on a high voltage platform.[42] This ECRIS is shown in Fig. 7. In order to minimize coil power consumption and work with a small plasma chamber for low microwave power levels, the axial magnetic field of ECR4 is partially made of FeNdB permanent magnets in a structure similar to the Neomafios 10 GHz ECRIS of Grenoble.[43] The axial structure is obtained with an uniformly magnetized ring, divided into 6 blocks, like the Grenoble source. A small magnet ring in the source midplane is used to shape the mirror ratio, and radially magnetized magnets placed at the ends increase the maximum mirror strength. While Neomafios 10 at Grenoble is entirely a permanent magnet structure, the ECR4 design includes coils, to allow tuning of the mirror field between .65 Tesla and 1.05 Tesla, and this should aid source optimization. The source assembly is expected to be complete by the middle of 1989.

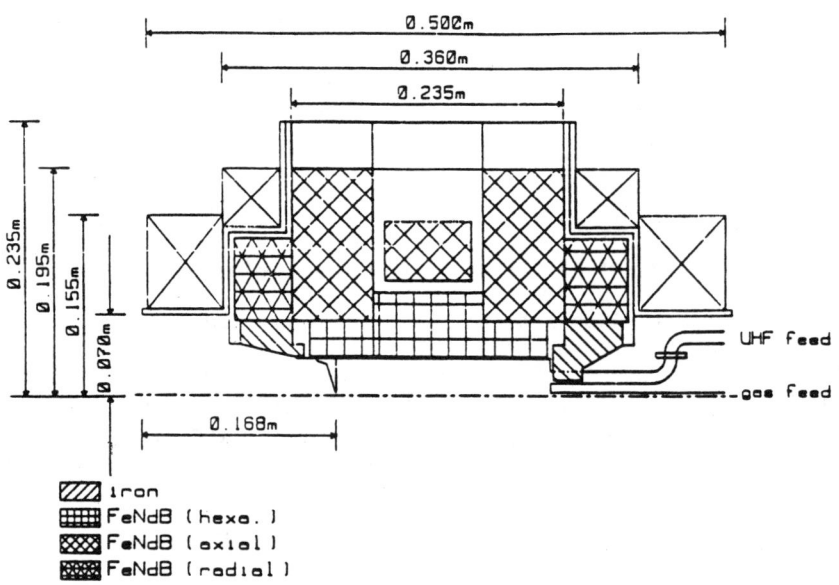

Fig. 7. Schematic diagram of the GANIL ECR4 ion source.

In summary, the trend in ECRIS design for highly charged ions is moving toward simple tandem mirror geometries with gradient first stages, operating at higher frequencies. The 10^{11} pps level may be pushed beyond 40+ for heavy mass beams produced in these sources.

REFERENCES

1. G. Roy, et.al., Proc.Int.Conf.ECRIS MSU-CP 47, 466 (1987).
2. D. Darquennes, et.al. Proc.Int.Conf.ECRIS MSU-CP 47, 491 (1987).
3. T. Clegg, private communication.
4. V. Bechtold, et.a.l, IEEE 84CH1996-3, 118 (1984).

5. G. H. Ryckewaert, et.al., IEEE 84CH1996-3, 226 (1984).
6. J. M. Loiseaux, et.al., IEEE 84CH1996-3, 188 (1984).
7. V. K. VanAsselt, et.al., IEEE 84CH1996-3, 177 (1984).
8. C.M. Lyneis, IEEE Trans. Nucl. SCI. NS-32, 1745 (1985).
9. J. Parker, ed., Proc. Int. Conf. ECRIS MSUCP 47 (1987).
10. H.G. Blosser, et.al., Proc. 11th Conf. Cyclotrons (ICONICS, Tokyo, 1986), p. 157.
11. N. Angert, et.al., 6th. Int. ECR Workshop (LBL, Berkeley, 1985), p. 94.
12. R. Geller, et. al., Proc. Int. Conf. ECRIS MSUCP 47, 1 (1987).
13. H. Haseroth, 1988 Int. Workshop ECRIS (Les Editions de Physique, France, 1989) to be publ.
14. R. Pardo, et.al., Proc. 7th Int. ECR Workshop (KFA Juelich, FRG, 1986), p. 223.
15. R. Geller, private comm.
16. F.W. Meyer, 6th. Int. ECR Workshop (LBL, Berkeley, 1985), p. 37.
17. G. Mank, et.al., Proc. 7th Int. ECR Workshop (KFA Juelich, FRG, 1986), p. 203.
18. M.H. Prior, et.al., 6th. Int. ECR Workshop (LBL, Berkeley, 1985), p. 225.
19. S.J. deZwart, et.al., 6th. Int. ECR Workshop (LBL, Berkeley, 1985), p. 83.
20. Y. Jongen, et.al., 6th. Int. ECR Workshop (LBL, Berkeley, 1985), p. 28.
21. T.A. Antaya, et.al., Proc. 11th Conf. Cyclotrons (ICONICS, Tokyo, 1986), p. 721.
22. H. Beuscher, 1988 Int. Workshop ECRIS (Les Editions de Physique, France, 1989) to be publ.
23. R. Geller, et.al., Proc. Int. Conf. ECRIS MSUCP 47, 1 (1987).
24. B. Jacquot, et.al., Proc. Int. Conf. ECRIS MSUCP 47, 254 (1987).
25. G. Melin, et.al., 1988 Int. Workshop ECRIS (Les Editions de Physique, France, 1989) to be publ.
26. C.M. Lyneis, Proc. Int. Conf. ECRIS MSUCP 47, 42 (1987).
27. F. Bourg, et.al., NIM A254, 13 (1987).
28. R. Geller, Proc. 11th Conf. Cyclotrons (ICONICS, Tokyo, 1986), p. 699.
29. F. Meyer, Proc. Int. Conf. ECRIS MSUCP 47, 522 (1987).
30. C. Lyneis, private comm.
31. T. Antaya, et.al., Proc. Int. Conf. ECRIS MSUCP 47, 86 (1987).
32. F. Meyer, private comm.
33. D.J. Clark, et.al., 1988 Int. Workshop ECRIS (Les Editions de Physique, France, 1989) to be publ.
34. A. Drentje, KVI Groningen Annual Report, 79 (1983).
35. M. Mack, et.al., Proc. 7th Int. ECR Workshop (KFA Juelich, FRG, 1986), p. 152.
36. T.A. Antaya, 1988 Int. Workshop ECRIS (Les Editions de Physique, France, 1989) to be publ.
37. R. Geller, et.al., 6th. Int. ECR Workshop (LBL, Berkeley, 1985), p. 1.
38. R. Geller, et.al., NIM A243, 244 (1987).
39. R. Geller, et.al. Proc. 7th Int. ECR Workshop (KFA Juelich, FRG, 1986), p. 187.

40. T.A. Antaya, et.al., Proc. Int. Conf. ECRIS MSUCP 47, 312 (1987).
41. C. Lyneis, 1988 Int. Workshop ECRIS (Les Editions de Physique, France, 1989) to be publ.
42. P.Sortais, et.al., 1988 Int. Workshop ECRIS (Les Editions de Physique, France, 1989) to be publ.
43. P. Sortais, et.al., Proc. Int. Conf. ECRIS MSUCP 47, 334 (1987).

DISCUSSION

<u>Schmieder</u>: Could you elaborate on the emittance of your source, and sources in general?

<u>Antaya</u>: You know the emittance is just the transverse canonical momentum divided by the longitudinal momentum. There are two terms: one coming from the intrinsic energy of the ions, and one coming from the magnetic field. At energies per charge of 5 to 10 eV, the magnetic term in the emittance dominates. So to first order, the radius of the extraction electrode times the magnetic field at that radius sets the emittance. Those numbers for ECR sources give upper bounds on the emittance of about 300 π mm-mrad unnormalized. An unusual characteristic of emittance measurements is that the emittance that you get is highly dependent upon the emittance measuring apparatus. For example, if the apparatus is misaligned, you will get a better emittance. So I think that to really compare the sources, you not only have to look at the intrinsic emittance but also the way that the emittance is measured. The numbers that I showed had analysis slits that limits the 10 mm. Also, there is a second divergence to about 40 mrad. From that system we get an acceptance of 100 π mm-mrad unnormalized, which is what we measure. For really intense beams from an ECR source, for example, a 1 mA He^+ beam, the space charge force in the beam is of the order of the strength of transport magnets in the system, so the emittance is much higher, with a strong contribution from the space charge. For such a beam, we get about 300 π mm-mrad at 10 kV. Normalized that is roughly 5 mm-mrad. So the normalized emittances (in mm-mrad) for typical ECR beams would be of the order 1 to 2. That, I think, in your system of units would make it about 10^{-6} m-rad. I saw numbers this morning that were 10^{-7} or better for an EBIS source. Again, in an EBIS source it is the effective radius of the beam and the magnetic field that are important because the energy spread is low. Since the effective radius of the beam is much smaller than for an ECR, it looks like the emittance from EBIS sources should also be smaller.

<u>Becker</u>: Geller started, as you pointed out, with triple MAFIOS. From our understanding today, is there any reason to go to a third stage in order to make the ratio of ion density to neutral density more favorable?

Antaya: There are disadvantages to two-stage sources. Maybe the other ECR guys would disagree, but I look around and do not see any, so I will give you my comments. Two-stage sources are less stable and the reason is that now you have two stages to balance. If you added a third stage, I think this would become more difficult. Single-stage sources have the highest stability. We now are building two-mirror sources, essentially single-stage, with performance like two-stage sources, but which are far simpler to operate. The new generation of ECR sources are all two-mirror, single-stage devices. They are: the 14 GHz source in Berkeley, the permanent magnet source, ECRY, in Ganile, and the superconducting source at MSU. With better mirror confinement we only need a single-stage. So I think the three-stage source would be a terrible animal to tame and I would not build one.

Keller: What fraction of the beam do your emittance plot border points include?

Antaya: That is 100% of the beam. There are quoted emittances from ECR sources that include a much smaller fraction of the beam. This is the total emittance from our source. It is roughly the value that we get by assuming that the magnetic field dominates the emittance.

EXPERIMENTS WITH A SYNCHROTRON X-RAY SOURCE AND CONVENTIONAL, ECR, AND STORAGE-RING ION SOURCES*

K. W. Jones, B. M. Johnson, and M. Meron
Brookhaven National Laboratory, Upton, NY 11973

ABSTRACT

The present intensities of photon beams produced by synchrotron-radiation x-ray sources and of ion beams from conventional ion sources, electron-cyclotron resonance ion sources (ECRIS), and cooled heavy-ion storage rings (CHISR) make possible investigations of photoionization and photoexcitation processes that have not previously been feasible. An evaluation of the signal and background rates for experiments that employ the different types of ion sources is given here.

INTRODUCTION

The performance of crossed- or merged-beam experiments in atomic physics is a difficult task in most cases. The ion densities that are produced are not high compared to the density of residual gas in the vacuum chamber and generally the signal of interest must be extracted from a sea of background events. The comparatively recent development of new types of ion sources now makes it possible to consider many new types of experiments. One of the most interesting areas is the study of the interaction of high-energy photon beams with ion beams.

Along with the new types of ion sources it is also necessary to have a suitable photon source. Lasers are appropriate for very low energies, but the regions of interest here are the hard x-ray energies. A brief calculation suffices to show that conventional x-ray sources can not produce sufficient flux to make the experiment feasible. It is necessary to use a synchrotron-radiation source (SRS). The purpose of the present discussion is to present estimates of signal and background rates for experiments that combine the use of the SRS with the three different types of ion sources: conventional sources of singly-charged ions, electron-cyclotron resonance ion source (ECRIS), and the cooled heavy-ion storage ring (CHISR).

SCIENTIFIC BACKGROUND

The reasons that measurements of photon-ion interactions are of interest to atomic physicists have been discussed by Manson[1] and by Jones et al.[2] Further discussions are given in a recent Brookhaven

---

*Work supported by the Fundamental Interactions Branch, Division of Chemical Sciences, Office of Basic Energy Sciences, US Department of Energy under Contract No. DE-AC02-76CH00016 and by the Brookhaven National Laboratory Exploratory Research Fund.

National Laboratory Conceptual Design Report for a National Atomic Physics Facility[3] and in the proceedings of a Workshop on Photon-Ion Interactions[4] held to study directions for ion physics experiments at the European Synchrotron Radiation Facility at Grenoble, France. The ideas that are listed can be grouped in three different categories: fundamental issues, poorly understood areas, and applications. The applications include the use of atomic physics in astronomy, plasmas, and Earth and planetary atmospheres.

One example of the type of experiment that might be done would be a measurement of the photoionization cross sections for a single element as a function of the ion charge state. This type of measurement has already been employed to study the importance of many-body correlations on atomic structure, but the scope was restricted because of the lack of a multi-purpose ion source.[5]

The conclusions drawn from this look at the scientific needs is that the ion source to be used in these experiments must be able to deliver beams of all elements for charge states that correspond to neutral to fully-stripped atoms. The needs for ion intensity for the performance of the experiments and associated questions of backgrounds will be considered below.

GENERAL CONSIDERATIONS

The usefulness of a particular type of ion source for a colliding- or merged-beam experiment is conveniently assessed in terms of the luminosity. The luminosity is defined as the number of interactions taking place per second per unit cross section. The luminosity can be evaluated for both the signal rate and for the background rate. Some, but not all, details of a particular experimental arrangement are contained within the luminosity value. For instance, the form factor that is used to define the overlap of the two beams is considered, but the detector efficiency and solid angle are not.

Adequate signal rates demand both high photon and high ion-beam intensities. The background rate is generally determined by interactions of the ion beam with residual gas molecules in the photon-ion interaction region. The signal-to-background ratio is then improved by increases in the photon flux or a reduction in the operating pressure. An increase in the ion current helps to improve the data acquisition rate, but will not improve the signal-to-background ratio.

For photoionization or excitation measurements of ion beams it is interesting to compare the values of ion densities produced by the various types of sources. This is done in Tables I-IV for the three types of ion sources that are under consideration here. For sake of comparison the density for a gas target is also included. It can be seen that the ion densities are always low compared to the gas target so that the crossed beam experiments are not easy to do.

The ion beams that are produced by the three sources are at radically different energies. The singly charged sources can be used at energies of a few keV. The ECRIS typically operates at a few tens of keV, while the operating energies of the heavy ion storage ring are in the 0.1-10 MeV/u region. The differences in velocities imply different ion densities for a given ion current. The effect is quite large as is evidenced by comparison of the currents produced by an ECRIS and a CHISR shown in Figure 1 with the ion densities given in Tables I-IV. Even though the CHISR produced currents many orders of magnitude larger in some cases, the ion density for the two types of sources are pushed towards equality by the great difference in the ion velocities.

Table I Beam Density from a Surface Ionization Ion Source[6]
 (energy = 2 keV)

Element	Beam Diameter cm^2	Ion Density cm^{-3}
Ba^+	0.2 x 0.3	2.0×10^5

Table II Density of Gas in Conventional Target Operated at 10^{-6} T

Density = 3.5×10^{10} cm^{-3}

Low pressure operation required for investigations of multiple-ionization processes following photoionization.

Table III Beam Densities from an ECRIS
 (energy = 15 keV)

Element	Beam Diameter cm^2	Ion Density
O^{6+}	0.5 x 0.5	2.4(7)
Ar^{8+}	0.5 x 0.5	2.3(7)
Ar^{16+}	0.5 x 0.5	1.8(3)
Xe^{14+}	0.5 x 0.5	5.4(6)
Xe^{22+}	0.5 x 0.5	4.3(5)

Table IV Beam Densities from CHISR

No Acceleration

Ion	One Stripping Target		Two Stripping Targets	
	a* cm	Ion Density cm^{-3}	a* cm	Ion Density cm^{-3}
C^{5+}	.76	2.60(6)	.87	1.80(6)
S^{9+}	1.15	1.34(6)	1.13	5.35(5)
Cu^{11+}	1.26	1.34(6)	1.25	2.97(5)
I^{13+}	1.36	8.85(5)	1.36	1.81(5)
Au^{13+}	1.48	9.04(5)	1.55	1.39(5)

Acceleration to 2.2 Tesla-meter

Ion	One Stripping Target		Two Stripping Targets	
	a* cm	Ion Density cm^{-3}	a* cm	Ion Density cm^{-3}
C^{6+}	.36	1.16(7)	.35	1.18(7)
S^{14+}	.64	4.34(6)	.40	4.27(6)
Cu^{21+}	.88	2.21(6)	.45	2.29(6)
I^{29+}	1.23	1.08(6)	.55	1.11(6)
Au^{33+}	1.67	7.10(5)	.69	7.03(5)

*a is the beam radius and is related to the width of the gaussian distribution of the beam density by $a = \sqrt{\sigma \cdot \sigma}$, where σ is the standard deviation.

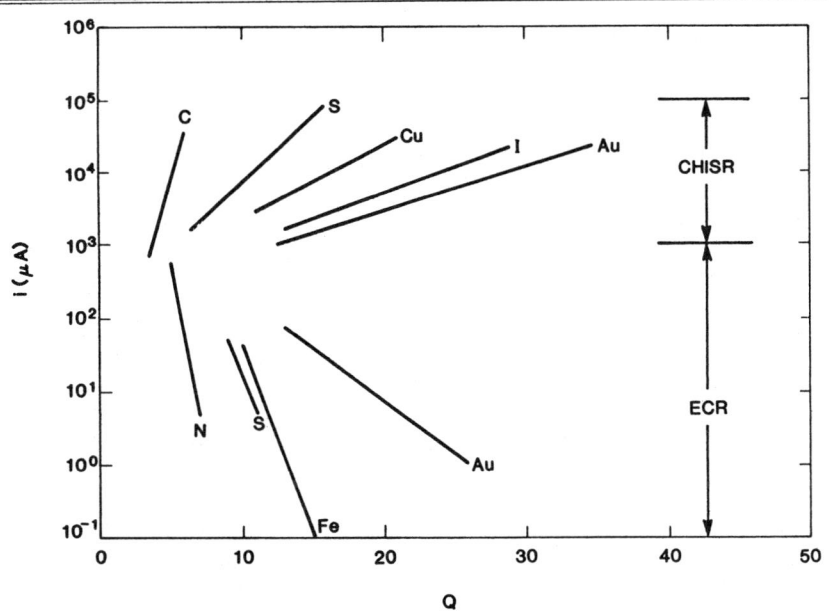

Fig. 1. Intensities of beams produced by ECRIS and CHISR.

Background effects that produce charge-changed ions are not the same for the three sources. At the low energies electron capture is the dominant process, while at the high energies of the CHISR the dominance of capture or loss mechanisms depends on the ion and the energy under consideration. An effect that is appreciable at the high energy end is direct Coulomb ionization of the inner-shells of the beam. In this case the center-of-mass energy for an ion-atom (e.g., hydrogen gas) encounter is of the order of several MeV and appreciable ionization results. In the case of the storage ring the capture and loss reactions also work to limit the lifetime of the stored ions in the ring.

THE PHOTON SOURCE

The photon source used for the only SR experiment to date, that by Lyon et al.,[6] was the ring located at Daresbury in England. The use of an ECRIS or CHISR has not yet been attempted at a SRS. For these cases, it has been assumed that the SRS is a superconducting wiggler device located at the National Synchrotron Light Source (NSLS) at Brookhaven. The use of a superconducting wiggler makes it feasible to consider ionization of K-shells of all elements. The spectrum produced by the wiggler is given in Figure 2 and typical photon beams that can be delivered to the location of the ion beam are given in Table V.

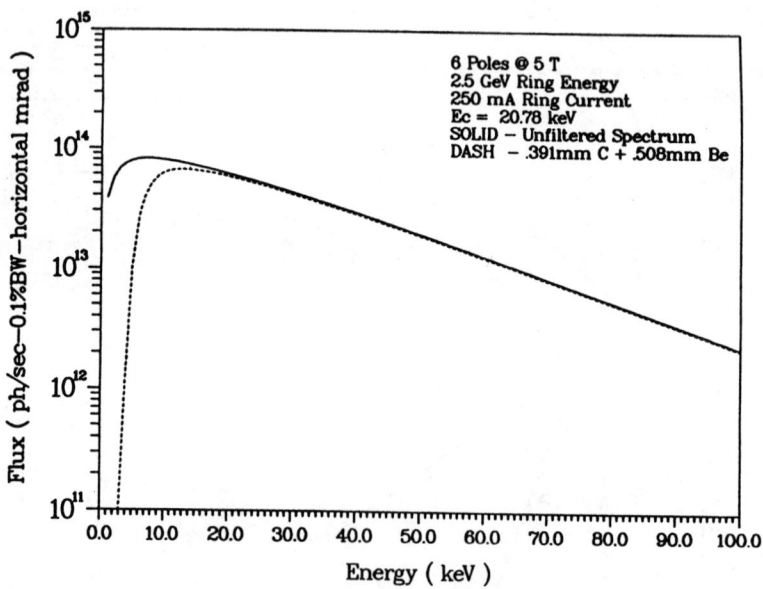

Fig. 2. Energy spectrum of photons produced by superconducting wiggler at the NSLS.

Table V Typical Photon Beams from the Superconducting Wiggler at the CHISR Ring

Beam Energy keV	ΔE and $\Delta E/E$	Horizontal Size (2σ) mm	Vertical Size (2σ) mm	Flux photons/(s-mr)
A. Unfocussed White Beam				
5		60	25	1×10^{17}
20		60	14	5×10^{16}
100		60	7	
B. Focussed White Beam				
5		10	25	1×10^{17}
20		10	14	
(High Energy Cut-Off at 20 keV.				
C. Silicon Monochromator				
5	1.0×10^{-3}			1.2×10^{13}
10	1.7×10^{-3}			1.1×10^{13}
20	2.7×10^{-3}			7.9×10^{12}
50	5.2×10^{-3}			2.2×10^{12}
100	9.0×10^{-3}			2.5×10^{11}

It must be remembered that the SRS is a pulsed machine. The NSLS produces bursts of photons with a width of 0.6-1. ns (4σ) that occur at intervals of 18.9 ns. The microstructure of the beam makes measurements of the background in the beam-beam experiments fairly easy.

PHOTOIONIZATION WITH A CONVENTIONAL ION SOURCE

The only experiments on the photoionization of an ion beam to date[6] used a surface ionization type source to produce beams such as Ba+ and K+. Currents of about 10 nA were obtained at an energy of 2 keV and a beam size of 2 mm x 3 mm. The photon fluxes incident on the ion beam varied from 10^9 to 6×10^{10} photons/s depending on the energy resolution of the monochromator employed. A merging length of 12.5 cm was used to achieve a signal rate of 100 events/s with the high resolution beam and a cross section of about 10^{-15} cm^2. By definition the luminosity is of the order of 10^{17} cm^{-2} s^{-1} for the long path length of 12.5 cm. This can be verified by direct calculation of the expected rate for the conditions mentioned above. The background rate was about 1/2 to 1/5 the signal at the point of highest cross section. The background luminosity is then around 2-5 $\times 10^{16}$ cm^{-2} s^{-1}. Extracting the signal from this background ultimately limits the size of cross section which can be investigated. Lyon et al.[6] cite a value of about 10^{20} cm^2 for that number.

EXPERIMENTS WITH AN ECRIS

The use of an ECRIS is a natural extension of the work done with the source of singly-charged ions. The variety of ion beams that can be produced are essential for systematic investigation of many problems in atomic physics. The luminosity of an experiment with an ECRIS and the NSLS can be easily calculated from the ion density and the assumption that the experiment is done with a photon beam of 10^{14} photons/s. If a 15-keV energy is assumed for a typical beam, such as Ar^{8+} confined to a region of about 5 mm x 5 mm in transverse dimension with a beam of 1.5 x 10^{14} particle/s, the luminosity is found to be 2.3 x 10^{21} $cm^{-2}s^{-1}$. The luminosity has increased by a factor of 2 x 10^4 compared to the Daresbury experiment and the signal-to-background ratio has also been improved since the photon flux is increased by 4-5 orders of magnitude. That, is the signal will be approximately equal to the background (assuming that the background rates do not change appreciably from the barium case) at about 10^{-20} cm^2. Under these conditions it might sometimes be feasible to do experiments at the level of 1 b cross sections. It should be noted that with the crossed beam geometry it will be possible to use a much smaller interaction length and the background will be reduced by at least an order of magnitude.

Another improvement in the ion currents may be effected by use of bunching as suggested by Andrä.[7] Klystron bunching could be used to bunch the beam by a factor of 20 or more and gain a further increment in current. This would correspond to producing ion bunches at the beam of 1 ns in length every 19 ns. Space charge will ultimately set a limit to the actual gains.

Luminosity values for the ECRIS are shown in Table VI for a number of different elements and charge states. The values in the table do not include the possible use of bunching. The values for the beam currents were taken from the data of Bourg et al.[8] Similar values are cited by Jones et al.[2]

EXPERIMENTS WITH CHISR

A new approach to the production of very high currents of heavy ions is the use of the cooled heavy ion storage ring. In this case, a beam of ions is produced by the stripping of high-energy beams produced by a conventional accelerator or by use of an ECRIS and subsequent acceleration to higher energies. The beam is then captured in a synchrotron storage ring where it can be recirculated for long periods of time. The synchrotron can be used to accelerate or decelerate the beam through a broad expanse of energies. Electron cooling can be used to produce beams of high quality, superior to the emittance of the conventional source. The average currents provided in the ring are generally far-superior to those that can be made with the ECRIS or electron-beam ion source (EBIS).

Table VI Luminosity Values for ECRIS and CHISR Calculated for 10^{14} photons/s Incident on the Ion Beam in Crossed-Beam Geometry. The energy of the ions from ECRIS is 15 keV/u in all cases.

Ion	Source	Luminosity $cm^{-2} s^{-1}$
C^{5+}	CHISR	1.3(21)
C^{6+}	CHISR	1.3(21)
O^{6+}	ECRIS	2.3(21)
S^{9+}	CHISR	8.8(20)
S^{14+}	CHISR	5.7(20)
Ar^{8+}	ECRIS	2.3(21)
Ar^{16+}	ECRIS	1.8(17)
Cu^{11+}	CHISR	6.1(20)
Cu^{21+}	CHISR	3.2(20)
I^{13+}	CHISR	4.3(20)
I^{29+}	CHISR	1.9(20)
Xe^{14+}	ECRIS	5.4(20)
Xe^{22+}	ECRIS	9.6(18)
Au^{13+}	CHISR	3.7(20)
Au^{33+}	CHISR	1.5(20)

A number of these rings are in the commissioning or construction stage at the present time. In all cases they are designed for possible experiments with auxiliary beams of electrons, ions, or laser photons. At Brookhaven a unique possibility exists for use of a storage ring in association with the photon beams produced at the National Synchrotron Light Source.[2,3,4,9] A summary of typical ion beams and beam densities in CHISR are shown in Table IV. The effective beam currents are very high because the revolution time of an ion in the ring is of the order of 500 ns. The effective beam current is then found by multiplying the number of stored ions by their revolution frequency. The beam currents are then of the order of mA. A comparison of the beam currents predicted for CHISR at BNL is made with the ECRIS currents obtained at Grenoble by Bourg et al.[8] in Figure 1. The current values for CHISR are very much higher. A comparison of the two sources is more complex and applicability of the two sources should consider the luminosities for signal and background.

Many of the backgrounds of importance for experiments in a ring depend on the lifetimes of the stored ions. The lifetimes can be estimated by using semi-empirical relationships for capture and loss cross sections.[10] While far from being perfect, these relationships are considered accurate to within a factor of two for capture and certainly to better than an order of magnitude for loss (with the possible exception of ions with only a K-shell electron left). In this context, our predictions for the lifetime of C are in agreement with recent measurements at the Heidelberg Test Storage Ring where a value of ~ 60 s was found for the lifetime of C^{+6} ions at an energy of 6.1 Mev/u and a pressure of 8×10^{-10} Torr.[11]

We emphasize that the superconducting wiggler used for the preliminary design of CHISR gives complete flexibility in doing photoionization of K-shells of all elements. This choice costs about a factor of 4 in flux when compared with predictions for a permanent magnet wiggler. This loss seems to be an acceptable price to pay to for the great versatility of the superconducting device.

An example of a photon-ion experiment is the measurement of photoionization cross sections. This is a very broad topic with many nuances. To show the practicality we consider a specific ion, Cu^{11+} and show that both K-shell and outer-shell photoionization measurements are feasible.

Figure 3 shows that the total photoionization cross section for Cu around the K-edge is 3×10^4 b/atom. The signal rate is found using the luminosity value of 4.3×10^{20} $cm^{-2}\text{-}s^{-1}$ for Cu^{11+} cited in Table VI. The signal rate is then around 13 Hz using a photon energy resolution better than 8 eV and a photon flux of 10^{14} Hz (E/E = 0.1%). (Outer shell cross sections are much larger and the rates will be in excess of 100 Hz).

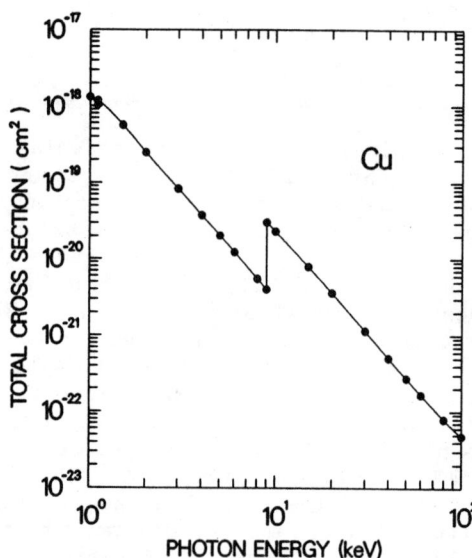

Fig. 3. Total cross section for the interaction of photons with copper as a function of energy.

The background rates can be easily calculated with the following approach. For the K-shell case, several electrons are removed in the photoionization process. Therefore, single charge-changing events are not a problem. The rate for direct Coulomb ionization of the K-shell can be found using a luminosity for interactions with the background gas in the experimental straight section of 2×10^{25} cm^2 s^{-1} and a K-shell vacancy production cross section of about 200 b for copper on hydrogen at about 2.9 MeV/u beam energy. The background rate is then found to be about 4000 Hz. This rate can be handled by conventional detectors with no difficulty.

The signal-to-background ratio is better than 13/4000 because advantage can be taken of the pulsed nature of synchrotron radiation by accepting events only when the photon beam is on. The correction for duty cycle improves the signal-to-background value. Normally, the NSLS operates with 25 x-ray bunches which have a width of 0.6-1.0 ns (4σ) and a revolution frequency of 567.7 ns. The duty factor is then 0.026 to 0.044 at the 4σ width. The duty factor is taken as about 0.03 for illustrative purposes. Therefore, we find: Signal/Background = 13/(4000 x .03) = .11.

Accurate measurements of the signal can be made in the course of a few minutes under these conditions. The uncertainty in the measurement, δ, is given by:

$$\delta^2 \approx \alpha \times b/(s^2 \times t_\delta)$$

where s is the signal counting rate, b the background counting rate, and α is the NSLS duty cycle, 0.03. Evaluation of this relationship for the calculated rates gives an uncertainty in the measurement of ± 10% for a measurement time of ~ 70 s and 5% for a measurement time of less than 5 min.

Now, look at the outer shell case. The signal rate is found as before, but now using a cross section of 1.2 Mb. The rate calculated using the above luminosity is 516 Hz.

The background in the one-electron loss detector is simply calculated. The electron loss lifetime[3] is 5000 s at a beam energy of 2.9 MeV/u and about 14,100 s at a beam energy of 7.1 MeV/u. The rate in the corresponding detector is then about around 5.5×10^5 Hz. This is still a rate which can be handled with a conventional detector. Or, the problem can be alleviated by running at reduced ion currents or using an aperture to reduce the fraction of beam accepted, etc.

Signal-to-background is then: Signal/Background = $516/(5.5 \times 10^5 \times .03)$ = .031 at the low beam energy and: Signal/Background = $516/(2.0 \times 10^5 \times .03)$ = .086 at the high beam energy.

The values are similar to the K-shell case calculated above and thus accurate measurements are also feasible within a matter of minutes. Using the formula cited above, we find that the time needed for a measurement with an uncertainty of 5% is 24 s at a beam energy of 2.9 MeV/u. Even if the ion beam current is reduced by an order of magnitude, to limit the count rate, the measurement time will still be below 5 min.

The above discussion shows that photoionization experiments in CHISR will be generally feasible without great experimental refinements. The photoionization experiments can be made more sophisticated by adding a detector at the target region to measure the photons/electrons emitted by the ionized atom. This additional specification makes it possible to measure the events following the ionization in much more detail and at the same time to considerably reduce background events due to much shorter interaction lengths. Detectors with large solid angles can be used so that the coincidence efficiency is very high. The luminosity can also be increased by using a broad-band of the synchrotron radiation spectrum. Improvements of up to 100-1000 could then be attained in the value of luminosity and signal-to-background ratio.

CONCLUSIONS

We have highlighted some aspects of photoionization or photoexcitation studies using three different types of ion sources. Conventional sources that produce beams of a few times ionized atoms were represented by a surface ionization source that was employed for the first such experiments. The extension of the first work to more highly ionized atoms using ECRIS and CHISR type devices was considered in terms of the signals and backgrounds involved. In a short discussion it is not possible to go into all the complexities of these experiments. Nevertheless, a good idea of the regions of applicability of each class source emerges from the data presented.

The conventional sources that can produce well-focussed beams from gaseous or solid materials are certainly best for cases of singly, and perhaps a few-times, ionized beams. The gains obtained by using the ECRIS are minimal. CHISR is not really at all suited to this work because the high-magnetic rigidity of the low-charge state ions makes it difficult or impossible to store them in the ring and because of short storage lifetimes at the low energies needed in the ring.

For light elements and ionization stages of heavier elements which are not too high the ECRIS is the simplest and most effective approach. The luminosities are comparable or better than those predicted for the CHISR approach. The beam quality in CHISR and other features associated with the higher energies could be helpful, but these points were not considered in this discussion. It can be seen, however, that the signal rate for Ar 14+ for the ECRIS beam is much less than for the beams produced by CHISR.

CHISR appears to be clearly superior for studies of beams of the highest charge states for elements from about sulfur through the rest of the periodic table. This assumes that there is a suitable source to produce the ions for injection of the ring. CHISR is also competitive in luminosity values with the ECRIS source for the lower charge states of the light elements and for many of the heavy elements. Again, it should be stated that the CHISR performance for lightly ionized atoms will be limited by magnetic rigidity and lifetime problems encountered in its operation.

It appears that the construction of a facility at a synchrotron light source that combines the features of these complementary devices would result in a superbly versatile and unique approach to the study of the interactions of the ion beams with photons from the synchrotron light source and also with beams of other ions or with beams of electrons. The extraordinary variety of fundamental experiments that could be done with the equipment would make it an extremely important tool for fundamental and applied atomic physics research.

REFERENCES

1. S. T. Manson, Nucl. Instrum. Methods B24/25, 429 (1987).
2. K. W. Jones, B. M. Johnson, M. Meron, B. Crasemann, Y. Hahn, V. O. Kostroun, S. T. Manson, and S. M. Younger, Comments At. Mol. Phys. 20, No. 1, 1 (1987).
3. Conceptual Design Report for a National Atomic Physics Facility. BNL Informal Report 42340, Dec. 1988.
4. Proc. Workshop on Photon-Ion Interactions, Grenoble, France, December 7-8, 1988, to be published.
5. J. A. R. Samson, Y. Shefer, and G. C. Angel, A Critical Test of Many-Body Theory: The Photoionization Cross Section of Cl as an Example of an Open-Shell Atom. Phys. Rev. Letts. 56, 2020-2023 (1986).
6. I. C. Lyon, B. Peart, J. B. West, and K. Dolder, J. Phys. B: At. Mol.Phys. 19, 4137 (1986).
7. H. J. Andrä, in Proc. Workshop on Photon-Ion Interactions, Grenoble, France, December 7-8, 1988.
8. F. Bourg, R. Geller, and B. Jacquot, Nucl. Instrum. Methods A254, 13 (1987).
9. K. W. Jones, B. M. Johnson, M. Meron, Y. Y. Lee, P. Thieberger, and W. C. Thomlinson. Nucl. Instrum. Methods B24/25, 381 (1987).
10. H. Halama, B. M. Johnson, K. W. Jones, M. Meron, and J. Schuchman, in American Vacuum Society Series 5, AIP Conf. Proc. No. 171, G. Lucovsky, Editor, pp. 275-282, American Institute of Physics, New York, 1988.
11. R. Schuch, private communication.

ATOMIC PHYSICS

ATOMIC PHYSICS RESEARCH USING HIGHLY CHARGED IONS FROM EBIS

H. Tawara
Institute of Plasma Physics, Nagoya University,
Nagoya 464-01, Japan

ABSTRACT
In this paper we review some topics of low energy atomic physics researches involving highly charged ions produced in EBIS and other ion sources : 1) collisions with electrons, 2) collisions with atoms(ions) and 3) collisions with solids.

I. INTRODUCTION
In the past ten years, a lot of information has become available on collisions of highly charged ions with atoms and solids. Among many thrusts toward such activities, the most intense came from high temperature fusion plasma research programs in various countries. Indeed, a number of investigations on highly charged ions have been supported by the fusion communities. On the other hand, the timely development of very powerful ion sources capable of producing highly charged ions such as electron beam ion sources (EBIS) or electron cyclotron resonance ion sources (ECR) has contributed significantly to obtaining reliable information on various collision processes.

II. COLLISIONS WITH ELECTRONS
A lot of experimental and theoretical data of collision processes of atoms by electron impact have been accumulated[1]. However, it is quite recent that precise measurements of the cross sections for ionization and excitation of ions, in particular highly charged ions, have become possible. Some nice reviews on this subject are available[2].

1. Experimental techniques
The most reliable technique for determining the cross sections of excitation/recombination/ionization of ions by electrons is the ion-electron crossed beam method developed by Dolder et al.[3] and most widely used presently. Trapped ion techniques (such as EBIS or EBIT) are sometimes used to estimate collision cross sections[4].

2. Ionization processes
2.1 Single and double ionization
The ionization of atoms or ions can occur through various processes :
a) single direct (knock-out) ionization :
$$e + A^{q+} \rightarrow e + A^{(q+1)+} + e$$
is generally believed to be dominant over multiple ionization. Though this is the case for light ions, a number of other processes are known to contribute significantly to ionization of many electron ions :
b) innershell excitation-autoionization :
$$e + A^{q+} \rightarrow e + A^{q+*} \rightarrow e + A^{(q+1)+} + e$$
b') innershell excitation-double autoionization :
$$e + A^{q+} \rightarrow e + A^{q+*} \rightarrow e + A^{(q+1)+*} + e \rightarrow e + A^{(q+2)+} + e + e$$
b'') innershell excitation-auto double ionization :
$$e + A^{q+} \rightarrow e + A^{q+*} \rightarrow e + A^{(q+2)+} + 2e$$
c) resonant recombination-double autoionization :
$$e + A^{q+} \rightarrow A^{(q-1)+**} \rightarrow A^{q+*} + e \rightarrow A^{(q+1)+} + e + e$$
c') resonant recombination-auto double ionization :

© 1989 American Institute of Physics

$$e + A^{q+} \rightarrow A^{(q-1)+*} \rightarrow A^{(q+1)+} + 2e$$

d) innershell ionization-autoionization :
$$e + A^{q+} \rightarrow e + A^{(q+1)+*} + e \rightarrow e + A^{(q+2)+} + e + e.$$

The contribution of processes b)-d), called as "indirect" processes, becomes significant with increasing the ionic charge of ions and is dominant over the direct process a) for a number of heavy ions with many electrons. Of course, it is not easy to separate the contribution from various processes mentioned above.

An experiment involving many electron alkaline earth metal systems[5] shows remarkable increase of ionization cross sections. Note that for low z ions (Be^+, Mg^+) no enhancement is seen, meanwhile for the largest z ions (Ba^+) the enhancement is strongest (see Fig.1).

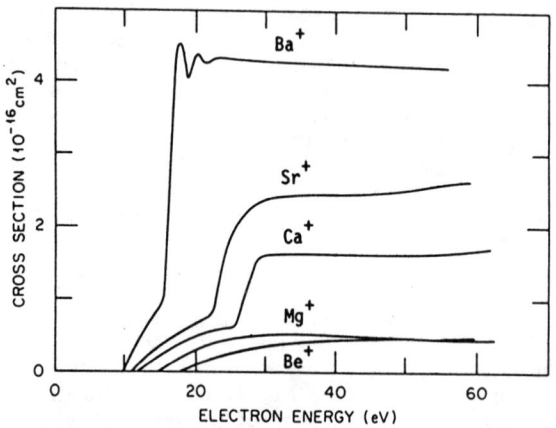

Fig.1 Cross sections for single electron ionization of Be^+, Mg^+, Ca^+, Sr^+ and Ba^+ ions by electrons[5].

For Xe^{q+}(q=2-6) ions, 4d-nl(=4f,5d,5f) excitation-autoionization can roughly account for the observed results[6] ($Xe^{6+}=4d^{10}5s^2$). The dominant contribution of 2p-shell ionization-autoionization is observed[2] in double ionization processes for Ar ions. However, there are examples which can not be explained only through such "indirect processes". Indeed there are discrepancies in ionization of, for example, Fe^{15+} ions which are expected to have intense resonant excitation-double autoionization at around 760 eV[7] (see Fig.2). On the other hand, the contribution of resonant recombination-autoionization processes has been confirmed for C^{3+} ions, though it is weak ($\cong$1% of direct ionization)[8]. **We need our extensive studies of ionization of various ions with different electronic configurations.**

2.2 Multiple ionization

Multiple ionization is generally weak, relative to single ionization. It should be noted that deeper innershell ionization (excitation) processes, followed by successive autoionization or Auger electron emissions, contribute dominantly to multiple ionization. In fact, the cross sections for L- and K-shell ionization of Ar atoms by electrons become comparable to those for Ar^{2+} and Ar^{5+} ion production from Ar atoms[9] (see Fig.3). **Systematic measurements of cross sections for multiple ionization have to be made for various ions.**

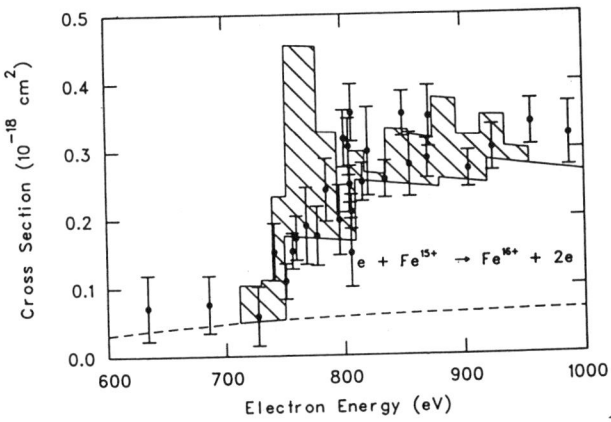

Fig.2 Cross sections for single electron ionization of Fe^{15+} ions by electrons[7].

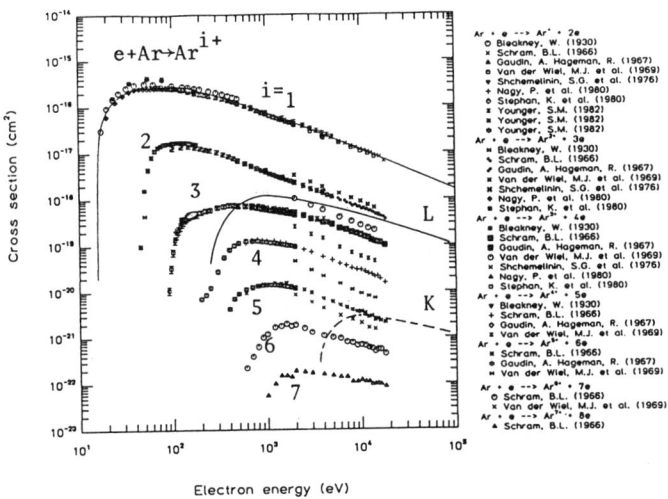

Fig.3 Comparison of cross sections for (i=1-7) multiple ionization and (L- and K-)innershell ionization of Ar atoms[9].

2.3 Threshold behaviors

The cross sections for ionization of ions are known to be zero at the threshold (significantly different from excitation processes) and, then, increase with increasing the impact electron energy just above threshold. The following power law for cross sections is proposed[10] :

$\sigma \equiv E^a$ ($a_1 \equiv 1$ for single ionization : $a_n \equiv n$ for n-times ionization).

Systematic studies are needed on the threshold behaviors in ionization of ions, in particular multiple ionization processes of highly charged ions. Such experiments are often hindered through the presence of the metastable ions in parent beams. Even a very small fraction of such metastable beams excludes the possibilities in working at threshold region where the cross sections should be extremely small.

2.4 Empirical formulas for ionization cross sections

Lotz formulas, based upon Born approximation for direct ionization, are most widely used[11]. If plotted as $\sigma * I_j^2$ versus E/I_j (σ : the ionization cross section ; I_j : the ionization potential of electron in the j-th shell and E is electron impact energy), all data must fall on a single line. For H-like ions, the ionization cross sections scale as z^{-4}. There are many proposed formulas which have their limited validity. However, no simple scaling formulas which include indirect processes are available.

3. Excitation processes

3.1 Threshold behaviors and resonances

Excitation of ions by electron impact has its features whose cross sections are finite and generally become maximum at thresholds,[2,12]. Another important feature near threshold region is the occurrence of a series of resonances where the cross sections show significant peaks. The resonances are a consequence of formation of some intermediate (compound doubly excited) autoionizing states during the incident electron-ion collisions which decay through emission of electrons. Although their widths are narrow, their contribution to total cross sections is significant. For H-like ions, the resonance contributions to the 1s->2s and 1s->2p excitation are estimated to be 20 and 10 %, respectively, with less contribution to the excitation into higher n-states[12]. On the other hand, at higher energies, reliable excitation cross sections can be estimated using some asymptotic behaviors. Some empirical formulas such as Gaunt-factor formula could sometimes provide reasonable excitation cross sections[2].

3.2 Experiments and data

The ion-electron crossed-beam technique is still the most reliable method for determining the excitation cross sections since the first experiment by Dance et al.[13] through observing photons. The obtained cross sections are generally less accurate (15-20 %), with the most serious errors being due to absolute calibration of photon detection efficiencies, compared with those of ionization ($\cong$2%). Up to now, most of experimental studies have concentrated on low charge ions because of the limited availabilities of ions.

A unique technique has recently been developed to determine electron impact excitation of highly charged ions through X-ray spectroscopy : by observing L X-rays from Ba^{46+} ions, the highest charge state ions ever studied, trapped in EBIT, the excitation cross sections at 5.69 and 8.2 keV electrons have been determined[14].

The excitation cross sections can also be estimated through the observed distinct structures in ionization cross section curves of highly charged ions, though the accuracy may not be so good as those for total ionization.

To get more insight in ion excitation by electrons, (relative) angular distributions (up to 20°) of the ns->np excitation cross sections, for example of Cd^+ or Zn^+, have been determined through observing the energy loss spectrum of scattered electrons, instead of photons, and found to be in agreement with the close-coupling calculations[15].

Experimental confirmation of calculations of resonances in excitation of highly charged ions should be pursued systematically.

4. Dielectronic recombination(DR)

$$e + A^{q+} \rightarrow A^{(q-1)+**} \rightarrow A^{(q-1)+*} + h\nu$$

is another form of decays of the doubly excited states formed through recombination of electron with target ion and the exact inverse process of photo-innershell excitation followed by autoionization. In order to investigate DR experimentally, the coincidence between photon and product ion or some resonance structures in the cross section curves or their combination is often used. However, there are serious difficulties : $A^{(q-1)+}$ ion formed has an electron in high Rydberg state which is easily influenced by stray electro-magnetic fields, resulting a significant loss of the product. In one of the first electron-ion crossed beam experiments of DR on Mg^+ ions[16] (see Fig.4) :

$$e + Mg^+(3s) \rightarrow Mg^{**}(3p,nl) \rightarrow Mg^*(3s,nl) + h\nu,$$

the observed results are found to be too much larger than theoretical calculation. This discrepancy, however, can be removed by taking into account the field ionization and mixing correctly.

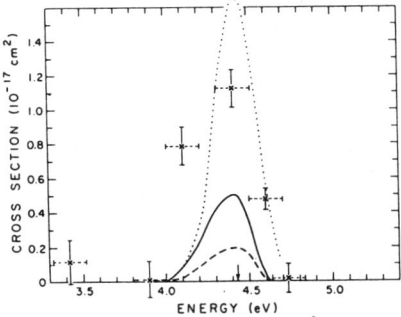

Fig.4 DR cross sections for $e + Mg^+$ collisions[16].

In well-collimated MeV ion-high density electron beam merging method, such environments are also found to influence the observations.
For DR of very high charge ions, a different technique is used : 10-100 MeV ion collides with an electron of atom which can be assumed to be quasi-free. The coincidence rates between X-ray and product ion show broad peaks which correspond to DR with an electron of Compton energy distribution profile, instead of a free electron. As expected, the detailed structures of DR can not be resolved but their gross cross sections are known with this method.
Another interesting method is the use of EBIS itself. Directly looking into EBIS and observing X-rays as a function of the confining electron energy, DR cross sections of Ar ions are estimated to be of the order of 10^{-20} cm^2, roughly agreeing with theoretical prediction for Ar^{14+} KLL DR. However, large electron energy spread ($\cong$15 eV mainly due to the space charge potential of the confining electron beams) prohibited studies of the detailed structures of DR in Ar ions[17].

III. COLLISIONS WITH ATOMS (IONS) AT LOW ENERGIES

Generally we can write down our collision processes between highly charged ion A^{q+} and neutral atom B as follows :

$$A^{q+} + B \rightarrow A^{r+}(M) + B^{i+}(N) + (r + i - q)*e + \Delta E$$

where M,N : quantum states (n,l,m) of projectile and target atom. Both the projectile and target may form multiply (M,N) excited states and, then, they may be autoionized.

The most important parameters here are 1) collision velocity, 2) projectile charge, 3) electron binding energy of target atom and 4) the number of electrons involved. At our slow collisions (≃ev/amu-keV/amu), the dominant process is capture of electrons into projectile ions, but ionization process of projectile ions is usually of minor importance. In such processes, electrons are usually captured into projectile excited states which, after emitting photons or electrons, get relaxed. Many of the observed features can be explained with the assumption that quasimolecules are formed during such slow collisions.

1. Theoretical aspects

A number of theories have been developed to treat electron capture processes at low energies. Some nice reviews are available[18]. Among theories and models, the classical over-barrier model is most convenient for understanding and explaining the expected and observed phenomena[19] and predicts total cross sections and the electron-capturing n-state can be estimated. The Landau-Zener model is also useful for qualitative as well as quantitative (not accurate) understanding and easily extended for multi-crossings[20]. Further modification of LZ formula can explain the l-distributions[21].

2. Total electron capture cross sections

At the intermediate energy range, total cross sections do not strongly depend on the collision energy[22], except for very light ions. This is due to the presence of a number of curve-crossings at proper impact parameter region (reaction window[21]) for highly charged ions. Based on available data[23], a number of useful empirical formulas[18] to estimate total cross sections for various ion-atom combinations are proposed[22]. One of the most convenient scaling formulas predicts total electron capture cross sections s for multi-electron capture:

$$\sigma_{q,q-k} = A*q^a*I^b \quad (k=1-4)$$

where I is the ionization potential(eV) of target atom, q the incident ion charge, A, a, b constants depending on k, the number of electrons

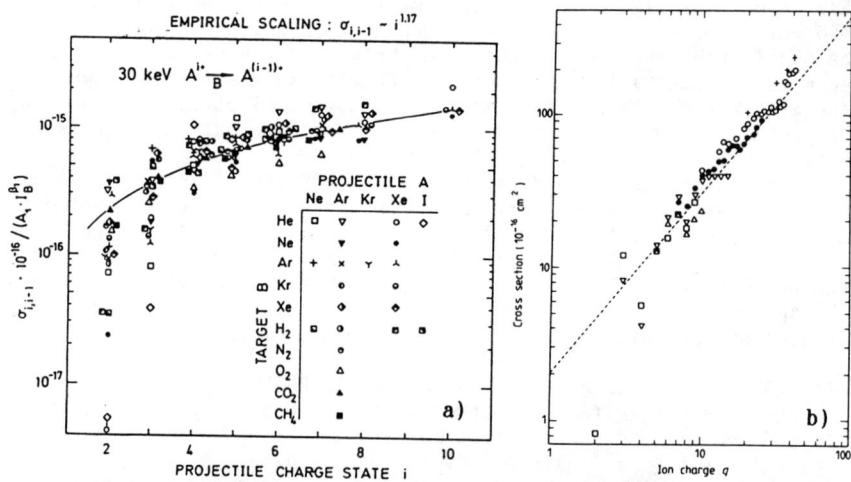

Fig.5 Scaled total cross section curve in comparison with experimental data. a) low q ions[22] ; b) high q ions[24].

captured[22]. For single electron capture (k=1), $A=1.43*10^{-12}(cm^2)$, a=1.17 and b=-2.76. b does not depend much on k. As an example, in Fig.5 is shown a comparison between a fitting curve and experimental data for single electron capture processes in various ion-target combinations. Note that there are scatterings of data at low charge states. In fact, for low q ions, sharp variations of cross sections are observed because a limited number of the levels are available for light elements[19]. Data up to q = 42 from He target are also found to follow with this empirical formula.[24] It should here pointed out that, in determining total cross sections for multi-electron capture, the angular distributions have to be known properly (see 4).
Here we should note that **even total cross sections can not be scaled anymore at the energy less than 0.1 keV/amu because of strong core electron effects**, though not confirmed experimentally yet (see 3.2). For highly charged projectiles, in particular in collisions with multi-electron targets, not only single electron capture process but also multiple electron capture process play an important role[25]. In some cases (such as C^{4+} + He system), two-electron capture is dominant over single-electron capture process at low energies. In multiple electron capture processes, electron capture into projectile from target atom simultaneously accompanied with target ionization(transfer ionization) can be one of the important processes in production of multiply charged recoil ions. In triple electron capture from neutral Xe target atoms by Xe^{8+} ions, for example, the most intense recoil Xe ions have the charge 5+ which means two more electrons are lost from target through (auto)ionization[26](see Fig.6).
Systematic measurements have to be made for total cross sections for very high charge state ions, in particular, for many electron targets including molecules and at very low energies of meV/amu - 1 eV/amu[27].

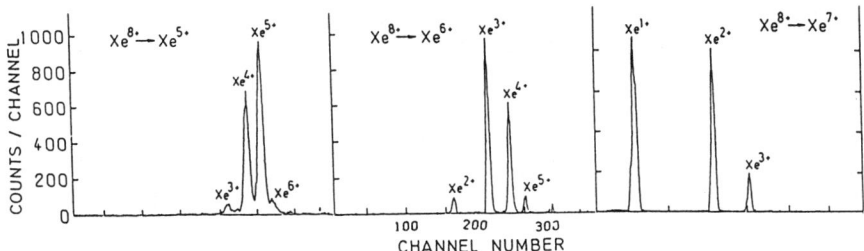

Fig.6 Correlation of projectile ions and secondary (recoil) ions produced in 80 keV Xe^{8+} + Xe collisions[26].

3. (n,l,m) partial cross sections
3.1 n-distributions
Through a series of observation of translational energy spectroscopy, a scaling law for estimating the most probable n-state has been introduced for light atoms such as He[24] :
$$n_0 = 0.76*q^{0.818}$$
which can be compared with the classical over-barrier model :
$$n_0 = 1.414*(q/z_2)^{0.75}$$
where z_2 is the effective charge of target atom.
We should stress that no scaling law can be valid at low energies to estimate n_0 (see 3.2). For many-electron targets, multi-electron capture becomes comparable to single electron capture and **systematic**

investigations on the correlated (M,N) states have to be performed.

3.2 (n,l)-distributions

The l-distributions should be influenced through Stark mixing among projectile sublevels by the electric field of residual target ions. For partially ionized ions (l-sublevels are not degenerate anymore but separate), intrashell Stark mixing among l-subshells due to target ion becomes diminished and the final (l,m) distributions are determined by the interaction of electron captured into the outer-shell with the core electrons in projectile (core effect). Indeed significant core effects are predicted in electron capture of highly charged ions. At low energies, even total cross sections are strongly dependent upon projectile ions themselves[28] (see Fig.7).

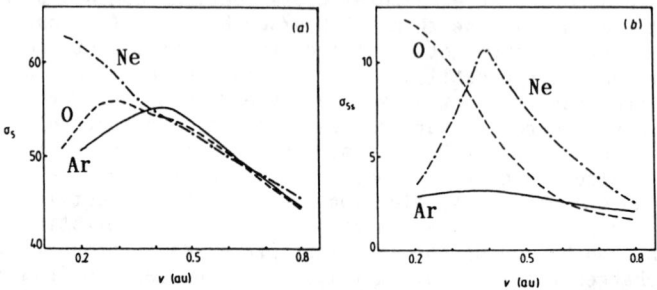

Fig.7 Partial cross sections for electron capture into 5l states of A^{7+} ions in collisions with atomic hydrogens as a function of collision velocity (A=O,Ne,Ar). a) total for n=5, b) 5s state.

Thus, there seems almost no way to infer or find any empirical formula to estimate the partial (n,l) cross sections at very low energies. The detailed studies for individual cases should be performed to get reliable information on l-distributions. On the other hand, only a limited experimental data on (n,l) distributions is available. The l-distributions are found to be strongly dependent on the velocity. The observed results confirm most theoretical calculations only for dominant channels in (n,l) distributions, meanwhile those for less dominant channels are different from each other[28,29] (see Fig.8). Observations in Ne^{q+} (q=1-10) + $Na(n_0 \cong 10)$ collisions at relatively low energies ($\cong$250 eV/amu) suggest that electron is captured mostly into lower l-states as the cross sections for visible lines are only 1/30 of total electron capture cross sections[30].

3.3 m-distributions

For bare ion + H(1s) collisions, rotational mixing among m-states occurs at our energy range, with the final m-substate distributions of varied degrees of alignment[31] :
a) at low velocities ($v_i \ll v_e$), m=0.
b) at intermediate velocities (finite rotational velocities), some rotational mixing among m-substates occurs with high degree of alignment, m=0,$\pm$1 (not statistical).

However, polarization measurements to determine m-distributions are still very limited (only for p states).
In Ne^{q+} + Na collisions, experimental values of polarization for dn=-1 (for example n=10->n=9 for q=10) can be reproduced through calculation for q=10-8, meanwhile those for q$\leq$7 are much smaller than calculation,

suggesting that the observed m-distributions are much wider than calculation due to core electron effect[30].
For multiple electron transfer, the influence of multiple-electron stripped residual ions to (n,l,m) distributions should become larger than that in single electron capture. This remains to be confirmed.

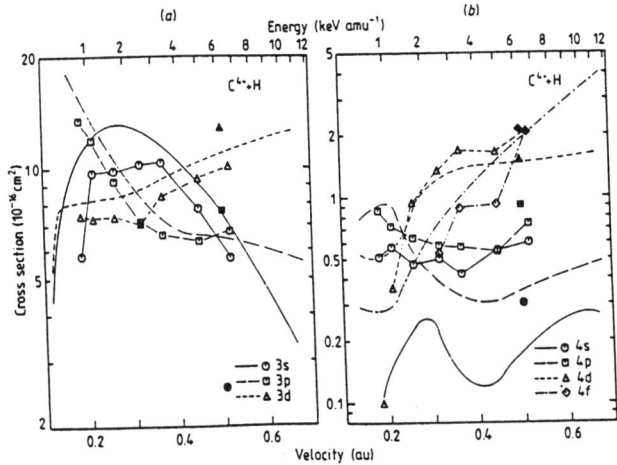

Fig.8 Comparison of theoretical and experimental cross sections for electron capture in C^{4+} + H collisions[29].

3.4 Autoionization states[32]
Doubly or multiply excited states formed through electron capture can be autoionized. The autoionizing electron spectra in N^{7+} + He -> Ne^{5+} (3l3l') + He^{2+} collisions reveal that electrons are dominantly captured (≡ 50 %) into highest l-state $^2G(2,0)$ (which means that two electrons captured sit on the opposite sides of nucleus). **Some electron spectroscopic work should be made for multi-electron (>3) capture processes.** On the other hand, as the autoionizing states emit electrons, then, the final product such as N^{6+} from N^{5+}(3l,3l') state can also be studied with translational energy spectroscopy. It should be noted that the products of multiple electron capture processes are scattered into large angles (see 4).

4. Angle-differential cross sections (see Fig.9)
Only few cases have been studied of their angular distributions[33]. In multiple electron capture processes, projectile ions are scattered into larger angles because of strong Coulomb repulsion between collision partners[34]. Angular distributions of charge-selected projectiles in coincidence with charge-selected recoil ions provide much detailed information of ion-atom collisions. Single electron capture process at 90 keV Ne^{7+}+Ne collisions is found to be strongly forward-peaked, meanwhile, with increasing the number of the captured electrons, large angle scatterings become important and also autoionization of the captured ions plays a role, increasing the ionic charge. Indeed, recoil Ne^{6+} ions are found to be produced, in addition to 5-4 electron transfer into projectiles, through single and double autoionization processes but not through 6-electron capture. The measured charge distributions of both collision partners are found to

be sharper than expected from binomial distributions and shifted
toward higher charge, suggesting significant autoionization. At large-
angle scatterings (>10 mrad), the mean charges of projectile and
recoil ions are nearly the same, indicating that both particles form a
quasimolecule and share equally the remaining electrons (in L-shell).

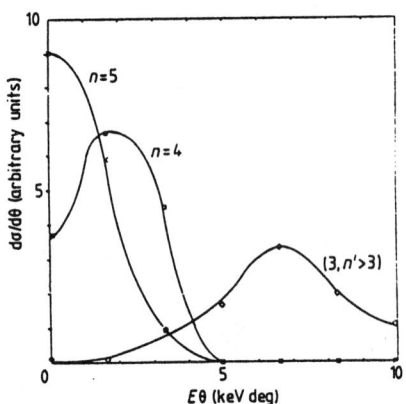

Fig.9 Relative angular scattering cross sections for 16 keV Ne^{9+} + He collisions[34].

IV. COLLISIONS WITH SOLIDS AT LOW ENERGIES
1. Scattered particles from surface
The scattering of projectile ions from surfaces and their mechanisms
depend on various processes and, among them, the electron transfer
processes between projectile and surface determine the fractions of
the scattered neutral and charged particles[35].

1.1 Charge distributions and electronic states of scattered particles
The observed data suggest that scattered particles are mostly neutral
and a very small fraction of singly and doubly charged ions are
included in the scattered particles. A far small fraction of trebly
charged ions are seen only in relatively high charge ion incidence.
In Fig.10 are shown the fractions of Ne, Ar and Kr ions taken at 15°
under 20 keV impact on clean tungsten surface at 15° incidence. From
these figures, the followings can be summarized[36] (see Fig.10) :
a) the fraction of singly charged ions in the scattered ions is nearly
independent of the incident ionic charge, b) the fraction of doubly
charged ions is increased and that of trebly charged ions is much
drastically enhanced when projectiles carry innershell vacancies, c)
this is because the time requiring to fill the innershell vacancies is
much longer than that for low charge ions. Thus, even after violent
collisions, a fraction of highly charged incident ions can survive
throughout collisions with surface. The scattered projectiles should
be in highly excited states and when they leave the surface, they can
decay by autoionization, resulting in the increase of the ion charge.
For example, most of trebly charged ions observed are due to doubly
charged ions autoionized on the way out, d) the energy distributions
of the doubly and trebly charged scattered particles have no low
energy tails, meanwhile those of singly charged ions have a large
portion of low energy tails, which are believed to result from a

series of collisions of the incident ions with target atoms in solid,
e) **the electronic states of the scattered ions, in particular of singly or doubly charged ions from ions with very high charge colliding with surfaces have to be investigated.** Under some conditions, the projectile ions have innershell vacancies as observed through Auger electrons or X-ray spectra (see 2.4).

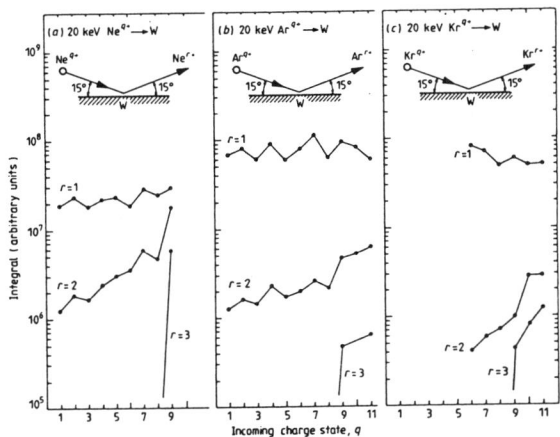

Fig.10 Yields of scattered Ar^{r+} ions with the charge of r=1, 2 and 3

1.2 Near-zero-angle or grazing incidence collision processes
In grazing incidence conditions, the projectile ion energy (E_x) parallel to surface is almost the same as their initial energy (E_0), meanwhile that normal to surface is roughly given as $E_y=E_0*\sin^2\theta$ (for $\theta\approx0.2°$, $E_y/E_x \approx 10^{-5}$)[37]. Thus, it becomes possible to study collision processes of highly charged ions with extremely small energy with surface using finely controllable high energy ions.
From a series of measurements[38], the charge distributions of the scattered ions after grazing incidence have been found to be different from those after passing foils. Such difference is strongly dependent on ion-surface atom combination and can be qualitatively understood from dynamic variation of their potential energy (due to image potential of the approaching ion inside solid) as a function of the distance between the incident ion and surface. For Li^+ + Cu system, for example, the outershell electron (2s) of Li projectile ions can be easily resonant-ionized through electron transfer into the empty states above the Fermi level of Cu, as, with the ions approaching surface, the potential energy level of the outershell electron of Li goes over the Fermi level of Cu target. Then, the resonant-ionization probabilities of Li 2s electrons are enhanced. Therefore, the fraction of neutral beams decreases significantly, compared with those from foils. On the contrary, for N^+ + Cu system, the resonant-electron capture is dominant. Thus, the fraction of neutral component of N ions is enhanced. **Systematic investigations on grazing incidence of highly charged ions should be performed.**

2. Emitted electrons
2.1 Total electron emission rate g
Total electron emission g is believed to consist of two components :

$g=g(PE)+g(KE)$, where $g(PE)$ and $g(KE)$ are due to the potential emission and kinetic emission, respectively.[39]
The potential emission term $g(PE)$ has the following features[40]:
a) the electron emission is due to interactions of vacant states of projectiles with surface valence-band states (no kinetic projectile energy is necessary) and dominant at low energies ($< 1*10^7$ cm/s) and decreases with increasing the collision energy (due to the decreasing time for resonance neutralization and autoionization), b) at low velocities, $g(PE)$ is proportional to total potential energy W available from the incident ions but nearly independent of ionic charge, c) roughly 100 eV of the energy is required for single electron emission at low energies ($< 1*10^7$ cm/s) due to potential emission mechanism.

On the other hand, the kinetic emission term $g(KE)$ has the following features : a) electron emission occurs through close collisions between projectile and target atom and thus need some minimum kinetic energy which roughly corresponds to about $1*10^7$ cm/s and increases with the collision energy and finally becomes dominant over the potential emission at higher energies, b) the emission rate is independent of the ionic charge but dependent only on W[40,41].

At higher energies and also for higher charge ions, g levels off at higher W. This is due to the increased time necessary for successive neutralization for higher charge ions. Then only a partial neutralization can be achieved before highly charged ions hit the surface (see Fig.11).

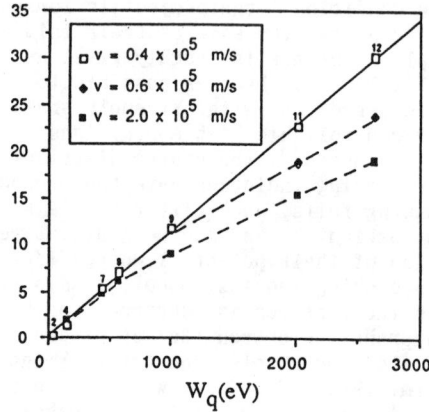

Fig.11 Dependence of total electron emission yield on potential energy of the incident Ar^{i+} (i=2-12) ions at various collision velocities[40].

The neutralization time for multiply charged ions can be estimated as follows : the number of collisions for complete neutralization is given as $N=W/W_0$ where W is total potential energy and W_0 is the average energy required for neutralization ($\cong 15$ eV). Thus, the complete neutralization time is $N*t$ where t is the average time for a single resonant capture + Auger decay ($\cong 10^{-15}$ s). On the other hand, the electron capturing state n_c for hydrogenic ions can be given as

$E_n=(1/2)(q/n_c)^2$, and the electron classical orbit radius of this state, $r_e=n_c^2 a_0/q$, the distance occurring resonance electron transfer is assumed to be a few times the electron orbit radius, $d \approx 5*r_e$ and the time of passage of ions is estimated to be d/v (v : ion velocity). At the velocities below 10^7 cm/s, complete neutralization can be achieved for intermediate charge ions as follows : For Ar^{12+} ions with the potential energy=2650 eV, N=177 and the total neutralization time is $177*10^{-15} \approx 1.8*10^{-13}$ s. On the other hand, $n_c=20$, $r=18$ Å, and then d=100 Å. Thus, the transit time is about 10^{-13} s (for 10^7 cm/s). However, for very heavy ions such as U^{92+} ions (total potential energy =740 keV), even at the velocity of 10^6 cm/s, only 10 % of the incident beams is neutralized before they hit the target surface (time for complete neutralization=$5*10^{-11}$ s : time of passage=$6*10^{-12}$ s) and, thus, they should still have a plenty of the innershell vacancies at the time of arrival on the surface (we should take into account very strong image effect and thus the transit time become small, resulting in less neutralization), f) **how large g(PE) is for very high charge ions of very low energies(PE>>KE), in particular at the lowest energies ? f) is g(PE) independent of the incident angle to surface ?**

2.2 Number distributions of emitted electrons

The distributions of the number of emitted electrons per incident ion can be known from, for example, the collected charge. Because of the limited energy resolution, up to 12 electrons could be resolved in passage of 2-5 MeV alpha particles[42] (see Fig.12). At our interest of relatively low energies, the solid must be at the ground potential, meanwhile the detector is at high voltage, tending to accompany serious noises. As about 30 electrons are emitted in low energy Ar^{12+} ion impact(see 2.1) a considerable fraction of ion collisions are expected, with a Poisson distribution, expected to emit as much as 60 electrons under a single ion impact.

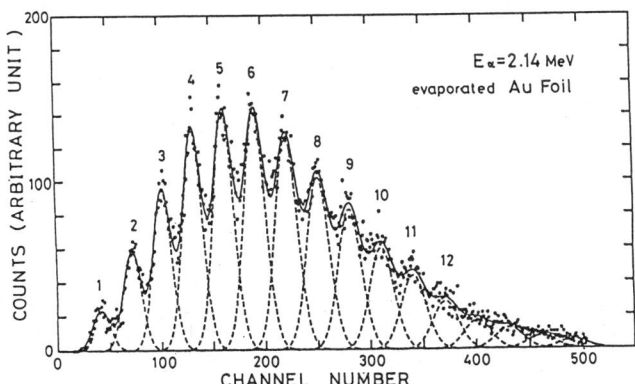

Fig.12 Distributions of electron number emitted from Au foils passing 2 MeV alpha particles[42].

2.3 Electron energy distributions (see Fig.13)

The energy distributions of electrons emitted from solids, as observed by Delaunay et al. using 15-70 keV Ar^{9+} ions normal incident upon Au, have the following features[43] : a) the mean energy is around 6 eV, b) the width (FWHM) is about 15 eV, c) the mean energy, peak position and width are weakly dependent on velocity but increase with increasing the ionic charge ; in the case g≡W, they are roughly constant ; in the case g≠W, the peak becomes broad for higher charge ions, suggesting the variation of the ionic states during electron emission.

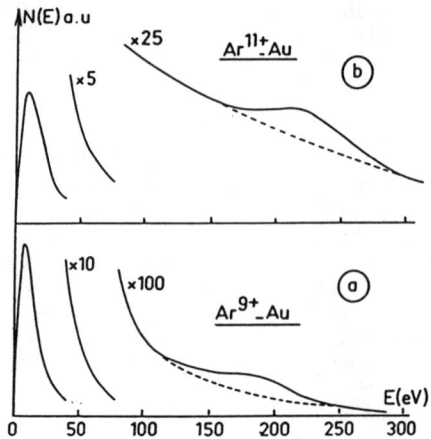

Fig.13 Energy distributions of electrons emitted from Au in Ar^{9+} and Ar^{11+} ions[43].

2.4 Auger electrons

Even though weak (only a few % of total emissions), they contain information on innershell electron processes in ion collisions with solids[43] :
a) the Auger electron emission is enhanced significantly if incident ion has innershell vacancies. For example, 500 eV Ar^{9+} ions with a single 2p vacancy produce significant L_{23}MM Auger electron peaks at 120-240 eV which is found to be very different from the expected value of direct Auger neutralization to Ar^{9+} ions (450 eV : no peak is seen there) but similar to that from neutral Ar atom targets or Ar^{+}+solid collisions, b) this fact suggests that, at low velocities, the outer M-shell vacancies are all filled prior to Auger decay and, then, the innershell vacancies decay before ions hit the surface, which is supported by the fact that LMM Auger electrons are sharp and Doppler-shifted, c) with increasing collision energy, Auger electron yields from Ar^{9+} impact decrease due to decrease of the time near surface and thus ions reach surface before deexcitation occurs ; at higher energies, ions with L-shell vacancies penetrate into solid before complete neutralization. Thus the level structures and accordingly the electron energy spectrum may changes, e) for higher charge Ar^{11+} collisions, Auger peak at 150-300 eV is energy-shifted due to some still-survived 2p vacancies because of the longer neutralization times (see 2.1), f)X-ray spectroscopy might also provide information on the electronic states of ions before/during/after collisions with solids.

3. Sputtering

Sputtering of solids by ions is due to kinetic collision cascades and should not strongly depend on the electronic or ionic states of the incident particles[44]. On the other hand, experimental data[45] show that sputtering yields from NaCl or ZnS increase with the ion charge over

q=1-6, meanwhile those from metals are independent of it, suggesting another mechanism for sputtering for non-metallic solids : electronic sputtering or Coulomb explosion[46]. Recent measurements of sputtering yields in 20 keV Ar^{q+} (q=1-9) on Si in 6° incidence to surface normal under much better surface conditions reveal[47] (see Fig.14):
a) yields of neutral Si atoms are independent of the incident charge from q=1-9, with sputtering yield of 1.3 atoms/ion, b) yields of secondary Si ions (about 1/200 of neutrals) are constant over q=1-6 ($4*10^{-3}$ ions/ion) and start to increase, reaching $12*10^{-3}$ ions/ion for Ar^{9+} ions, which are consistent with other measurements[48], c) because sputtered ion yields are a minor contribution to total sputtering, the major features of sputtering by highly charged ions are believed to be practically the same as those by low charged ions, d) this is also supported by similar energy spectra of sputtered ions observed in both Ar^+ and Ar^{9+} ion impact[47], e) **What happens in sputtering when the potential energy of projectiles becomes comparable to or larger than their kinetic energy ?**

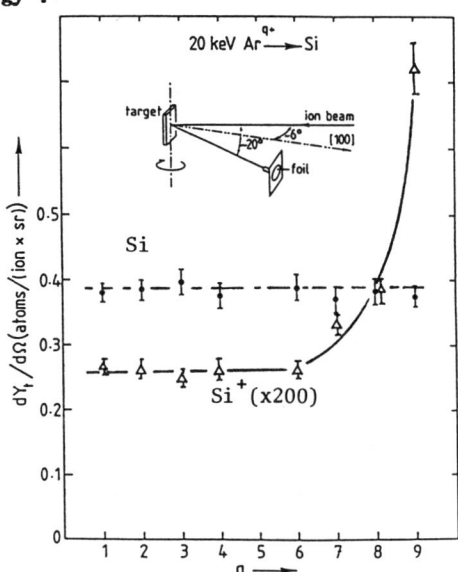

Fig.14 Sputtered neutral and ion yields of Si target by 20 keV Ar^{q+} ion impact as a function of q. Note that ion yields are multiply by a factor of 200[43].

V. CONCLUDING REMARKS

We have seen remarkable progresses on studies involving highly charged ions in collisions with electrons, atoms(ions) and solids. A great part of the success should owe to the development of powerful ion sources, such as EBIS and ECR. Although a number of the impressive work have been reported up to now, they are scattered and a lack of systematics. In particular, we are still short of information on the state-selected collision processes.
It is felt that we are almost ready for obtaining ions with much higher charge than before using EBIS and now we are surely at the stage of giving serious consideration in working with ultra-high

charge ions such U^{92+} ions. There should be very interesting topics to be pursued experimentally as well as theoretically. What happens when very slow U^{92+} ions approach solids ? Because of very large total potential energy ($\cong$750 keV), complete neutralization of surfaces, in particular of non-metallic surfaces, can take too much time, before which a part of the surface may Coulomb-explode. It is also wondered if their electron capture processes could be understood just as an extension of our present knowledge, because their life times become so short that they may decay before the collision partners separates.

REFERENCES
1. H.Tawara & T.Kato, At.Data and Nucl.Data Tables 36(1987)167 ; K.L. Bell,H.B.Gilbody,J.G.Hughes,A.E.Kingston & F.J.Smith, J.Phys.Chem. Ref.Data 12(1983)891 ; J.W.Gallagher & A.K.Pradhan,JILA(Univrsity of Colorado)report no.30(1985)
2. D.H.Crandall, Atomic Physics of Highly Ionized Atoms (Plenum,1982) p.399 ; R.A.Phaneuf, Atomic Processes in Electron-Ion and Ion-Ion Collisions (Plenum Press,1986)p.117
3. K.T.Dolder,M.F.Harrison & P.C.Thonemann,Proc.Roy.Soc.A26(1961)367
4. E.D.Donets & V.P.Ovsyannikov, Sov.Phys.JETP53(1981)466
5. B.Peart & K.Dolder, J.Phys.B8(1975)56
6. D.C.Griffin,C.Bottcher,M.S.Pindzola,S.M.Younger,D.C.Gregory & D.H. Crandall, Phys.Rev.A29(1984)1729
7. D.C.Gregory,L.J.Wang,F.W.Meyer & K.Rinn, Phys.Rev.A35(1987)3256
8. A.Muller,G.Hoffmann,K.Tinchert & E.Salzborn, Phys.Rev.Lett.61 (1988)1352
9. H.Tawara,private communication
10. A.R.P.Rau, Electronic and Atomic Collisions(North-Holland,1984)711
11. W.Lotz, Z.Phys. 216(1968)241 ; T.Kato, IPPJ-AM-2 (Institute of Plasma Physics, Nagoya Univ.,1977) ; M.S.Pindzola,D.C.Griffin, C. Bottcher, S.M.Younger & H.T.Hunter, Nucl.Fusion Suppl.(1987)21 ; M.Arnaud & R.Rothenflug, Astron.Astrophys.Suppl.Ser.60(1985)425 ; A.Burgess & M.C.Chidichimo, Mon.Not.R.astrophys.Soc.203(1983)1269 ; L.A.Vainshtein & V.P.Shevelko, Opt.Spectrosc.(USSR)63(1987)11
12. V.A.Bazylev & M.I.Chibisov, Sov.Phys.Usp.24(1981)278 ; Y.Itikawa, Phys.Rep.143(1986)69 ; M.A.Hayes & M.J.Seaton, J.Phys.B11(1978)79
13. D.F.Dance,M.F.A.Harrrison & A.C.H.Smith, Proc.Roy.Soc.A290(1966)73
14. R.E.Marrs,M.A.Levine,D.A.Knapp & J.R.Henderson, Phys.Rev.Lett.60 (1988)1715
15. A.Chutjian, Phys.Rev.A29(1984)64
16. D.S.Belic,G.H.Dunn,T.J.Morgan,D.W.Mueller and C.Timmer, Phys.Rev. Lett.50(1983)339 ; G.H.Dunn, Atomic Proceses in Electron-Ion and Ion-Ion Collisions(Plenum,1986)p.93 ; Y.Hahn and K.J.LaGattuta, Phys.Rep.166(1988)195
17. J.P.Briand,P.Charles,J.Arianer,H.Laurent,C.Goldstein,J.Dubau, M.Loulergue & F.Bely-Dubau, Phys.Rev.Lett.52(1984)617
18. R.K.Janev,L.P.Presnyakov & V.P.Shevelko, Physics of Highly Charged Ions (Springer,1985)
19. H.Ryufuku,K.Sasaki & T.Watanabe, Phys.Rev.A21(1980)745
20. A.Salop & R.E.Olson, Phys.Rev.A13(1976)1312
21. K.Taubjerg, J.Phys.B19(1986)L367
22. A.Muller & E.Salzborn, Phys.Lett.62A(1977)391 ; E.Salzborn & A. Muller, Electronic and Atomic Collisions (North-Holland,1980)p.407
23. J.W.Gallagher,B.H.Bransden & R.K.Janev, J.Phys.Chem.Ref.Data 12 (1983)873 ; H.Tawara,T.Kato & Y.Nakai, At.Data & Nucl.Data Tables

32(1985)235 ; W.K.Wu,B.A.Huber & K.Wiesemann, At.Data and Nucl.Data Tables 40(1988)
24. H.Tawara,T.Iwai,Y.Kaneko,M.Kimura,N.Kobayashi,S.Ohtani,K.Okuno, S.Takagi & S.Tsurubuchi, J.Phys.B18(1985)337
25. L.Liljeby,G.Astner,A.Barany,H.Cederquist,H.Danared,S.Huldt, P.Hvelplund,A.Johnson,H.Knudsen & K.G.Rensfelt, Phys.Scripta 33 (1986)310 ; D.H.Crandall, Phys.Rev.A16(1977)958
26. W.Groh,A.Muller,C.Achenbach,A.S.Schlachter & E.Salzborn,Phys.Lett. 85A(1981)77
27. H.Y.Wang & D.A.Church, Phys.Rev.A36(1987)4261
28. C.Harel & H.Jouin, J.Phys.B21(1988)859 ; M.Gargaud & M.McCarroll, J.Phys.B21(1988)513
29. D.Dijkkamp,D.Ciric,E.Vlieg,A.de Boer & F.J.de Heer, J.Phys.B18 (1985)4763 ; W.Fritsch & C.D.Lin, J.Phys.B17(1984)3271
30. L.J.Lembo,K.Danzmann,Ch.Stoller,W.E.Meyerhof & T.W.Hansch, Phys.Rev.A37(1988)1141
31. A.Salin, J.de Phys.45(1984)671
32. A.Bordenave-Montesquieu,P.Brenoit-Cattin,M.Boudjema,A.Gleizes & H.Bachaus, J.Phys.B20(1987)L695 ; S.Tsurubuchi,T.Iwai,Y.Kaneko, M.Kimura,N.Kobayashi,S.Ohtani,K.Okuno,S.Takagi & H.Tawara,J.Phys. B15(1982)L733
33. H.Danared,H.Andersson,G.Astner,A.Barany,P.Defrance & S.Rachafi, J.Phys.B20(1987)L165 ; Phys.Scripta 36(1987)756 : L.N.Tunnell, C.L.Cocke,J.P.Giese,E.Y.Kamber,S.L.Varghese & W.Waggoner, Phys. Rev.A35(1987)3299
34. H.Schmidt-Bocking,M.H.Prior,R.Dorner,H.Berg,J.O.K.Pedersen,C.L. Cocke,M.Stockli & A.S.Schlachter,Phys.Rev.A37(1988)4640 ; M.Barat, M.N.Gaboriaud,L.Guillemot,P.Roncin,H.Laurent & S.Andriamonje,J. Phys.B20(1987)5771 ;
35. N.H.Tolk,J.C.Tully,W.Heiland & C.W.White(ed.),Inelatic Ion-Surface Collisions(Academic Press,1977) ; H.Winter & R.Zimny, Coherence in Atomic Collision Physics (Plenum,1988)p.283
36. S.T.de Zwart,T.Fried,U.Jellen,A.L.Boers & A.G.Drentje, J.Phys.B18 (1985)L623 ; K.J.Snowdon, NIM B34(1988)309
37. J.Burgdorfer,E.Kupfer & H.Gabreil, Phys.Rev.A35(1987)4963
38. H.Winter,R.Zimny,A.Schirmacher,B.Becker & H.J.Andra, Z.Phys.A311 (1983)267
39. M.Fehringer,M.Delaunay,R.Geller,P.Varga & H.Winter,NIMB23(1987)245
40. M.Delaunay,M.Fehringer,R.Geller,D.Hitz,P.Varga & H.Winter, Phys. Rev.B35(1987)4232
41. K.Oda,A.Ichimiya,Y.Yamada,T.Yasue & S.Ohtani, NIM B33(1988)345
42. S.Mizugashira (private communication)
43. M.Delaunay,C.Benazeth,N.Benazeth,R.Geller & C.Mayoral, Surf.Sci. 195(1988)455 ; S.T.de Zwart, NIM B23(1987)239 ; F.W.Meyer,C.C. Havener,K.J.Snowdon,S.H.Overbury,D.M.Zehner & W.Heiland, Phys. Rev.A35(1987)317
44. R.Behrisch, Sputtering by Particle Bombardment 1 (Springer,1981)
45. Y.Yu.Arifov,E.K.Vasileva,D.D.Gruich,S.F.Kovalenko & S.N.Morozov, Izv.Akad.Nauk SSSR,ser.Fiz.40(1976)2621 ; Sh.S.Radzhabov,R.R. Rakhimov & D.Abdusalimov, Izv.Akad.Nauk SSSR,ser.Fiz.40(1976)2543
46. I.S.Bitenskii,M.N.Murakhmetov & E.S.Parilis, Sov.Phys.-Tech.Phys. 24(1979)818
47. S.T.deZwart,T.Fried,D.O.Boerma,R.Hoekstra,A.G.Drentje & A.L.Boers, Surf.Sci.177(1986)L939
48. K.Wittmaack, Sur.Sci.90(1979)557

DISCUSSION

<u>Marchetti</u>: Could you say something about what energy resolutions you would need to see the resonant recombination process in this cross beam?

<u>Tawara</u>: I think you should have the energy resolution better than 0.1 eV and hopefully 0.01 eV. Otherwise, it would all smear out.

ATOMIC PHYSICS MEASUREMENTS IN AN ELECTRON BEAM ION TRAP

R.E. Marrs, P. Beiersdorfer, C. Bennett, M.H. Chen, T. Cowan,
D. Dietrich, J.R. Henderson, D.A. Knapp,
A. Osterheld, M.B. Schneider, and J.H. Scofield
Lawrence Livermore National Laboratory
P.O. Box 808, L-296, Livermore, CA 94550, USA

M.A. Levine
Lawrence Berkeley Laboratory, Berkeley, CA 94720

ABSTRACT

An electron Beam Ion Trap at Lawrence Livermore National Laboratory is being used to produce and trap very-highly-charged ions ($q \leq 70+$) for x-ray spectroscopy measurements. Recent measurements of transition energies and electron excitation cross sections for x-ray line emission are summarized.

INTRODUCTION

A new device called an Electron Beam Ion Trap (EBIT) is operating at Lawrence Livermore National Laboratory for the purpose of studying very-highly charged ions using x-ray spectroscopy. The EBIT device, shown in Fig. 1, is basically a short EBIS optimized for x-ray spectroscopy measurements of the confined ions, for production of extremely high charge states, and for very long confinement times (up to several hours). The EBIT apparatus is described elsewhere in these proceedings by Levine.

The EBIT at LLNL is being used for precision wavelength measurements of x-ray transitions in highly charged ions, and also for the study of electron–ion collision cross sections and excitation mechanisms for x-ray line emission. In all cases the EBIT device has demonstrated its ability to provide these measurements for ionization–stages which have not been accessible before.[1,2,3] The purpose of the present paper is to provide a sampling of these types of measurements.

PRECISION WAVELENGTH MEASUREMENTS

The small diameter of the electron beam (70μm) as it passes through the trapped ions in EBIT produces an ideal source for high resolution wavelength measurements using Bragg diffraction crystals. Two different types of spectrometers are in use. One is a Johann spectrometer, shown in Fig. 2, which can be configured for either soft (< 1 keV) or hard (> 3 keV) x-rays. It operates in a scanning mode and uses a conventional x-ray tube for wavelength calibration. The other type of spectrometer uses the von Hamos geometry, in which the crystal is flat in the diffraction plane but curved out of the diffraction plane to provide focusing.[4] A position sensitive proportional counter is used for x-ray detection. The wavelength calibration for this spectrometer comes from x-ray transitions of known energy produced in EBIT itself.

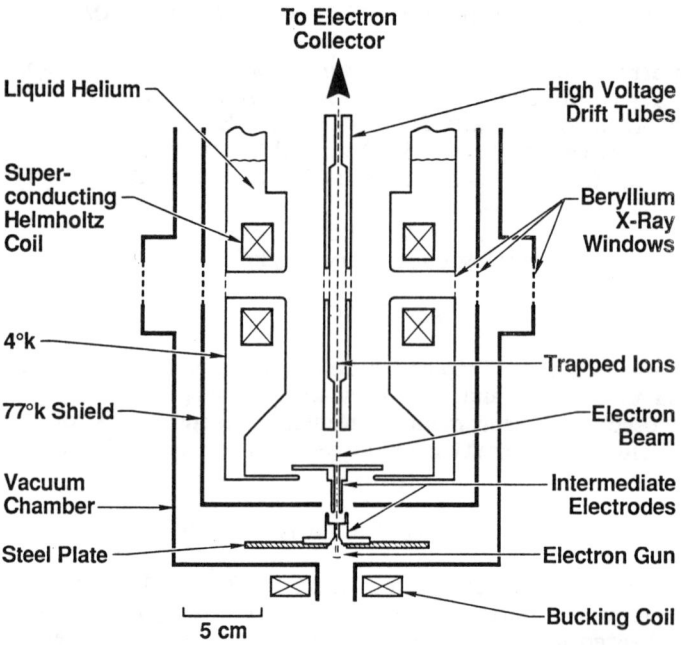

Fig 1. – The electron beam ion trap. The beryllium windows can be removed for soft x-ray or VUV spectroscopy. Two different drift-tube geometries have been used. In the one shown here the drift tube structure operates at a single electrical potential. In the other version (not shown), three electrically isolated drift tubes can be biased at different potentials to form a higher axial barrier.

The Johann spectrometer is being used for Lamb shift measurements in hydrogenlike and heliumlike high-Z ions. These measurements are motivated by the desire to test QED predictions in the strong field limit. We have performed a preliminary experiment on H-like nickel, in which we measured the 2p to 1s transition energies, in order to compare the measured Lamb shift with theory. Fig. 3 displays a partial spectrum for H-like nickel. Our preliminary measurements are in agreement with theory, and promise eventual precision much better than has been previously obtained from high-Z ion beams produced at accelerators.

The von Hamos spectrometer has been used to observe x-ray spectra from highly charged neonlike ions. For example, we have measured the $(2s_{1/2}^{-1} \; 3p_{3/2})_{J=1} \rightarrow$ ground state transition in Ne-like ytterbium (Z=70). This transition is of interest because it allows studying the screening of the self energy in a multielectron ion. In order to determine the energy of this transition, the spectrum was calibrated by recording the position of the Lyman-alpha lines of hydrogenic zinc, as shown in Fig. 4.

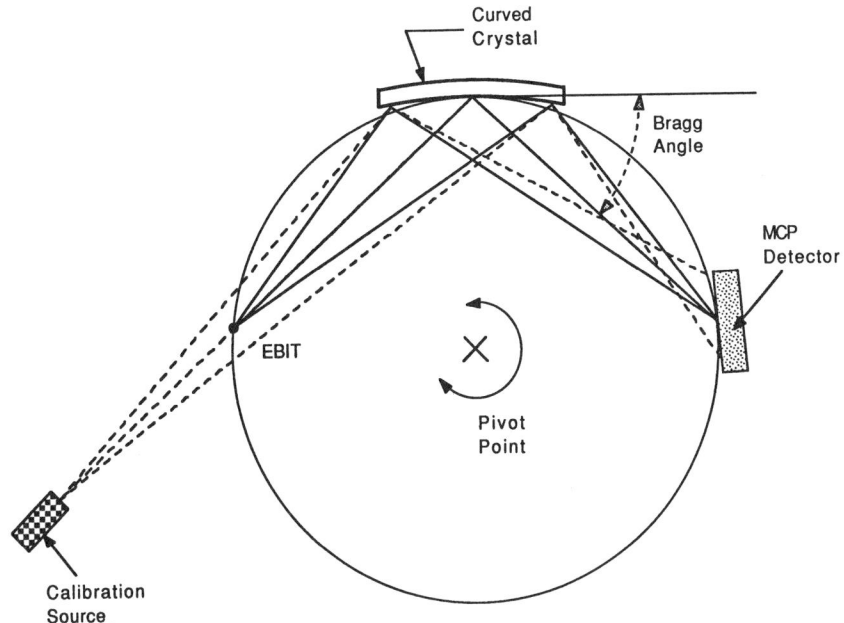

Fig. 2 – Johann spectrometer arrangement. For soft x-ray measurements a position-sensitive microchannel-plate detector is located tangent to the Rowland circle, and the spectrometer operates in ultrahigh vacuum. For higher x-ray energies a position sensitive proportional counter is placed on the Rowland circle facing the crystal, and a vacuum chamber is unnecessary. For scanning operation the crystal and detector move together while EBIT and the calibration source remain fixed.

A preliminary analysis indicates that the measured transition energy is 4 eV less than theoretical calculations. This agrees with differences found earlier between experimental and calculated values in neonlike silver (Z=47), xenon (Z=54), lanthanum (Z=57)[5], gold (Z=79) and bismuth (Z=83).[6]

DIELECTRONIC RECOMBINATION MEASUREMENTS

Dielectronic recombination (DR) is the resonant capture of an incident electron into a doubly excited state followed by x-ray emission:

$$A^{q+} + e^- \rightarrow [A^{(q-1)+}]^{**} \rightarrow A^{(q-1)+} + \gamma_{DR}$$

The related processes of resonant excitation (RE) can also produce x-ray emission at similar energies:

$$A^{q+} + e^- \rightarrow [A^{(q-1)+}]^{**} \rightarrow [A^{q+}]^* + e^- \rightarrow A^{q+} + e^- + \gamma_{RE}$$

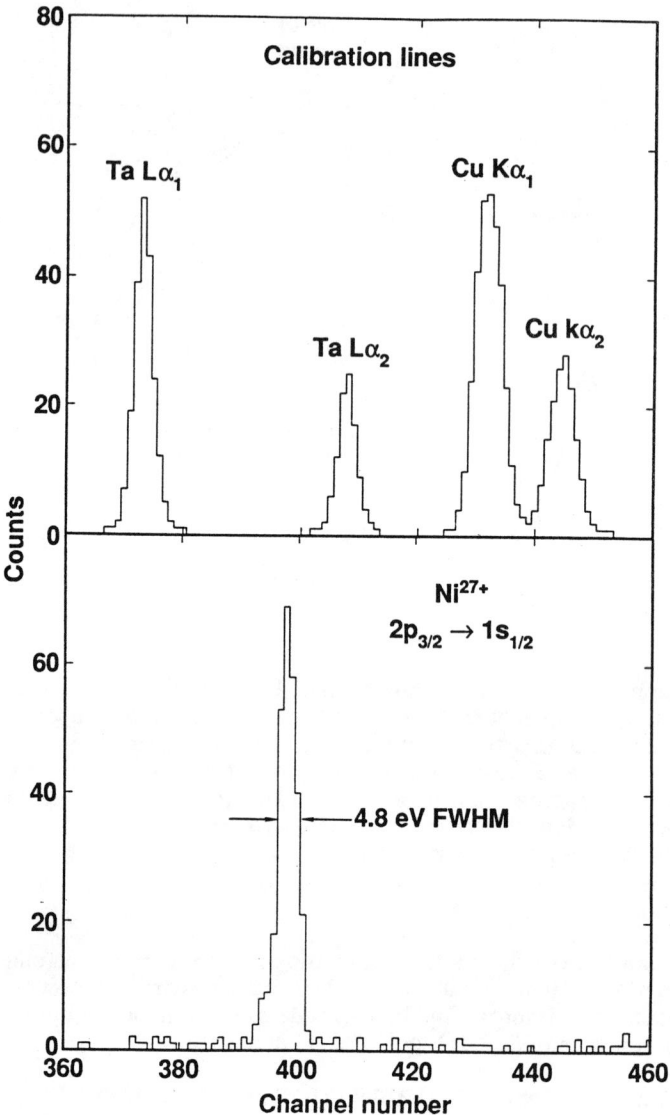

Fig. 3 – Data from the Johann spectrometer. Top: The different calibration lines shown were measured in separate scans with different exposure times. Bottom: The $2p_{3/2} \to 1s_{1/2}$ line from Ni^{27+} ions in EBIT. The x-ray energy range shown is approximately 8000–8170 eV.

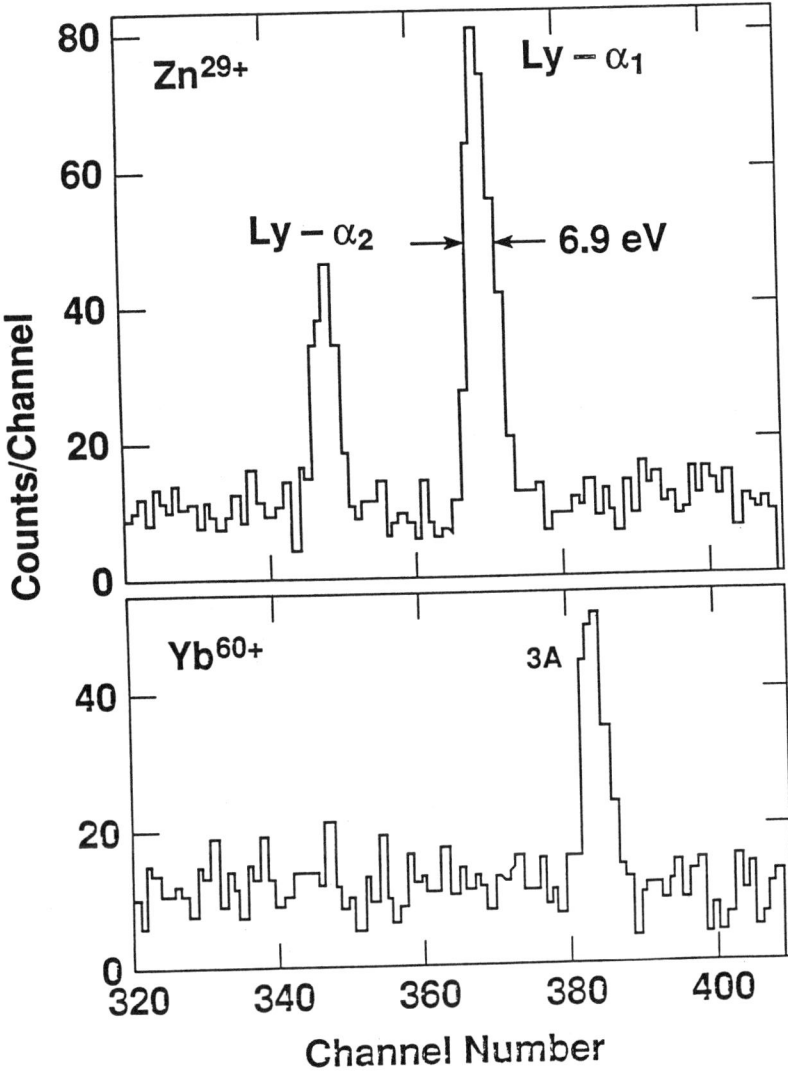

Fig. 4 – Spectra from the von Hamos spectrometer. Top: Lyman–alpha calibration lines from H–like zinc. Bottom: The $(2s_{1/2}^{-1}\,3p_{3/2})J=1 \to$ g.s. transition, denoted 3A, in Ne–like ytterbium.

The DR process is important in hot plasmas, whose x-ray emission spectra contain satellite lines characteristic of DR. However there are no detailed measurements of DR cross sections for the high ionization stages (such as He-like Fe^{24+} or Ni^{26+}) which are produced in tokamaks[7,8] and solar flares.[9] Because of the precise control of the electron beam energy in EBIT and the good energy resolution, EBIT is a powerful tool for measuring a resonant process such as DR.

Since the DR cross sections being measured are quite large, setting the electron beam energy on a DR resonance quickly destroys the charge state being measured. We overcome this problem by rapidly switching the electron energy between the resonant energy and a higher nonresonant energy which restores the ionization balance. Typically only ~ 10% of the time is spent on resonance. An excitation function is obtained by taking successive data runs in which the lower electron energy (i.e. the DR energy) is changed by a small amount (typically 20-eV steps) and normalizing each run to a common nonresonant upper energy.

We have measured a DR excitation function for He-like Ni^{26+} by detecting the K x-rays from the recombined Ni^{25+} ions. For He-like target ions DR produces one and only one K x-ray photon; so the number of K x-rays produced at a given incident electron energy is proportional to the DR cross section. There is one previous observation of DR x-rays from an EBIS. In that case an x-ray signal from DR of Ar^{14+} ions was observed.[10]

Examples of EBIT spectra obtained in a Ge detector at three different electron energies are shown in Fig. 5. We have obtained cross sections for the KLL DR resonant feature by normalizing the K x-rays from DR to theoretical radiative recombination cross sections, whose characteristic x-rays are also clearly observed in our spectra. The results are displayed in Fig. 6 and compared to a multiconfiguration Dirac Fock calculation.[11]

ELECTRON IMPACT EXCITATION MEASUREMENTS

Electron impact excitation (IE) cross sections may be obtained in a manner similar to the DR measurements described above. In this case, however, the electron energies are carefully chosen to avoid resonances, since they can influence the observed line intensities. To date, we have obtained impact excitation data for two Ne-like ions in different regions of the periodic table, Ba^{46+} and Au^{69+}. This should help provide an understanding of the role of relativistic and QED effects in electron-ion collisions. Since some of the fine structure splittings in these ions are unresolved in a Si(Li) or Ge detector, we have also obtained spectra in a Bragg diffraction spectrometer. This spectrometer employed a flat crystal (for broad spectral coverage) and a position sensitive proportional counter.

Our Ba^{46+} IE cross section measurements have been published elsewhere.[1] Briefly, cross section information was obtained for five n=3 excited states in Ne-like barium. The results generally support existing theoretical collision strengths.[12]

Electron impact excitation of Ne-like Au^{69+} was studied at an excitation energy of 18 keV in the same manner as for Ba^{46+}. Gold x-ray spectra from a Ge detector and a Bragg diffraction crystal are shown in Figs. 7 and 8, respectively. Analysis of these data is still in progress.

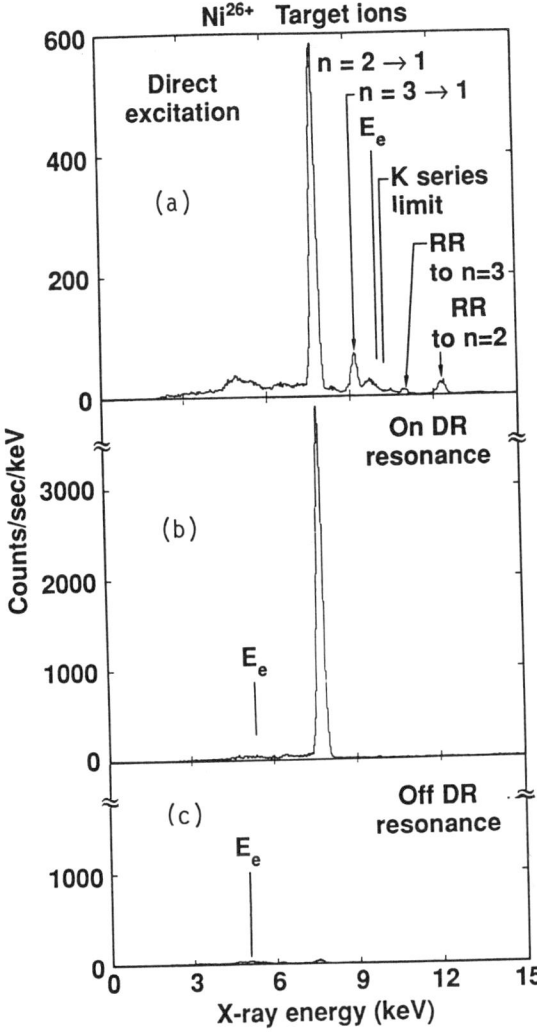

Fig. 5 – X-ray spectra from Ni^{26+} target ions excited at three different electron beam energies. (a) Above the n=2 direct-excitation threshold, E_e = 10.0 keV; (b) On resonance, E_e = 5.34 keV; (c) Off resonance, E_e = 5.05 keV.

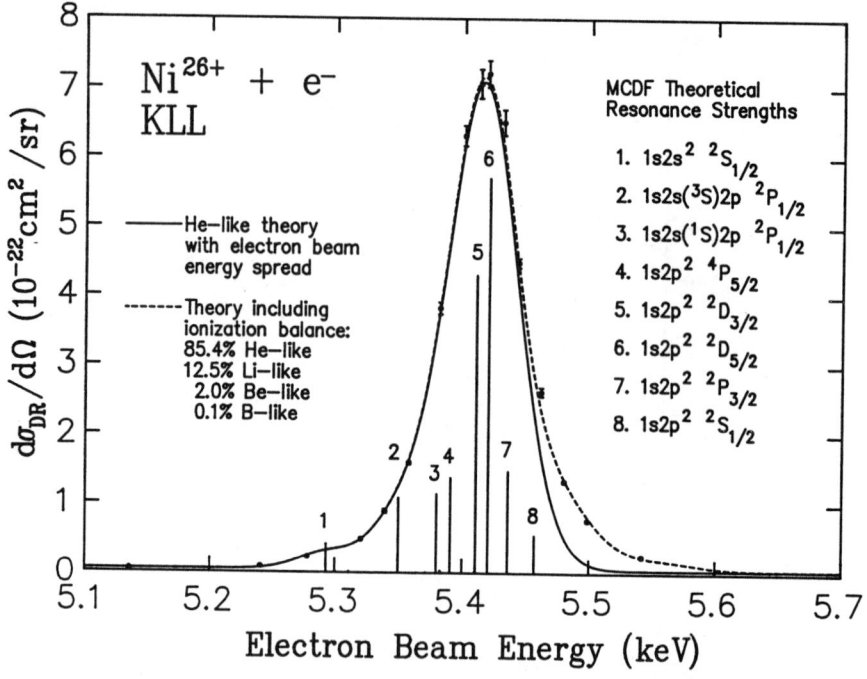

Fig. 6 – Structure of the KLL DR feature compared to theory for He-like Ni target ions. The vertical bars are the theoretical resonance strengths. The curve is the theoretical cross section obtained by folding the calculated resonance strengths with the experimental electron energy resolution function. The data points at electron energies $\gtrsim 5.4$ keV show a contribution from DR on lower-charge-state Ni ions, which are also present in the trap.

However some interesting observations can already be made. The $(2p_{3/2}3p_{1/2})_{J=2}$ transition shows a relative intensity stronger than expected from observations of lower Z ions such as Ba^{46+}. Model calculations predict that this line is blended with a magnetic dipole transition $(2p_{3/2}\ 3p_{1/2})_{J=1}$. The M1 line is absent in the spectra of ions below $Z \approx 74$ due to unfavorable branching ratios. However, in the case of gold it is predicted to have an intensity which is 30% the size of the $(2p_{3/2}\ 3p_{1/2})_2$ transition.

X-RAY LINE EXCITATION MECHANISMS

The intensity ratios of x-ray lines from highly charged ions are often used to infer the properties of the medium containing the emitting ions.

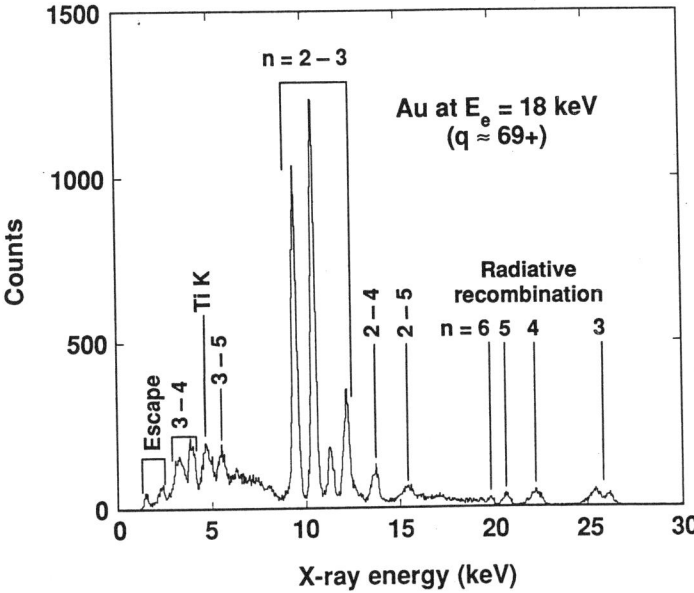

Fig. 7 – X-ray spectrum from highly ionized gold (mostly q = 69+ and 68+) obtained in a Ge detector at an incident electron energy of 18.0 keV.

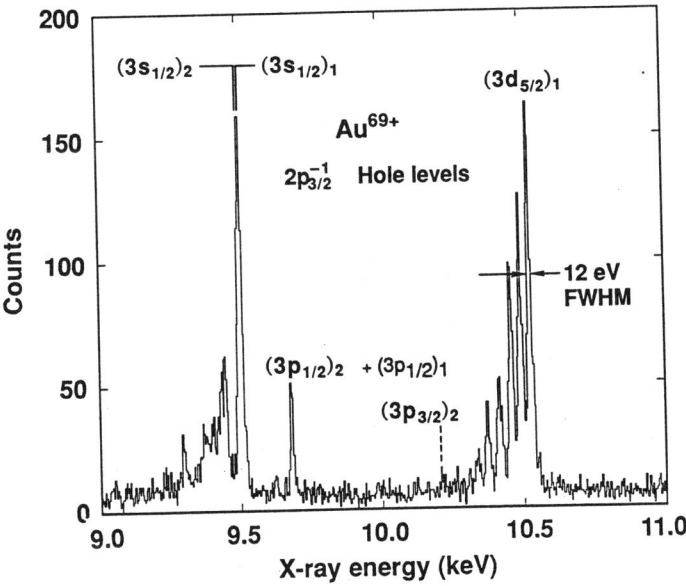

Fig. 8 – Spectrum of the $2p_{3/2}$ hole transitions in Ne-like Au^{69+} obtained with a flat LiF 220 Bragg diffraction crystal. The unlabeled satellite lines are transitions in Na- and Mg-like gold.

This is especially true in the case of astrophysical and laser-produced plasmas where alternative methods for determining quantities such as electron density or electron temperature are unavailable. However, the interpretation of the x-ray line emission is often model dependent, and uncertainties in the atomic models translate into uncertainties in the inferred quantities.

We have begun to investigate the intensity ratios of the x-ray line emission of Ba^{46+} as a function of electron beam energy in order to study the effect of cascades on the x-ray line intensities. In Fig. 9 we have plotted the ratio of the intensity of the $(2p_{1/2}^{-1} 3s_{1/2})_1$ transition (labeled 3F) and the $(2p_{3/2}^{-1} 3d_{5/2})_1$ transition (labeled 3D) versus beam energy. Also shown in the figure are theoretical predictions for the line ratio from a collisional–radiative model which includes all collisions and radiative decays involving levels of the type $2\ell\, n\ell'$. The collision strengths, energy levels, and radiative transition rates used in the model are from an atomic physics package developed by Klapisch et al.[13] The figure shows that a significantly higher theoretical line ratio is obtained when the number of levels in the calculation is increased. The 37-level model, which includes the ground state and the 36 levels of the type $2\ell\, 3\ell'$ predicts a line ratio which is 30% lower than the experimental result at 8 keV. Much better agreement with the data is found if the calculations include all 89 levels with excited electrons in the n=3,4 shells, or if the calculations include all 157 levels with

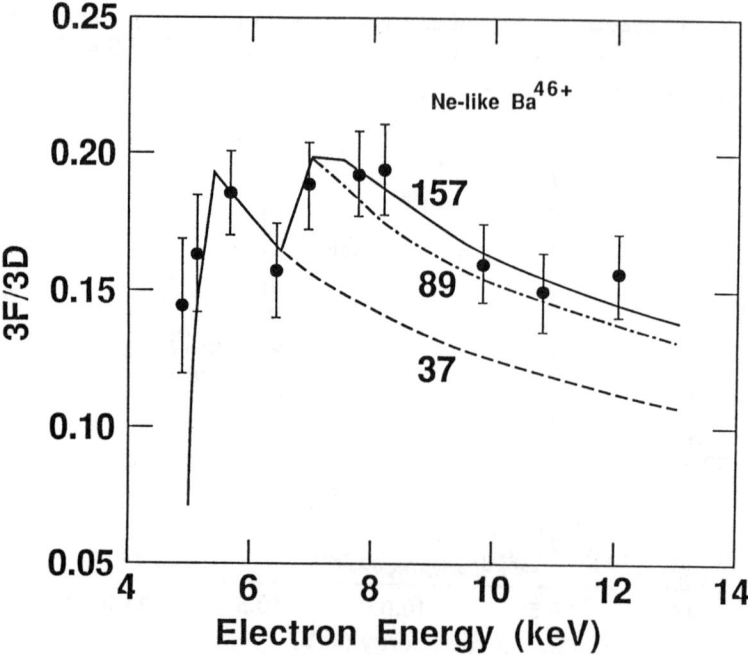

Fig. 9 – Measured ratio of the 3F and 3D emission lines from Ne–like Ba^{46+}. The curves show the results from models which include 37, 89, or 157 levels.

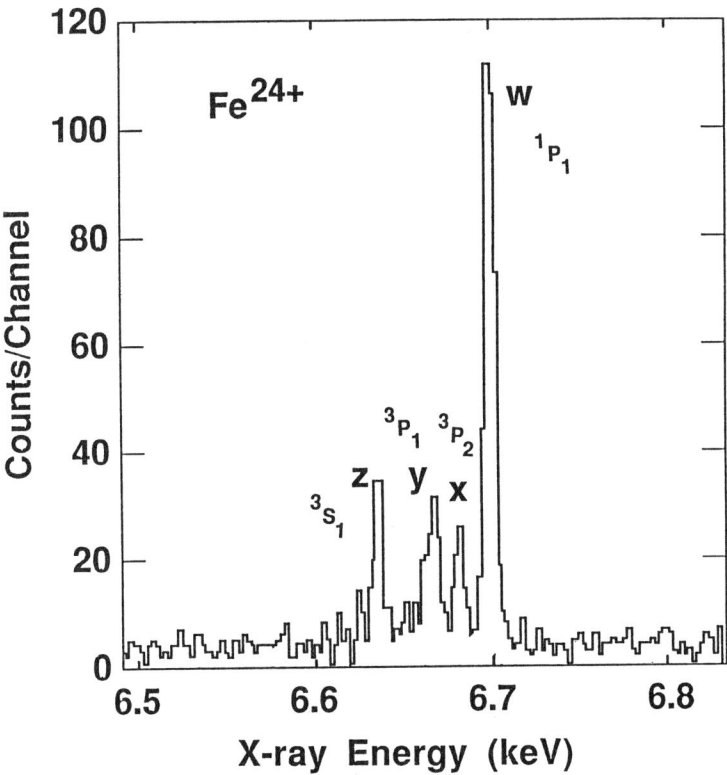

Fig. 10 – Spectrum of n=2 → 1 transitions im He–like Fe^{24+} excited at an electron energy of 8.2 keV.

excited electrons in the n=3,4,5 shells. The comparison between the calculations and the data shows that cascade feeding from high-n levels is a significant population mechanism for neonlike ions, which cannot be ignored in theoretical models.

We are also studying excitation mechanisms in H-like and He-like systems. A typical x-ray spectrum of the n=2 → 1 transitions in He-like iron is shown in Fig. 10. A series of such spectra, obtained at different electron energies, will be used to remove model dependent uncertainties which affect the use of these line ratios as a plasma diagnostic.

SUMMARY

The electron beam ion trap, with its capability for x-ray spectroscopy, is now established as a unique tool for the study of highly charged ions. In this paper we have given examples of electron collision cross section measurements and precision transition-energy measurements performed with the EBIT at LLNL. The ionization stages available for x-ray spectroscopy with EBIT already exceed those observed in the most advanced tokamaks, and essentially match the highest charge states studied in beam-foil spectroscopy.

ACKNOWLEDGMENTS

Work performed under the auspices of the U.S. Department of Energy by the Lawrence Livermore National Laboratory under contract No. W-7405-ENG-48.

REFERENCES

1. R.E. Marrs, M.A. Levine, D.A. Knapp, and J.R. Henderson, Phys. Rev. Lett. 60, 1715 (1988).
2. M.A. Levine, R.E. Marrs, J.R. Henderson, D.A. Knapp, and M.B. Schneider, Phys. Scr. T22, 157 (1988).
3. R.E. Marrs, M.A. Levine, D.A. Knapp, and J.R. Henderson, in Electronic and Atomic Collisions, edited by H.B. Gilbody, W.R. Newell, F.H. Read, and A.C.H. Smith (North-Holland, Amsterdam, 1988), p. 209.
4. L. von Hamos, Ann. Phys. 17, 716 (1933).
5. P. Beiersdorfer et al. Phys. Rev A 34, 1297 (1986); P. Beiersdorfer et al., Phys. Rev. A 37, 4153 (1988).
6. G.A. Chandler, et al., Phys. Rev. A 39, 565, (1989); D.D. Dietrich, et al., Nucl. Inst. Meth. B 24/25, 301 (1987).
7. F. Bombarda, R. Giannella, E. Kallne, G.J. Tallents, F. Bely-Dubau, P. Faucher, M. Cornille, J. Dubau, and A.H. Gabriel, Phys. Rev. A 37, 504 (1988).
8. H. Hsuan, M. Bitter, K.W. Hill, S. von Goeler, B. Grek, D. Johnson, L.C. Johnson, S. Sesnic, C.P. Bhalla, K.R. Karim, F. Bely-Dubau, and P. Faucher, Phys. Rev. A 35, 4280 (1987).
9. K. Tanaka, T. Watanabe, K. Nishi, and K. Akita, Ap. J., 254 L59 (1982).
10. J.P. Briand, P. Charles, J. Arianer, H. Laurent, C. Goldstein, J. Dubau, M. Loulergue, and F. Bely-Dubau, Phys. Rev. Lett. 52, 617 (1984).
11. M.H. Chen, Phys. Rev. A 33, 994 (1986).
12. K.J. Reed, Phys. Rev. A 37, 1791 (1988).
13. M. Klapisch, private communications; A. Bar-Shalom, M. Klapisch, and J. Oreg, Phys. Rev. A 38, 1773 (1988).

DISCUSSION

Johnson: Ross, most of your studies so far seem to be limited to closed shell configurations. How easy do you think it will be to extend your work to partially filled shells?

Marrs: Well, we have already started to do that but we are still staying close to the closed shells. We have been trying to measure $\Delta n=0$ transitions in sodium and magnesium-like ions. You can do that. The price you pay is that the ionization balance is not as pure and the atomic structure is more complicated.

Knapp: There are certain regimes where you can get ion balances off the shell by using dielectronic recombination to drive a transition much faster than the excitations. So, near the helium-like shell, for example, you can get basically pure non-helium like charge-states.

Stockli: How long did it take to accumulate the nickel Lyman spectrum?

Marrs: I do not remember but it was small compared to the time it took to scan through and convince ourselves that we understood everything. A couple of hours, maybe. Most of the time was spent understanding the systematics, making sure we were on the Rowland circle, and trying to see how the peak shifted as a function of temperature; things like that.

Stockli: In such an experiment, the ions could capture secondary electrons into high Rydberg states, which could cause satellites observed on the low energy side of your lines. Do you have another explanation for this low energy tail?

Marrs: I think that it is probably the line shape being asymmetric in the spectrometer. It is almost just statistics. Anyway, I think we are in a fortunate position in this sense, compared to other people. If you tried to do these measurements using the beam foil technique, then certainly there are other electrons around. In our case the target ions just sit there in an electron beam that is only 10^{12} or so electrons per cubic centimeter. Hence, there just are not going to be spectator electrons in high Rydberg levels. I think we actually have an advantage with respect to that.

Stockli: Do you know what causes the line width to be 4 eV?

Marrs: Yes, the natural width is much smaller. The biggest contribution to the width comes from the detector resolution.

Stockli: The other question I have is you had an improvement on EBIT. I did not understand that. You must have changed your drift tubes.

Marrs: Well, a second EBIT is being built by engineers and they are going to float a whole CAMAC crate in a rack with electronics in it and bias the end drift tubes just a few hundred volts relative to the middle one.

Beebe: What is the switching range right now for the voltage?

Marrs: We have trouble getting much above 20 kV.

Beebe: And is that down the tubes?

Marrs: Yes. We are using a better high voltage regulator tube for the upgrade; then we will see how high we can go. It may be that the drift tubes will not go to 50 kV.

Beebe: For most of your work you could get away with floating and switching through smaller voltages?

Marrs: Yes. Of all the data I showed, I think the highest voltage was 18 kV, which was used to make neon-like gold.

Marchetti: Can you say anything about the effect of the magnetic field and the tracking field on the rates? I know the Zeeman shift would be almost nothing, but at those rates you could break all the symmetries, or at least the spherical symmetry.

Marrs: Well, I think you said it right when you said the Zeeman energy was negligible. The electric field is quite small too compared to the fields in a highly charged ion.

Marchetti: But in terms of transitions, don't you open up new channels because you break the selection rules by having the magnetic field.

Marrs: You mean something like hyperfine splitting?

Marchetti: For example, in all your radiative rates wouldn't you have the possibility of getting quadrupole contributions? I am asking whether you can break the dipole selection rules through a radiative transition just because you have a magnetic field.

Marrs: I do not think that is important.

CONCLUDING REMARKS

R. Becker

Many things have to be said. When we started on Monday, it was really exciting, especially for those of us who came from the very crowded Denton Conference to enjoy this nice environment of 50 people (sometimes less), speaking the same tongue and having one major subject. I suppose the major subject of this meeting has undoubtedly been cooling. The Livermore group has shown in EBIT how wonderful cooling can work, and now all EBIS people probably have hope that they can make use of cooling in order to improve the performance of their sources.

Now if I may just rush through the program, I see that one veteran of the field, Jean Faure, gave a good overview of all the EBISs and their branches of development; the discussion after his talk merged into the main topic of cooling.

We had Michael Kleinod indicating that we should make things simpler, especially for those who want to have EBIS sources for atomic physics.

Leif Liljeby had to be welcomed in the EBIS club, showing up with A^{18+}.

Kazuhiko Okuno surprised us with the fancy work of making a cheap EBIS with liquid nitrogen cooled normal conducting copper coils.

Bob Schmieder has a really rich uncle who provided him with isotope enriched xenon and he could show up with Xe^{38+}. His cooling calculations were interesting and convincing in this respect.

Val Kostroun gave us an excellent lecture on electron focussing and injection and wonderful measurements about them. I think this careful work is very impressive.

Finally Mort Levine has raised a question, "Is EBIT an EBIS?" This impressive work allows him to indulge in semantics. I remember that Geller said that if physics can not help, semantics must do it. I would say in your case, you are allowed to do semantics since physics has helped you.

Jean Faure showed details about DIONE with which some of us are already quite familiar, and he made a statement that sometimes he has an ion traffic jam.

Martin Stockli impressed us with his careful work of installing an EBIS on a high platform, although he is not in the EBIS club yet. Anyhow, he has created a TREBIS, or whatever you would like to call a Travel EBIS mounted on wheels.

Timothy Antaya was quite impressive in reviewing the field of ECR sources and showing that with cooling (this was a positive feedback on the ECR meeting in Michigan), the obscure gas mixing occurring in ECR is understood. You can really see how to improve the emittance of an ECR source by cooling, and this is a hope as well for EBIS sources.

Carl Herrlander showed us the beauty of symmetry on CRYRING which will probably result in very good focussing conditions.

In the afternoon of Tuesday we had posters. I suppose the most remarkable thing which everyone of us will keep in mind, besides the content of the posters, is that it is the first time I have realized it is possible to mail posters. You no longer have to attend conferences, you just mail a poster.

On Wednesday, Dick Thomas showed us results of careful cathode measurements, impregnated cathodes of all kinds. I got very upset about our cathodes. Do we really believe that we can make a compression by a factor of thousand? Seeing these patchy cathodes, one has to think seriously about them and make measurements.

Dick True showed us about electron focussing; it may be favorable to have at least a partially immersed beam.

Kurt Amboss then made the statement that transverse fields can be surprisingly damaging. I suppose all EBIS people know it.

Val Kostroun invented something which seems strange at the moment, but it is just proof of the criterion of invariance of systems. It is possible to move a vacuum system instead of moving the magnet. His finding that alignment should be accurate to one beam diameter is important.

Bob Harper showed us wonderful three-dimensional beam profiles on the EBIT gun and it was quite impressive to see how things evolve.

We had a panel discussion on beam plasma interactions and EBIS physics. It was difficult for me to reach a conclusion on this because I have my own views. It is, perhaps, more important that a definition was tried to give EBIS a special flavor and you have been successful in semantics again proposing that it should be quiescent.

In the afternoon we had this wonderful boat trip and dinner which everyone of us enjoyed. Of course this is something that you can only do in New York.

On Thursday we came back to work and it was hard. Jeremy Good told us about precision winding. He was quite successful in doing some commercials.

Martin Stockli explained to us that for the same reasons it is unreasonable to build two or three story parking lots in Kansas, it also would be unreasonable to have a vertical EBIS.

Kimo Welch explained sealography and it was amazing to see what careful work must be done in order to understand all of these things. I believe most people now spend money and take conflat flanges. He convinced us that this is at least not unreasonable.

Roderich Keller showed us interesting results on his ion source and gave us hope that we can get more ions than we need.

Richard True really impressed us with his magnetron-type injection gun, but we have to carefully consider how this fits into EBIS work.

John Staples gave an excellent overview on RFQ accelerators; RFQ's in pulse mode fit quite well to EBISs for injecting into accelerators.

Carl Herrlander showed how CRYSIS would work on 50 kV, for an RFQ of Alvin Schempp's design, and something we will remember for a long time, he invented the RF Murphy.

Tom Ludlam explained the RHIC proposal. His statement that he was looking for the largest chunks of nuclear matter available was impressive. Remember the nice viewgraph where he showed the collision of two chunks of nucleons forming a quark burger!

Krsto Prelec explained the source requirements to RHIC. It is very demanding for EBIS sources to do it with a single beam. One should remember the early days of American gun development: one should have something like a revolving gun to do it better.

John Staples did the specifics of RFQs for RHIC and, if I got it correctly, he said RFQ design is easy for this proposal; if the source does not do well, the RFQ will do it. I doubt it.

We had another panel discussion, this time on the future prospects of EBIS. Some of the homework has been done: we have now reached current densities exceeding 1000 A/cm^2. We have exceeded the charge states available from ECRs, but we do not have sufficiently high ion yields. Also, sometimes 50% of space charge compensation really has got out so we really have to show that we can use multi ampere beams with reasonable high voltage and can feed accelerators in short pulses with reasonable ions. This must be done. I suppose one important thing in this direction is that we have to get rid of secondaries.

On this last day, everything is fresh in your memory and perhaps I should pass through quite rapidly.

Hiro Tawara showed what we can all do with this NICE ion source, which was the first one devoted to atomic physics only; very nice.

Ross Marrs showed us what kind of x-ray spectroscopy is possible in an EBIT to study DR, RC, RE, and how charged states can be readily selected just by the electron energy.

Ed Beebe convinced us of the evidence of ion cooling in the Cornell EBIS, and this is really about work he has done. I hope he can continue with the thesis.

Marilyn Schneider really got heated up and again we learned successful semantics from the Livermore crew: "evaporation cooling" is a very good picture of what is going on there.

So I suppose this was a very successful symposium and we wish to meet again in two or three years at another place in the world. Then we will see what has been done in the meantime. For myself, I want to thank all of you because I feel it was a good symposium and we have learned a lot. As a veteran, I have the feeling that the field is now progressing faster.

Thank you.

APPENDICES

APPENDIX I: List of Participants

Dr. James G. Alessi
AGS Department, Bldg. 911B
Brookhaven National Laboratory
Upton, NY 11973

Dr. Kurt Amboss
Hughes Electron Dynamics Division
3100 W. Lomita Boulevard
P.O. Box 2999
Torrance, CA 90509-2999

Dr. Timothy Antaya
National Superconducting
 Cyclotron Laboratory
Michigan State University
East Lansing, MI 48824

Dr. Lahsen Assoufid
Ward Laboratory
Cornell University
Ithaca, NY 14853

Dr. John Bailey
Yale University
c/o Bldg. 911C
Brookhaven National Laboratory
Upton, NY 11973

Dr. Reinard Becker
Institut für Angewandt Physik
 der Universitat Frankfurt a.M.
Robert-Mayer-Strasse 2-4
D-6000 Frankfurt a.M. 1
FEDERAL REPUBLIC OF GERMANY

Dr. Edward Beebe
Ward Laboratory
Cornell University
Ithaca, NY 14853

Dr. John Benjamin
AGS Department, Bldg. 901A
Brookhaven National Laboratory
Upton, NY 11973

Dr. Ilan Ben-Zvi
The Weizmann Institute of
 Science
Rehovot 76100
ISRAEL

Mr. Basil DeVito
AGS Department, Bldg. 911A
Brookhaven National Laboratory
Upton, NY 11973

Dr. Jean Faure
CEN-SACLAY
Laboratoire National Saturne
91191 GIF SUR YVETTE CEDEX
FRANCE

Dr. Eric Forsyth
ADD, Building 1005S-4
Brookhaven National Laboratory
Upton, NY 11973

Dr. Jeremy A. Good
Cryogenic Consultants Ltd.
Metrostone Building
231 The Vale
London W3 7QS
UNITED KINGDOM

Dr. Harald Hahn
ADD, Building 1005S
Brookhaven National Laboratory
Upton, NY 11973

Dr. Robert Harper
Tube Division
Raytheon Company
190 Willow Street
Waltham, MA 02254

Dr. Carl Johan Herrlander
Manne Siegbahn Institute
Frescativagen 24
S-104 05 Stockholm
SWEDEN

Dr. Ady Hershcovitch
AGS Department, Bldg. 911B
Brookhaven National Laboratory
Upton, NY 11973

Dr. Brant Johnson
Department of Applied Science
Building 815
Brookhaven National Laboratory
Upton, NY 11973

Dr. Keith Jones
Department of Applied Science
Building 815
Brookhaven National Laboratory
Upton, NY 11973

Dr. Roderich Keller
Mail Stop 47-112
1 Cyclotron Road
Lawrence Berkeley Laboratory
Berkeley, CA 94720

Dr. Michael Kleinod
Institut fur Angewandt Physik
   der Universitat Frankfurt a.M.
Robert-Mayer-Strasse 204
D-6000 Frankfurt a.M. 1
FEDERAL REPUBLIC OF GERMANY

Dr. David Knapp
Mail Stop L-296
P.O. Box 808
Lawrence Livermore National
   Laboratory
Livermore, CA 94550

Dr. Val O. Kostroun
Ward Laboratory
Cornell University
Ithaca, NY 14853

Mr. Vincent Kovarik
AGS Department, Bldg. 911B
Brookhaven National Laboratory
Upton, NY 11973

Dr. Ahovi Kponou
AGS Department, Bldg. 911B
Brookhaven National Laboratory
Upton, NY 11973

Dr. Frank Krienen
Boston University
c/o Bldg. 911C
Brookhaven National Laboratory
Upton, NY 11973

Dr. Y.Y. Lee
AGS Department, Bldg. 911B
Brookhaven National Laboratory
Upton, NY 11973

Dr. Morton A. Levine
Mail Stop 4/230
1 Cyclotron Road
Lawrence Berkeley Laboratory
Berkeley, CA 94720

Dr. Leif Liljeby
Manne Siegbahn Institute
Frescativagen 24
S-104 05 Stockholm
SWEDEN

Dr. Derek Lowenstein
AGS Department, Bldg. 911B
Brookhaven National Laboratory
Upton, NY 11973

Dr. Tom Ludlam
ADD, Building 510C
Brookhaven National Laboratory
Upton, NY 11973

Mr. Vincent Marchetti
Ward Laboratory
Cornell University
Ithaca, NY 14853

Dr. Roscoe E. Marrs
Mail Stop L-296
P.O. Box 808
Lawrence Livermore National
  Laboratory
Livermore, CA 94550

Dr. Charles Meitzler
AGS Department, Bldg. 930
Brookhaven National Laboratory
Upton, NY 11973

Dr. Mati Meron
Department of Applied Science
Building 815
Brookhaven National Laboratory
Upton, NY 11973

Dr. John Noe
Physics Department
The University at Stony Brook
Stony Brook, NY 11794

Dr. Kazuhiko Okuno
Department of Physics
Tokyo Metropolitan University
2-1-1 Fukasawa, Setagaya-ku
Tokyo 158
JAPAN

Dr. Bernie M. Penetrante
Mail Stop L-417
P.O. Box 808
Lawrence Livermore National
  Laboratory
Livermore, CA 94550

Dr. Krsto Prelec
AGS Department, Bldg. 911B
Brookhaven National Laboratory
Upton, NY 11973

Dr. Mark Rhoades-Brown
ADD, Bldg. 1005S
Brookhaven National Laboratory
Upton, NY 11973

Dr. Norman Rostoker
Physics Department
University of California
Irvine, CA 92717

Dr. Robert W. Schmieder
Div. 8347
Sandia National Laboratory
Livermore, CA 94551

Dr. Marilyn B. Schneider
Mail Stop L-45
P.O. Box 808
Lawrence Livermore National
  Laboratory
Livermore, CA 94550

Dr. Theodorus J. Sluyters
AGS Department, Bldg. 911B
Brookhaven National Laboratory
Upton, NY 11973

Dr. John Staples
MS-64-121
Bldg. 64, Room 224A
Lawrence Berkeley Laboratory
Berkeley, CA 94720

Dr. Martin Stockli
Department of Physics
Cardwell Hall, Room 117
Kansas State University
Manhattan, KS 66506

Dr. Hiroyuki Tawara
Institute of Plasma Physics
Nagoya University
Chikusa-ky
Nagoya 464
JAPAN

Dr. Richard E. Thomas
Code 6844
U.S. Naval Research Laboratory
Washington, D.C. 20375

Dr. Richard True
Electron Devices Division
Litton Systems, Inc.
960 Industrial Road
San Carlos, CA 94070-4194

Mr. John W. Walker
Cryogenic Consultants Ltd.
Box 416
Warwick, NY 10990

Ms. Joanne Wang
Ward Laboratory
Cornell University
Ithaca, NY 14853

Dr. Harvey Wegner
Physics Department
Building 901A
Brookhaven National Laboratory
Upton, NY 11973

Dr. Kimo Welch
AGS Department, Bldg. 911B
Brookhaven National Laboratory
Upton, NY 11973

APPENDIX II: List of Authors

A

Amboss, K.
Antaya, T.A.
Antoine, P.

B

Bardsley, J.N.
Becker, R.
Beebe, E.N.
Beiersdorfer P.
Bennett, C.L.
Bisson, C.L.

C

Chen, M.H.
Ciret, J.C.
Cocke, C.L.
Courtois, A.
Cowan, T.

D

Degueurce, L.
Deitinghoff, H.
Dietrich, D.

E

Engstrom, A.

F

Faure, J.

G

Gastineau, B.
Gobin R.
Good, J.A.
Gros, P.

H

Haney, S.
Harper, R.
Henderson, J.R.
Herrlander, C.J.
Hershcovitch, A.

J

Johnson, B.M.
Jones, K.W.

K

Kallberg, A.
Keller, R.
Klein, H.
Kleinod, M.
Knapp, D.A.
Kostroun, V.O.
Kovarik, V.

L

Leaux, P.
Leroy, P.A.
Levine, M.A.
Liljeby, L.
Ludlam, T.W.
Lyneis, C.M.

M

Marrs, R.E.
McIntyre, G.T.
Meron, M.

O

Okuno, K.
Osterheld, A.

P

Pate, D.J.
Penetrante, B.M.
Penicaud, J.P.
Prelec, K.
Puri, M.P.

R

Richard, P.

S

Schmieder, R.W.
Schneider, M.B.
Scofield, J.H.
Soltan, A.
Staples, J.
Stockli, M.P.

T

Tawara, H.
Thomas, R.E.
Toly, N.
True, R.
Tuozzolo, J.E.

V

Van Hook, A.R.

W

Weeks, J.
Welch, K.
Wilkins, P.

APPENDIX III: List of Sessions
INTERNATIONAL SYMPOSIUM ON ELECTRON BEAM ION SOURCES
AND THEIR APPLICATIONS

Brookhaven National Laboratory
November 14-18. 1988

Monday, November 14, 1988

9:00   Opening Remarks
        D. Lowenstein, Chairman, AGS Department
        A. Hershcovitch, Symposium Chairman

        Session I - EBIS Devices
        Chairperson: R. Becker

9:20   Review of EBIS Devices
        J. Faure
        Saclay, France

10:10   BREAK

10:40   Progress Report on the Frankfurt EBIS
        M. Kleinod, R. Becker, and H. Klein
        Institute Fur Angewandte Physik, Frankfurt, FRG

11:20   Status Report on the Stockholm Cryogenic Electron
        Beam Ion Source
        L. Liljeby
        Manne Siegbahn Institute of Physics,
        Stockholm, Sweden

12:00   LUNCH

(Monday continued)

Session II - EBIS Devices
Chairperson: C.J. Herrlander

2:00 Performance of MINI-EBIS Cooled by Liquid Nitrogen
K. Okuno
Tokyo Metropolitan University, Tokyo, Japan

2:40 The Sandia EBIS Program
R.W. Schmieder
Sandia National Laboratory, Livermore, CA, USA

3:20 BREAK

3:50 The Cornell Superconducting Solenoid Cryogenic EBIS
V.O. Kostroun
Cornell University, Ithaca, NY, USA

4:30 The EBIT Apparatus
M. Levine
Lawrence Berkeley Laboratory, Berkeley, CA, USA

5:30 – 7:30 Wine and Cheese Reception
Courtesy of AUI

Tuesday, November 15, 1988

Session III – Devices
Chairperson: L. Liljeby

9:00 The Status of Dione Heavy Ion Injector
      J. Faure, et al.
      Saclay, France

9:40 Report on the KSU-CRYEBIS
      M.P. Stockli
      Kansas State University, Manhattan, KS, USA

10:20 BREAK

10:50 ECR Ion Sources Status 1988
      T.A. Antaya
      Michigan State University, East Lansing, MI, USA
      C.M. Lyneis
      Lawrence Berkeley Laboratory, Berkeley, CA, USA

11:30 Status of CRYRING
      C.J. Herrlander
      Manne Siegbahn Institute of Physics,
      Stockholm, Sweden

12:10 LUNCH

(Tuesday continued)

Session IV - Posters/Site Tours
Chairperson: K. Prelec

2:00   High Brightness, Steady State Electron Beam Source
Utilizing Hollow Cathode Plasmas
    A. Hershcovitch, V. Kovarik, and K. Prelec
    Brookhaven National Laboratory, Upton, NY, USA

Results for KRION I, II, and III
    E.D. Donets
    JINR, Dubna, USSR

Uses of EBIS, ECRIS, and Storage Rings for Atomic
Physics Experiments
    K.W. Jones, B.M. Johnson, and M. Meron
    Brookhaven National Laboratory, Upton, NY, USA

The Reliable Operation of CRYEBIS I
    J. Faure, et al.
    Saclay, France

IEEE Transactions on Electron Devices, Part II
    R. True
    Litton Systems, Electron Devices Division,
    San Carlos, CA, USA

Studies of Collisional Ion Cooling in EBIT
    M.B. Schneider
    Lawrence Livermore Laboratory, Livermore, CA, USA

Dielectronic Recombination on He Like Nickel
    D.A. Knapp
    Lawrence Livermore Laboratory, Livermore, CA, USA

Multi-Ampere e-Beam
    F. Krienen
    Boston University, Boston, MA, USA

MEVVA Applications to EBIS
    I. Brown
    Lawrence Berkeley Laboratory, Berkeley, CA, USA

(Tuesday continued)

    Electron Beam Dynamics in EBIS Sources
      M. Tagger
      Saclay, France

    RFQ's and Linacs for EBIS Devices
      R. Hamm
      AccSys Technology Inc., Pleasanton, CA, USA

    The Feasibility of an EBIS for the CERN Lead Project
      R. Becker
      Institute Fur Angewandte Physik, Frankfurt, FRG

    Principles of Ion Collisional Cooling in EBIT
      B.M. Penetrante, et al.
      Lawrence Livermore Laboratory, Livermore, CA, USA

Wednesday, November 16, 1988

Session V - Electron Beams
Chairperson: J. Faure

8:30   e-Beam Emitters
       R. Thomas
       Naval Research Laboratory, Washington, DC, USA

9:10   Advances in e-Beam Formation, Focussing and Collection
       R. True
       Litton Systems, Electron Devices Division,
       San Carlos, CA, USA

9:50   BREAK

10:20   The Effect of Small Transverse Magnetic Fields on Electron Beam Transmission
       K. Amboss
       Hughes Aircraft Company, Electron Dynamics Division,
       Torrance, CA, USA

11:00   Electron Beam Alignment in an EBIS
       V.O. Kostroun
       Cornell University, Ithaca, NY, USA

11:30   Characteristics of Typical Pierce Guns for PPM Focussed TWTs
       R. Harper and M.P. Puri
       Ratheon Company, Waltham, MA, USA

11:55   Panel Session
       e-Beam Plasma Interactions in an EBIS Trap; EBIS Physics; Diagnostics
       Panel Members
       N. Rostoker, Chairperson, V. Kostroun, R. Marrs,
       R. Schmieder, and H. Tawara

12:55   LUNCH

3:30   Leave for Cruise and Banquet

Thursday, November 17, 1988

Session VI - Related Technologies; Primary Ions
Chairperson: H. Tawara

9:00   Precision Measurements and Winding Accuracy of Superconducting Magnets
       J.A. Good
       Cryogenic Consultants LTD., London, UK

9:40   The Magnetic Alignment and Mapping of Long Horizontal Solenoids for EBIS Devices
       M.P. Stockli
       Kansas State University, Manhattan, KS, USA

10:10   BREAK

10:40   Numerical Simulations Related to EBIS Design
       R. Becker
       Institute Fur Angewandte Physik, Frankfurt, FRG

11:10   Sealography and Cryopumping in EBIS Devices
       K. Welch
       Brookhaven National Laboratory, Upton, NY, USA

11:50   Primary Ion Sources for EBIS Devices
       R. Keller
       Lawrence Berkeley Laboratory, Berkeley, CA, USA

12:30   A Novel Electron Source for EBIS Machines
       R. True
       Litton Systems, Electron Devices Division, San Carlos, CA, USA

12:45   LUNCH

(Thursday continued)

Session VII - RFQ's, Preinjectors, etc.
Chairperson: M.P. Stockli

2:00   Heavy Ion RFQ's
            J. Staples
            Lawrence Berkeley Laboratory, Berkeley, CA, USA

2:30   RFQ Injector for CRYRING
            A. Schempp, H. Deitinghoff, and H. Klein
            Institute Fur Angewandte Physik, Frankfurt, FRG
            A. Kallberg, A. Soltan, and C.J. Herrlander
            Manne Siegbahn Institute of Physics,
            Stockholm, Sweden

3:00   BREAK

3:30   Heavy Ion Requirements for RHIC
            T. Ludlam
            Brookhaven National Laboratory, Upton, NY, USA

4:00   Source Options for RHIC
            K. Prelec
            Brookhaven National Laboratory, Upton, NY, USA

4:20   RFQ for an EBIS Based RHIC Preinjector
            J. Staples
            Lawrence Berkeley Laboratory, Berkeley, CA, USA

4:40   Panel Session
            Future Prospect and Limits
               Panel Members
            J. Faure, Chairperson, R. Becker, C. Herrlander,
            M. Kleinod, and M. Levine

Friday, November 18, 1988

Session VIII - Atomic Physics
Chairperson: M.A. Levine

9:00   Atomic Physics Research Using Highly Charged Ions from EBIS
        H. Tawara
        Institute of Plasma Physics, Nagoya University, Japan

9:40   Atomic Physics Measurements in EBIT
        R. Marrs
        Lawrence Livermore Laboratory, Livermore, CA, USA

10:20   BREAK

10:50   Atomic Physics Experiments with CEBIS I
        E.N. Beebe
        Cornell University, Ithaca, NY, USA

11:20   Collisional Cooling of Highly Charged Ions in EBIT - An Experimental Realization
        M.B. Schneider, M.A. Levine, J.R. Henderson, D.A. Knapp, R.E. Marrs, and B.M. Penetrante
        Lawrence Livermore Laboratory, Livermore, CA, USA

11:45   CONCLUSION

AIP Conference Proceedings

		L.C. Number	ISBN
No. 1	Feedback and Dynamic Control of Plasmas – 1970	70-141596	0-88318-100-2
No. 2	Particles and Fields – 1971 (Rochester)	71-184662	0-88318-101-0
No. 3	Thermal Expansion – 1971 (Corning)	72-76970	0-88318-102-9
No. 4	Superconductivity in d- and f-Band Metals (Rochester, 1971)	74-18879	0-88318-103-7
No. 5	Magnetism and Magnetic Materials – 1971 (2 parts) (Chicago)	59-2468	0-88318-104-5
No. 6	Particle Physics (Irvine, 1971)	72-81239	0-88318-105-3
No. 7	Exploring the History of Nuclear Physics – 1972	72-81883	0-88318-106-1
No. 8	Experimental Meson Spectroscopy –1972	72-88226	0-88318-107-X
No. 9	Cyclotrons – 1972 (Vancouver)	72-92798	0-88318-108-8
No. 10	Magnetism and Magnetic Materials – 1972	72-623469	0-88318-109-6
No. 11	Transport Phenomena – 1973 (Brown University Conference)	73-80682	0-88318-110-X
No. 12	Experiments on High Energy Particle Collisions – 1973 (Vanderbilt Conference)	73-81705	0-88318-111-8
No. 13	π-π Scattering – 1973 (Tallahassee Conference)	73-81704	0-88318-112-6
No. 14	Particles and Fields – 1973 (APS/DPF Berkeley)	73-91923	0-88318-113-4
No. 15	High Energy Collisions – 1973 (Stony Brook)	73-92324	0-88318-114-2
No. 16	Causality and Physical Theories (Wayne State University, 1973)	73-93420	0-88318-115-0
No. 17	Thermal Expansion – 1973 (Lake of the Ozarks)	73-94415	0-88318-116-9
No. 18	Magnetism and Magnetic Materials – 1973 (2 parts) (Boston)	59-2468	0-88318-117-7
No. 19	Physics and the Energy Problem – 1974 (APS Chicago)	73-94416	0-88318-118-5
No. 20	Tetrahedrally Bonded Amorphous Semiconductors (Yorktown Heights, 1974)	74-80145	0-88318-119-3
No. 21	Experimental Meson Spectroscopy – 1974 (Boston)	74-82628	0-88318-120-7
No. 22	Neutrinos – 1974 (Philadelphia)	74-82413	0-88318-121-5
No. 23	Particles and Fields – 1974 (APS/DPF Williamsburg)	74-27575	0-88318-122-3
No. 24	Magnetism and Magnetic Materials – 1974 (20th Annual Conference, San Francisco)	75-2647	0-88318-123-1
No. 25	Efficient Use of Energy (The APS Studies on the Technical Aspects of the More Efficient Use of Energy)	75-18227	0-88318-124-X

No. 26	High-Energy Physics and Nuclear Structure – 1975 (Santa Fe and Los Alamos)	75-26411	0-88318-125-8
No. 27	Topics in Statistical Mechanics and Biophysics: A Memorial to Julius L. Jackson (Wayne State University, 1975)	75-36309	0-88318-126-6
No. 28	Physics and Our World: A Symposium in Honor of Victor F. Weisskopf (M.I.T., 1974)	76-7207	0-88318-127-4
No. 29	Magnetism and Magnetic Materials – 1975 (21st Annual Conference, Philadelphia)	76-10931	0-88318-128-2
No. 30	Particle Searches and Discoveries – 1976 (Vanderbilt Conference)	76-19949	0-88318-129-0
No. 31	Structure and Excitations of Amorphous Solids (Williamsburg, VA, 1976)	76-22279	0-88318-130-4
No. 32	Materials Technology – 1976 (APS New York Meeting)	76-27967	0-88318-131-2
No. 33	Meson-Nuclear Physics – 1976 (Carnegie-Mellon Conference)	76-26811	0-88318-132-0
No. 34	Magnetism and Magnetic Materials – 1976 (Joint MMM-Intermag Conference, Pittsburgh)	76-47106	0-88318-133-9
No. 35	High Energy Physics with Polarized Beams and Targets (Argonne, 1976)	76-50181	0-88318-134-7
No. 36	Momentum Wave Functions – 1976 (Indiana University)	77-82145	0-88318-135-5
No. 37	Weak Interaction Physics – 1977 (Indiana University)	77-83344	0-88318-136-3
No. 38	Workshop on New Directions in Mossbauer Spectroscopy (Argonne, 1977)	77-90635	0-88318-137-1
No. 39	Physics Careers, Employment and Education (Penn State, 1977)	77-94053	0-88318-138-X
No. 40	Electrical Transport and Optical Properties of Inhomogeneous Media (Ohio State University, 1977)	78-54319	0-88318-139-8
No. 41	Nucleon-Nucleon Interactions – 1977 (Vancouver)	78-54249	0-88318-140-1
No. 42	Higher Energy Polarized Proton Beams (Ann Arbor, 1977)	78-55682	0-88318-141-X
No. 43	Particles and Fields – 1977 (APS/DPF, Argonne)	78-55683	0-88318-142-8
No. 44	Future Trends in Superconductive Electronics (Charlottesville, 1978)	77-9240	0-88318-143-6
No. 45	New Results in High Energy Physics – 1978 (Vanderbilt Conference)	78-67196	0-88318-144-4
No. 46	Topics in Nonlinear Dynamics (La Jolla Institute)	78-57870	0-88318-145-2
No. 47	Clustering Aspects of Nuclear Structure and Nuclear Reactions (Winnepeg, 1978)	78-64942	0-88318-146-0
No. 48	Current Trends in the Theory of Fields (Tallahassee, 1978)	78-72948	0-88318-147-9

No. 49	Cosmic Rays and Particle Physics – 1978 (Bartol Conference)	79-50489	0-88318-148-7
No. 50	Laser-Solid Interactions and Laser Processing – 1978 (Boston)	79-51564	0-88318-149-5
No. 51	High Energy Physics with Polarized Beams and Polarized Targets (Argonne, 1978)	79-64565	0-88318-150-9
No. 52	Long-Distance Neutrino Detection – 1978 (C.L. Cowan Memorial Symposium)	79-52078	0-88318-151-7
No. 53	Modulated Structures – 1979 (Kailua Kona, Hawaii)	79-53846	0-88318-152-5
No. 54	Meson-Nuclear Physics – 1979 (Houston)	79-53978	0-88318-153-3
No. 55	Quantum Chromodynamics (La Jolla, 1978)	79-54969	0-88318-154-1
No. 56	Particle Acceleration Mechanisms in Astrophysics (La Jolla, 1979)	79-55844	0-88318-155-X
No. 57	Nonlinear Dynamics and the Beam-Beam Interaction (Brookhaven, 1979)	79-57341	0-88318-156-8
No. 58	Inhomogeneous Superconductors – 1979 (Berkeley Springs, W.V.)	79-57620	0-88318-157-6
No. 59	Particles and Fields – 1979 (APS/DPF Montreal)	80-66631	0-88318-158-4
No. 60	History of the ZGS (Argonne, 1979)	80-67694	0-88318-159-2
No. 61	Aspects of the Kinetics and Dynamics of Surface Reactions (La Jolla Institute, 1979)	80-68004	0-88318-160-6
No. 62	High Energy e^+e^- Interactions (Vanderbilt, 1980)	80-53377	0-88318-161-4
No. 63	Supernovae Spectra (La Jolla, 1980)	80-70019	0-88318-162-2
No. 64	Laboratory EXAFS Facilities – 1980 (Univ. of Washington)	80-70579	0-88318-163-0
No. 65	Optics in Four Dimensions – 1980 (ICO, Ensenada)	80-70771	0-88318-164-9
No. 66	Physics in the Automotive Industry – 1980 (APS/AAPT Topical Conference)	80-70987	0-88318-165-7
No. 67	Experimental Meson Spectroscopy – 1980 (Sixth International Conference, Brookhaven)	80-71123	0-88318-166-5
No. 68	High Energy Physics – 1980 (XX International Conference, Madison)	81-65032	0-88318-167-3
No. 69	Polarization Phenomena in Nuclear Physics – 1980 (Fifth International Symposium, Santa Fe)	81-65107	0-88318-168-1
No. 70	Chemistry and Physics of Coal Utilization – 1980 (APS, Morgantown)	81-65106	0-88318-169-X
No. 71	Group Theory and its Applications in Physics – 1980 (Latin American School of Physics, Mexico City)	81-66132	0-88318-170-3
No. 72	Weak Interactions as a Probe of Unification (Virginia Polytechnic Institute – 1980)	81-67184	0-88318-171-1
No. 73	Tetrahedrally Bonded Amorphous Semiconductors (Carefree, Arizona, 1981)	81-67419	0-88318-172-X

No. 74	Perturbative Quantum Chromodynamics (Tallahassee, 1981)	81-70372	0-88318-173-8
No. 75	Low Energy X-Ray Diagnostics – 1981 (Monterey)	81-69841	0-88318-174-6
No. 76	Nonlinear Properties of Internal Waves (La Jolla Institute, 1981)	81-71062	0-88318-175-4
No. 77	Gamma Ray Transients and Related Astrophysical Phenomena (La Jolla Institute, 1981)	81-71543	0-88318-176-2
No. 78	Shock Waves in Condensed Matter – 1981 (Menlo Park)	82-70014	0-88318-177-0
No. 79	Pion Production and Absorption in Nuclei – 1981 (Indiana University Cyclotron Facility)	82-70678	0-88318-178-9
No. 80	Polarized Proton Ion Sources (Ann Arbor, 1981)	82-71025	0-88318-179-7
No. 81	Particles and Fields –1981: Testing the Standard Model (APS/DPF, Santa Cruz)	82-71156	0-88318-180-0
No. 82	Interpretation of Climate and Photochemical Models, Ozone and Temperature Measurements (La Jolla Institute, 1981)	82-71345	0-88318-181-9
No. 83	The Galactic Center (Cal. Inst. of Tech., 1982)	82-71635	0-88318-182-7
No. 84	Physics in the Steel Industry (APS/AISI, Lehigh University, 1981)	82-72033	0-88318-183-5
No. 85	Proton-Antiproton Collider Physics –1981 (Madison, Wisconsin)	82-72141	0-88318-184-3
No. 86	Momentum Wave Functions – 1982 (Adelaide, Australia)	82-72375	0-88318-185-1
No. 87	Physics of High Energy Particle Accelerators (Fermilab Summer School, 1981)	82-72421	0-88318-186-X
No. 88	Mathematical Methods in Hydrodynamics and Integrability in Dynamical Systems (La Jolla Institute, 1981)	82-72462	0-88318-187-8
No. 89	Neutron Scattering – 1981 (Argonne National Laboratory)	82-73094	0-88318-188-6
No. 90	Laser Techniques for Extreme Ultraviolt Spectroscopy (Boulder, 1982)	82-73205	0-88318-189-4
No. 91	Laser Acceleration of Particles (Los Alamos, 1982)	82-73361	0-88318-190-8
No. 92	The State of Particle Accelerators and High Energy Physics (Fermilab, 1981)	82-73861	0-88318-191-6
No. 93	Novel Results in Particle Physics (Vanderbilt, 1982)	82-73954	0-88318-192-4
No. 94	X-Ray and Atomic Inner-Shell Physics – 1982 (International Conference, U. of Oregon)	82-74075	0-88318-193-2
No. 95	High Energy Spin Physics – 1982 (Brookhaven National Laboratory)	83-70154	0-88318-194-0
No. 96	Science Underground (Los Alamos, 1982)	83-70377	0-88318-195-9

No. 97	The Interaction Between Medium Energy Nucleons in Nuclei – 1982 (Indiana University)	83-70649	0-88318-196-7
No. 98	Particles and Fields – 1982 (APS/DPF University of Maryland)	83-70807	0-88318-197-5
No. 99	Neutrino Mass and Gauge Structure of Weak Interactions (Telemark, 1982)	83-71072	0-88318-198-3
No. 100	Excimer Lasers – 1983 (OSA, Lake Tahoe, Nevada)	83-71437	0-88318-199-1
No. 101	Positron-Electron Pairs in Astrophysics (Goddard Space Flight Center, 1983)	83-71926	0-88318-200-9
No. 102	Intense Medium Energy Sources of Strangeness (UC-Sant Cruz, 1983)	83-72261	0-88318-201-7
No. 103	Quantum Fluids and Solids – 1983 (Sanibel Island, Florida)	83-72440	0-88318-202-5
No. 104	Physics, Technology and the Nuclear Arms Race (APS Baltimore –1983)	83-72533	0-88318-203-3
No. 105	Physics of High Energy Particle Accelerators (SLAC Summer School, 1982)	83-72986	0-88318-304-8
No. 106	Predictability of Fluid Motions (La Jolla Institute, 1983)	83-73641	0-88318-305-6
No. 107	Physics and Chemistry of Porous Media (Schlumberger-Doll Research, 1983)	83-73640	0-88318-306-4
No. 108	The Time Projection Chamber (TRIUMF, Vancouver, 1983)	83-83445	0-88318-307-2
No. 109	Random Walks and Their Applications in the Physical and Biological Sciences (NBS/La Jolla Institute, 1982)	84-70208	0-88318-308-0
No. 110	Hadron Substructure in Nuclear Physics (Indiana University, 1983)	84-70165	0-88318-309-9
No. 111	Production and Neutralization of Negative Ions and Beams (3rd Int'l Symposium, Brookhaven, 1983)	84-70379	0-88318-310-2
No. 112	Particles and Fields – 1983 (APS/DPF, Blacksburg, VA)	84-70378	0-88318-311-0
No. 113	Experimental Meson Spectroscopy – 1983 (Seventh International Conference, Brookhaven)	84-70910	0-88318-312-9
No. 114	Low Energy Tests of Conservation Laws in Particle Physics (Blacksburg, VA, 1983)	84-71157	0-88318-313-7
No. 115	High Energy Transients in Astrophysics (Santa Cruz, CA, 1983)	84-71205	0-88318-314-5
No. 116	Problems in Unification and Supergravity (La Jolla Institute, 1983)	84-71246	0-88318-315-3
No. 117	Polarized Proton Ion Sources (TRIUMF, Vancouver, 1983)	84-71235	0-88318-316-1

No. 118	Free Electron Generation of Extreme Ultraviolet Coherent Radiation (Brookhaven/OSA, 1983)	84-71539	0-88318-317-X
No. 119	Laser Techniques in the Extreme Ultraviolet (OSA, Boulder, Colorado, 1984)	84-72128	0-88318-318-8
No. 120	Optical Effects in Amorphous Semiconductors (Snowbird, Utah, 1984)	84-72419	0-88318-319-6
No. 121	High Energy e^+e^- Interactions (Vanderbilt, 1984)	84-72632	0-88318-320-X
No. 122	The Physics of VLSI (Xerox, Palo Alto, 1984)	84-72729	0-88318-321-8
No. 123	Intersections Between Particle and Nuclear Physics (Steamboat Springs, 1984)	84-72790	0-88318-322-6
No. 124	Neutron-Nucleus Collisions – A Probe of Nuclear Structure (Burr Oak State Park - 1984)	84-73216	0-88318-323-4
No. 125	Capture Gamma-Ray Spectroscopy and Related Topics – 1984 (Internat. Symposium, Knoxville)	84-73303	0-88318-324-2
No. 126	Solar Neutrinos and Neutrino Astronomy (Homestake, 1984)	84-63143	0-88318-325-0
No. 127	Physics of High Energy Particle Accelerators (BNL/SUNY Summer School, 1983)	85-70057	0-88318-326-9
No. 128	Nuclear Physics with Stored, Cooled Beams (McCormick's Creek State Park, Indiana, 1984)	85-71167	0-88318-327-7
No. 129	Radiofrequency Plasma Heating (Sixth Topical Conference, Callaway Gardens, GA, 1985)	85-48027	0-88318-328-5
No. 130	Laser Acceleration of Particles (Malibu, California, 1985)	85-48028	0-88318-329-3
No. 131	Workshop on Polarized ^{3}He Beams and Targets (Princeton, New Jersey, 1984)	85-48026	0-88318-330-7
No. 132	Hadron Spectroscopy–1985 (International Conference, Univ. of Maryland)	85-72537	0-88318-331-5
No. 133	Hadronic Probes and Nuclear Interactions (Arizona State University, 1985)	85-72638	0-88318-332-3
No. 134	The State of High Energy Physics (BNL/SUNY Summer School, 1983)	85-73170	0-88318-333-1
No. 135	Energy Sources: Conservation and Renewables (APS, Washington, DC, 1985)	85-73019	0-88318-334-X
No. 136	Atomic Theory Workshop on Relativistic and QED Effects in Heavy Atoms	85-73790	0-88318-335-8
No. 137	Polymer-Flow Interaction (La Jolla Institute, 1985)	85-73915	0-88318-336-6
No. 138	Frontiers in Electronic Materials and Processing (Houston, TX, 1985)	86-70108	0-88318-337-4
No. 139	High-Current, High-Brightness, and High-Duty Factor Ion Injectors (La Jolla Institute, 1985)	86-70245	0-88318-338-2

No. 140	Boron-Rich Solids (Albuquerque, NM, 1985)	86-70246	0-88318-339-0
No. 141	Gamma-Ray Bursts (Stanford, CA, 1984)	86-70761	0-88318-340-4
No. 142	Nuclear Structure at High Spin, Excitation, and Momentum Transfer (Indiana University, 1985)	86-70837	0-88318-341-2
No. 143	Mexican School of Particles and Fields (Oaxtepec, México, 1984)	86-81187	0-88318-342-0
No. 144	Magnetospheric Phenomena in Astrophysics (Los Alamos, 1984)	86-71149	0-88318-343-9
No. 145	Polarized Beams at SSC & Polarized Antiprotons (Ann Arbor, MI & Bodega Bay, CA, 1985)	86-71343	0-88318-344-7
No. 146	Advances in Laser Science–I (Dallas, TX, 1985)	86-71536	0-88318-345-5
No. 147	Short Wavelength Coherent Radiation: Generation and Applications (Monterey, CA, 1986)	86-71674	0-88318-346-3
No. 148	Space Colonization: Technology and The Liberal Arts (Geneva, NY, 1985)	86-71675	0-88318-347-1
No. 149	Physics and Chemistry of Protective Coatings (Universal City, CA, 1985)	86-72019	0-88318-348-X
No. 150	Intersections Between Particle and Nuclear Physics (Lake Louise, Canada, 1986)	86-72018	0-88318-349-8
No. 151	Neural Networks for Computing (Snowbird, UT, 1986)	86-72481	0-88318-351-X
No. 152	Heavy Ion Inertial Fusion (Washington, DC, 1986)	86-73185	0-88318-352-8
No. 153	Physics of Particle Accelerators (SLAC Summer School, 1985) (Fermilab Summer School, 1984)	87-70103	0-88318-353-6
No. 154	Physics and Chemistry of Porous Media—II (Ridge Field, CT, 1986)	83-73640	0-88318-354-4
No. 155	The Galactic Center: Proceedings of the Symposium Honoring C. H. Townes (Berkeley, CA, 1986)	86-73186	0-88318-355-2
No. 156	Advanced Accelerator Concepts (Madison, WI, 1986)	87-70635	0-88318-358-0
No. 157	Stability of Amorphous Silicon Alloy Materials and Devices (Palo Alto, CA, 1987)	87-70990	0-88318-359-9
No. 158	Production and Neutralization of Negative Ions and Beams (Brookhaven, NY, 1986)	87-71695	0-88318-358-7

No. 159	Applications of Radio-Frequency Power to Plasma: Seventh Topical Conference (Kissimmee, FL, 1987)	87-71812	0-88318-359-5
No. 160	Advances in Laser Science–II (Seattle, WA, 1986)	87-71962	0-88318-360-9
No. 161	Electron Scattering in Nuclear and Particle Science: In Commemoration of the 35th Anniversary of the Lyman-Hanson-Scott Experiment (Urbana, IL, 1986)	87-72403	0-88318-361-7
No. 162	Few-Body Systems and Multiparticle Dynamics (Crystal City, VA, 1987)	87-72594	0-88318-362-5
No. 163	Pion–Nucleus Physics: Future Directions and New Facilities at LAMPF (Los Alamos, NM, 1987)	87-72961	0-88318-363-3
No. 164	Nuclei Far from Stability: Fifth International Conference (Rosseau Lake, ON, 1987)	87-73214	0-88318-364-1
No. 165	Thin Film Processing and Characterization of High-Temperature Superconductors	87-73420	0-88318-365-X
No. 166	Photovoltaic Safety (Denver, CO, 1988)	88-42854	0-88318-366-8
No. 167	Deposition and Growth: Limits for Microelectronics (Anaheim, CA, 1987)	88-71432	0-88318-367-6
No. 168	Atomic Processes in Plasmas (Santa Fe, NM, 1987)	88-71273	0-88318-368-4
No. 169	Modern Physics in America: A Michelson-Morley Centennial Symposium (Cleveland, OH, 1987)	88-71348	0-88318-369-2
No. 170	Nuclear Spectroscopy of Astrophysical Sources (Washington, D.C., 1987)	88-71625	0-88318-370-6
No. 171	Vacuum Design of Advanced and Compact Synchrotron Light Sources (Upton, NY, 1988)	88-71824	0-88318-371-4
No. 172	Advances in Laser Science–III: Proceedings of the International Laser Science Conference (Atlantic City, NJ, 1987)	88-71879	0-88318-372-2
No. 173	Cooperative Networks in Physics Education (Oaxtepec, Mexico 1987)	88-72091	0-88318-373-0
No. 174	Radio Wave Scattering in the Interstellar Medium (San Diego, CA 1988)	88-72092	0-88318-374-9
No. 175	Non-neutral Plasma Physics (Washington, DC 1988)	88-72275	0-88318-375-7

No. 176	Intersections Between Particle Land Nuclear Physics (Third International Conference) (Rockport, ME 1988)	88-62535	0-88318-376-5
No. 177	Linear Accelerator and Beam Optics Codes (La Jolla, CA 1988)	88-46074	0-88318-377-3
No. 178	Nuclear Arms Technologies in the 1990s (Washington, DC 1988)	88-83262	0-88318-378-1
No. 179	The Michelson Era in American Science: 1870–1930 (Cleveland, OH 1987)	88-83369	0-88318-379-X
No. 180	Frontiers in Science: International Symposium (Urbana, IL, 1987)	88-83526	0-88318-380-3
No. 181	Muon-Catalyzed Fusion (Sanibel Island, FL, 1988)	88-83636	0-88318-381-1
No. 176	Intersections Between Particle and Nuclear Physics (Third International Conference) (Rockport, ME 1988)	88-62535	0-88318-376-5
No. 177	Linear Accelerator and Beam Optics Codes (La Jolla, CA 1988)	88-46074	0-88318-377-3
No. 178	Nuclear Arms Technologies in the 1990s (Washington, DC 1988)	88-83262	0-88318-378-1
No. 179	The Michelson Era in American Science: 1870–1930 (Cleveland, OH 1987)	88-83369	0-88318-379-X
No. 180	Frontiers in Science: International Symposium (Urbana, IL 1987)	88-83526	0-88318-380-3
No. 181	Muon-Catalyzed Fusion (Sanibel Island, FL 1988)	88-83636	0-88318-381-1
No. 182	High T_c Superconducting Thin Films, Devices, and Application (Atlanta, GA 1988)	88-03947	0-88318-382-X
No. 183	Cosmic Abundances of Matter (Minneapolis, MN 1988)	89-80147	0-88318-383-8
No. 184	Physics of Particle Accelerators (Ithaca, NY 1988)	87-07208	0-88318-384-6
No. 185	Glueballs, Hybrids, and Exotic Hadrons (Upton, NY 1988)	89-83513	0-88318-385-4
No. 186	High-Energy Radiation Background in Space (Sanibel Island, FL 1987)	89-083833	0-88318-386-2
No. 187	High-Energy Spin Physics (Minneapolis, MN 1988)	89-083948	0-88318-387-0

MAR 2 3 1990